FLUID FILTRATION: GAS
Volume I

A symposium
sponsored by
ASTM Committee F-21
on Filtration and The
American Program Committee
of The Filtration Society
Philadelphia, PA, 20–22 Oct. 1986

ASTM SPECIAL TECHNICAL PUBLICATION 975
Robert R. Raber, Farr Co., editor

ASTM Publication Code Number (PCN)
04-975001-39

 1916 Race Street, Philadelphia, PA 19103

7351-6302

CHEMISTRY

Library of Congress Cataloging-in-Publication Data

Fluid filtration.

(ASTM special technical publication; 975)
"ASTM publication code number (PCN) 04-975001-39."
"ASTM publication code number (PCN) 04-975002-39"—
V. 2, t.p.
Includes bibliographies and indexes.
Contents: v. 1. Gas/Robert R. Raber, editor;
v. 2. Liquid/Peter R. Johnston and Hans G. Schroeder,
editors.
   1. Filters and filtration—Congresses. 2. Fluids—
Congresses. I. Raber, Robert R. II. Johnston,
Peter R. III. Schroeder, Hans G. IV. ASTM Committee
F-21 on Filtration. V. Filtration Society (Great
Britain). American Program Committee. VI. Series.
TP156.F5F576 1986   660.2'84245   86-22237
ISBN 0-8031-0926-1 (set)
ISBN 0-8031-0945-8 (v. 1)
ISBN 0-8031-0946-6 (v. 2)

NOTE

The Society is not responsible, as a body,
for the statements and opinions advanced
in this publication.

Printed in Baltimore, MD
October 1986

Foreword

The symposium on Gas and Liquid Filtration was presented at Philadelphia, Pennsylvania, 20–22 October 1986. The symposium was sponsored by ASTM Committee F-21 on Filtration and The American Program Committee of The Filtration Society. Robert R. Raber, Farr Co., served as chairman of the Gas Filtration sessions, and Peter R. Johnston, Ametek, Inc., and Hans G. Schroeder, International Consultants Association, served as chairmen of the Liquid Filtration sessions. Mr. Raber is also the editor of this publication (Volume I—Gas), while Mr. Johnston and Mr. Schroeder are co-editors of Volume II—Liquid.

Related
ASTM Publications

Fluid Filtration: Liquid (Volume II), STP 975 (1986) 04-975002-39.

A Note of Appreciation
to Reviewers

The quality of the papers that appear in this publication reflects not only the obvious efforts of the authors but also the unheralded, though essential, work of the reviewers. On behalf of ASTM we acknowledge with appreciation their dedication to high professional standards and their sacrifice of time and effort.

ASTM Committee on Publications

Contents

Overview   ix

ANALYSIS OF FUNDAMENTAL FILTRATION MECHANISMS

Air Filtration by Fibrous Media—BENJAMIN Y.H. LIU AND
    KENNETH L. RUBOW   1

Numerical Modelling of Electrically Enhanced Fibrous
    Filtration—FRANK S. HENRY AND TEOMAN ARIMAN   13

Key Parameters Used in Modelling Pressure Loss of Fibrous Filter
    Media—DONALD R. MONSON   27

Particle Collection Mechanisms Pertinent to Granular Bed
    Filtration—KENNETH W. LEE   46

Effect of Particle Deposition on the Performance of Granular
    Filters—CHI TIEN   60

Characteristics of Membrane Filters for Particle
    Collection—KENNETH L. RUBOW AND BENJAMIN Y.H. LIU   74

APPLICATIONS AND TESTING: FLAT MEDIA TESTING

Quantifying the Porous Structure of Fabrics for Filtration
    Applications—BERNARD MILLER, ILYA TYOMKIN, AND
    JOHN A. WEHNER   97

The Dustron: An Automated Flat Sample Filtration Performance
    Testing System—JAMES E. MOULTON AND
    JAMES C. WILSON, SR.   110

Automated Systems for Filter Efficiency Measurements—
    RICHARD J. REMIARZ, JUGAL K. AGARWAL, AND
    BRIAN R. JOHNSON   127

A Standard Test Method for Initial Efficiency Measurements on
    Flatsheet Media—ROBERT M. NICHOLSON   141

Results of ASTM Round Robin on Filter Efficiency Using Latex
Spheres and on Airflow Resistance—DANIEL A. JAPUNTICH   152

APPLICATIONS AND TESTING: RESPIRATORS

Respirator Filtration Efficiency Testing—ERNEST S. MOYER   167

Quantitative Fit Testing of Respirators: Past, Present,
Future—WARREN R. MYERS   181

APPLICATIONS AND TESTING: FILTRATION FOR OCCUPIED SPACES

Filtration As a Method for Air Quality Control in Occupied
Spaces—JAMES E. WOODS AND BRIAN C. KRAFTHEFER   193

Air Filter Testing: Current Status and Future
Prospects—RICHARD D. RIVERS AND DAVID J. MURPHY   214

Development of an Artificial Dust Spot Technique for Use with
ASHRAE 52-76—ROBERT R. RABER   229

APPLICATIONS AND TESTING: PROTECTION OF EQUIPMENT

Time Resolved Measurements of Industrial Pulse-Cleaned Cartridge
Dust Collectors—BRUCE McDONALD, ROBIN E. SCHALLER,
MARK R. ENGEL, BENJAMIN Y.H. LIU, DAVID Y.H. PUI,
AND TODD W. JOHNSON   241

Filter Tests to Support Gas Turbine Applications—R. BRUCE TATGE   257

A Review of the Use of SAE Standard J726 in Heavy Duty Engine Air
Cleaner Testing—ROBERT M. NICHOLSON AND
LLOYD E. WEISERT   266

APPLICATIONS AND TESTING: PROTECTION OF THE ENVIRONMENT

Industrial Fabric Filter Bag Test Methods: Usefulness and
Limitations—JOHN D. McKENNA AND LEIGHTON H. HALEY, JR.   277

The Need for Standard Test Methods for Fiberglass Fabrics Used on Utility Coal-Fired Boilers—LARRY G. FELIX, DAN V. GIOVANNI, ROBERT H. GHERI, AND WALLACE B. SMITH   292

Fabric Filter Design: The Case of a Missing Parameter— WAYNE T. DAVIS   302

Electrostatic Enhancement of Fabric Filtration— DOUGLAS W. VAN OSDELL AND ROBERT P. DONOVAN   316

Progress in Fabric Filtration Technology for Utility Applications—DUANE H. PONTIUS AND WALLACE B. SMITH   332

New Filter Efficiency Tests Being Developed for the Department of Energy—RONALD C. SCRIPSICK   345

Standards for Pressure Drop Testing of Filters as Applied to HEPA Filters—DOUGLAS E. FAIN   364

APPLICATIONS AND TESTING: PROTECTION OF PROCESSES

Air Cleanliness Requirements for Clean Rooms—J. GORDON KING   383

Air Cleanliness Validation for Cleanrooms—JAMES L. FLANNERY AND JAMES P. WALCROFT   390

Mechanisms and Devices for Filtration of Critical Process Gases— MAURO A. ACCOMAZZO AND DONALD C. GRANT   402

Index   421

Overview

1986 marks the tenth anniversary of ASTM Committee F-21 on Filtration. Organized on 10-11 March 1976 in Philadelphia, Committee F-21 has worked within the following scope: "The development of test methods, performance specifications, practices, definitions, and classifications, and the stimulation of research to support performance standards for filtration systems, media, and equipment."

In the decade since the committee's founding, significant advances have been made in particle measurement technology and in the overall state of the art in filtration. These facts led the Executive Subcommittee of F-21 to conclude that a symposium on Gas and Liquid Filtration would be most timely. This Special Technical Publication (STP 975) contains 45 refereed technical papers that were reviewed and revised prior to their presentation at the 20-22 October 1986 symposium. STP 975 is organized into Volume I on Gas Filtration with 29 papers and Volume II on Liquid Filtration with 16 papers.

Volume I was sponsored by Subcommittee F21.20 on Gas Filtration. From its beginnings in 1976, F21.20 has had strong research interests and brought together many, diverse technical experts twice each year to develop full consensus methods. The first six years of research and development resulted in the publication in 1982 of ASTM F 778, Standard Methods for Gas Flow Resistance Testing of Filtration Media. The next principal effort of the subcommittee was directed toward developing a standard for measuring filtration media efficiency versus size using polystyrene latex spheres. This standard is now in its final draft stages. In 1982, F 21.20 also undertook the task of converting Federal Standards FF-300 and FF-310 on air conditioning filters into consensus standards. The ASTM version of FF-300 was published as ASTM F 872 Standard Specification for Filter Units, Air Conditioning: Viscous-Impingement Type, Cleanable in 1984; the FF-310 equivalent is now in committee ballot.

The papers in this volume are divided into two sections: Section I, Analysis of Fundamental Filtration Mechanisms; and Section II, Applications and Testing. Section I has been organized by Dr. Kenneth Rubow of the University of Minnesota. It consists of six papers: three on fibrous filtration, two on granular bed filtration, and one on membrane filtration. The topics covered include clean filter efficiency, pressure drop, electrostatic enhancement, and effects of particle loading.

Section II consists of 23 papers. It is divided into six subsections on flat media testing, respirators, filtration for occupied spaces, protection of equipment, protection of the environment, and protection of processes. These topical subsections have been organized by Mr. James C. Wilson of Hollingsworth & Vose Co. (Flat Media Testing); Mr. Daniel A. Japuntich of 3M Corp. (Respirators); Dr. James E. Woods of Honeywell, Inc. and Mr. Richard D. Rivers of EQS, Inc. (Filtration for Occupied Spaces); Mr. Robert M. Nicholson of Donaldson, Inc. (Protection of Equipment); Dr. Wayne T. Davis of the University of Tennessee (Protection of the Environment); and Dr. Alvin Lieberman of Hiac-Royco, Inc. (Protection of Processes). Dr. Vern Bergman of Lawrence Livermore Laboratory, University of California also assisted in the early organizational phases.

One of the goals of this publication has been to cast theory in a clear and concise format; another goal is to provide a forum for future directions in filter testing. I hope it will provoke an ongoing interest in improving not only test methods but also the means by which filtration equipment is specified.

> Robert R. Raber
> Farr Company
> El Segundo, CA
> Symposium chairman and
> editor

Analysis of Fundamental Filtration Mechanisms

Benjamin Y. H. Liu and Kenneth L. Rubow

AIR FILTRATION BY FIBROUS MEDIA

REFERENCE: Liu, B.Y.H., and Rubow, K. L., "Air Filtration by Fibrous Media," Fluid Filtration: Gas, Volume I, ASTM STP 975, R. R. Raber, American Society for Testing and Materials, Philadelphia, 1986

ABSTRACT: Theories of air filtration by fibrous media have been reviewed and the pertinent equations for calculating filter pressure drop and efficiency summarized. Calculation results have been presented showing the single fiber efficiency as a function of particle size for various fiber sizes and packing densities. The existence of a most penetrating particle size for fibrous media has been demonstrated and shown to vary between 0.015 to 0.9 μm for fiber diameters of 1 to 30 μm, packing densities of 0.01 to 0.3 and filtration velocities of 1 to 100 cm/s.

KEYWORDS: GAS FILTRATION, FIBROUS FILTERS, AEROSOL FILTRATION

Fibrous filters are widely used for the collection and removal of airborne particles. The efficiency and pressure drop of a fibrous filter are affected by the size of the filter fibers, their packing density in the filter media and the velocity of gas flow through the filter. In addition, the filter efficiency also depends on the size of the particles being collected and, to a somewhat lesser extent, particle shape and density.

Numerous authors have contributed to the development of modern aerosol filtration theories. No attempt has been made in this paper to review the different theories in detail. Those interested in an in-depth review may find the book by Davies [1] and the review articles by Kirsch and Stechkina [2] and Pich [3] useful. The purpose of this paper is to summarize some of the useful equations in filtration theory for filter design, and present the calculation results on filter performance based on the use of these equations.

Drs. Liu and Rubow are respectively Professor and Director, and Research Associate and Manager, Particle Technology Laboratory, Mechanical Engineering Department, University of Minnesota, Minneapolis, MN 55455

1

FILTRATION THEORY FOR FIBROUS FILTERS

Of primary importance in the theory of air filtration is the theory for pressure drop and filter efficiency. Both the pressure drop and filter efficiency are important because they combine to determine the overall performance of the filter for practical applications.

Consider a fiber of unit length inside a filter matrix in which it is embeded. The fiber experiences a drag force, F_d per unit fiber length in proportion to the gas velocity, U, and viscosity, μ,

$$F_d = F* \mu U \tag{1}$$

where F* is the dimensionless fiber drag parameter. Equation 1 holds in the so-called Darcy law regime, which is valid for low velocity, small Reynolds number flows such as those encountered in high efficiency particular air (HEPA) filters used in respirators and for clean room air filtration. The overall pressure drop of a flat-sheet filter media is then related to F_d and l_f,

$$\Delta p = F_d \ l_f = F* \mu U \ l_f \tag{2}$$

where l_f is the specific fiber length, or the length of the filter fiber per unit filter area. It is useful to define a filter area factor

$$S = D_f \ l_f \tag{3}$$

in terms of which

$$\Delta p = \frac{F* \mu U S}{D_f} \tag{4}$$

The area factor for a specific filter can be determined by measuring the specific filter mass, i.e. the mass of the filter per unit filter area, the fiber diameter, D_f, and the fiber density, ρ_f, and calculating S using the following equation,

$$S = \frac{4 \ m_f}{\pi \ D_f \ \rho_f} \tag{5}$$

Theory of fibrous filters shows that for flow in the continuum regime where the mean free path of the gas molecules is small compared to the fiber size, F* is a function of the packing density, α, of the filter only, the packing density being related to the specific filter mass, m_f, thickness, L, and density, ρ_f by

$$\alpha = \frac{m_f}{\rho_f L} \tag{6}$$

According to the Kuwabara theory for flow around a cylindrical fiber,

$$F* = \frac{4 \pi}{K} \tag{7}$$

in which K is the hydrodynamic factor

$$K = - \frac{1}{2} \ln \alpha + \alpha - \frac{1}{4} \alpha^2 - \frac{3}{4} \tag{8}$$

The experimental pressure drop for a filter is usually found to be smaller then that predicted by Eq.7. The ratio

$$\epsilon = \frac{(\Delta p)_{theoretical}}{(\Delta p)_{experimental}} \tag{9}$$

is known as the inhomogeneity factor or the effective fiber length factor. Yeh and Liu [4,5] and Lee and Liu [6,7] have shown that $\epsilon \approx 1.6$ for the experimental polyester filters they studied. The factor may differ for other filters depending on the material of the fiber and method of manufacture. Figure 1 shows the relationship between F* and α determined by Monson [8] for filters over a wide range of values of α and the comparison of the data with a "mitered cylinder" pressure drop model develped by him.

The efficiency of filter fiber for particle collection can be described by a single fiber efficiency, η_s, defined as the fraction of particles collected by the fiber from the volume of air geometrically swept out by the fiber. In terms of this single fiber efficiency, the overall efficiency of the filter is

$$\eta = 1 - e^{-\eta_s S} \tag{10}$$

here S is the filter area factor defined earlier. Using the Kuwabara low and a boundary layer analysis, Lee and Liu [6,7] have shown that he theoretical single fiber efficiency for diffusion is

$$\eta_D = 2.58 \left(\frac{1 - \alpha}{K}\right)^{1/3} Pe^{-2/3} \tag{11}$$

ıd that for interception is

$$\eta_R = \left(\frac{1 - \alpha}{K}\right) \frac{R^2}{1 + R} \tag{12}$$

ere Pe is the Peclet number

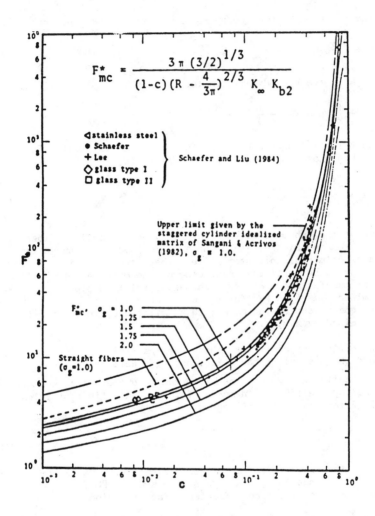

Figure 1.   Dimensionless fiber drag parameter (Monson, 1985)

$$Pe = U D_f / D \tag{13}$$

R is the interception parameter

$$R = D_p / D_f \tag{14}$$

In the above equations, D_p and D_f are respectively the particle and fiber diameters and D is the diffusion coefficient of the particle. In the regime where particle collection by inertial impaction is unimportant, as is in the case of HEPA and high efficiency respirator air filters, one can write as a first approximation

$$\eta_s = \eta_D + \eta_R \tag{15}$$

Using the theoretical expressions for η_D and η_R in Eq. 11 and 12 and multiplying both sides of Eq. 15 by $Pe\, R/(1 + R)^{1/2}$ we have

$$\eta_s\, Pe\, R\, /(1 + R)^{1/2} = 2.58\, (\frac{1 - \alpha}{K})^{1/3}\, Pe^{1/3}\, R/(1 + R)^{1/2}$$

$$+ \left[(\frac{1 - \alpha}{K})^{1/3}\, Pe^{1/3}\, R/(1 + R)^{1/2} \right]^3 \tag{16}$$

Equation 16 suggests that the quantity, $\eta_s\, Pe\, R/(1 + R)^{1/2}$, on the left hand side of the equation should be a single valued function of the quantity, $[(1-\alpha)/K]^{1/3}\, Pe^{1/3}\, R/(1 + R)^{1/2}$, on the right hand side of the equation. The actual relationship between these two quantities for a real filter can thus be obtained by plotting the experimental single fiber efficiency data with the use of these two dimensionless quantities. Figure 2 shows the result of such a plot by Lee and Liu [6,7] for a Dacron polyester filter ($D_f = 11$ μm) they studied. The correlation covers a particle diameter range of 0.034 to 1.3 μm, a velocity range of 1 to 30 cm/s and a packing density range of 0.0086 to 0.1513. Further, the actual relationship for the filter can be represented by the equation

$$\eta_s = \frac{2.58}{\epsilon} (\frac{1 - \alpha}{K})^{1/3}\, Pe^{-2/3} + \frac{1}{\epsilon} (\frac{1 - \alpha}{K})\, \frac{R^2}{1 + R} \tag{17}$$

where $\epsilon = 1.6$ is the inhomogeneous factor mentioned previously. In the absence of data for a real filter, the above equation can be used for predictive purposes.

PERFORMANCE OF FIBROUS FILTERS

The influence of filter fiber size, filter packing density and filtration velocity on filter efficiency has been calculated by means

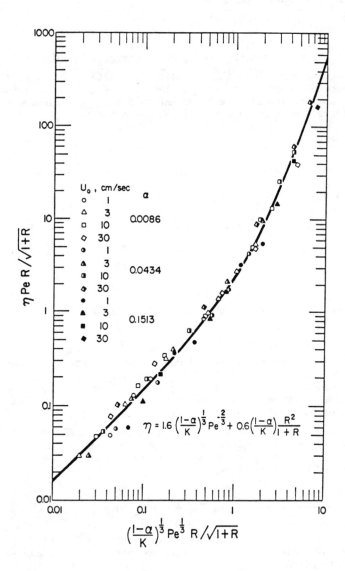

Figure 2. Lee and Liu correlation for filter efficiency

of Eq. 17. The results are shown in the series of figures presented below.

Figure 3 to 6 shows the single fiber efficiency as a function of particle size for various filter fiber sizes. The velocity is assumed to be constant at 10 cm/s and the packing density α has been set to 0.01, 0.03, 0.1 and 0.3 respectively. The result shows that the single fiber efficiency first decreases with increasing particle size in the diffusion regime and increases with increasing particle size in the interception regime. The efficiency reaches a minimum in the intermediate particle size range and this minimum generally occurs at a particle size of 0.15 to 0.9 μm depending on the size of the filter fiber and the packing density. In the case of $\alpha = 0.1$, for instance, the minimum efficiency is reached at a particle size of approximately 0.15 μm for $D_f = 1$ μm, increasing to 0.65 μm for $D_f = 30$ μm. In the case of $\alpha = 0.01$, this most penetrating particle size, $\hat{D}_p$ varies from approximately 0.2 μm to 0.9 μm for filter fiber sizes of 1 to 30 μm.

The influence of filtration velocity on a filter efficiency is shown more clearly in Figure 7 in which D_f and α have been kept constant at 3 μm and 0.03 respectively. Increasing the velocity is seen to cause a decrease in both the efficiency and the most penetrating particle size. This is consistant with experimental observations. Similar results for $D_f = 30$ μm and $\alpha = 0.03$ are shown in Figure 8.

The influence of filter packing density, α, on filter efficiency is shown in Figure 9, the filter fiber size and velocity having been kept constant at 3 μm and 10 cm/s respectively. Increasing the packing density is seen to cause the efficiency to increase both in the diffusion and the interception regimes. Further, increasing α causes the most penetrating particle size to decrease. However, the latter effect is not large. Increasing α from 0.01 to 0.3 causes $\hat{D}_p$ to decrease only from 0.35 to 0.2 μm approximately.

SUMMARY AND CONCLUSIONS

While a comprehensive theory for the quantitative prediction of filter efficiency from first principles does not yet exist, the advances in filtration theory and data correlation procedure during the past ten years have reached the point where the influence of filter parameters on filter performance can be predicted with a reasonable degree of accuracy. The pertinent equations for filter performance prediction have been summarized in this paper and results presented showing the performance of filter as a function of filter packing density and filter fiber size for various filtration velocities and particle sizes. The result shows that a most penetrating particle size exists for fibrous filters and this most penetrating size varies from 0.15 to 0.9 μm for filter fiber sizes of 1 to 30 μm, packing densities of 0.01 to 0.3 and velocities of 1 to 100 cm/s.

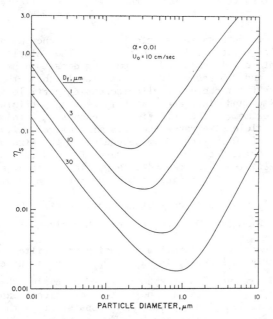

Figure 3. Single fiber efficiency for α =0.01 and U_0=10 cm/s

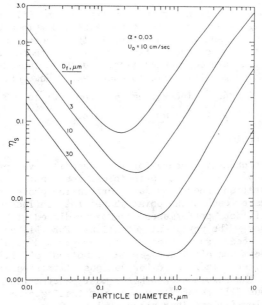

Figure4. Single fiber efficiency for α=0.03 and U_0=10 cm/s

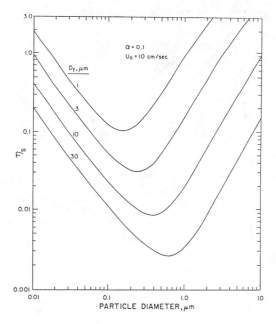

Figure 5. Single fiber efficiency for α =0.1 and U_o= 10 cm/s

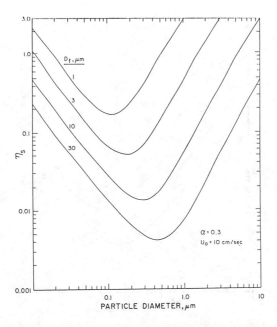

Figure 6. Single fiber efficiency for α=0.3 and U_o=10 cm/s

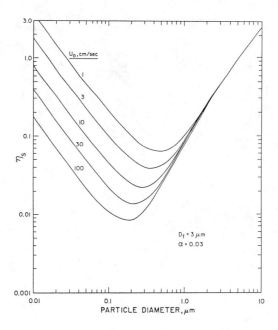

Figure 7. Single fiber efficiency for D_f=3 μm and α =0.03

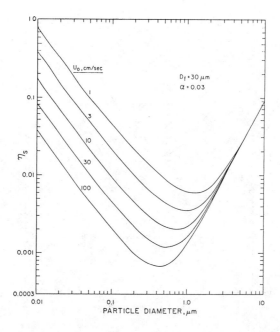

Figure 8. Single fiber efficiency for D_f=30 μm and α =0.03

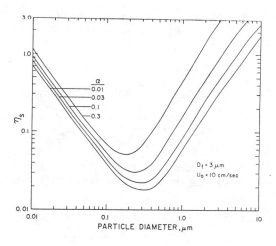

Figure 9. Single fiber efficiency for D_f=3 μm and U_o=10 cm/s

ACKNOWLEDGEMENT

This research is supported by a grant from the Center for Disease Control, No. 5 R01 OH01485. The support of the research by the Center is gratefully acknowledged.

REFERENCE

[1] Davies, C.N., *Air Filtration*, Academic Press, London, 1973

[2] Kirsch, A.A. and Stechkina, I.B., "The Theory of Aerosol Filtration with Fibrous Filters," in *Fundamentals of Aerosol Science* (D. T. Shaw, Ed.), Wiley, New York, 1978

[3] Pich, J., "Theory of Aerosol Filtration by Fibrous and Membrane Filters," in *Aerosol Science* (C.N. Davies, Ed.), Academic Press, London, 1966.

[4] Yeh, H.C., and Liu, B.Y.H., "Aerosol Filtration by Fibrous Filters--Part I. Theoretical," *J. Aerosol Sci*. 5:191-204, 1973.

[5] Yeh, H.C., and Liu, B.Y.H., "Aerosol Filtration by Fibrous Filters--Part II. Experimental," *J. Aerosol Sci*. 5:205-217, 1973.

[6] Lee, K.W., and Liu, B.Y.H., "Experimental Study of Aerosol Filtration by Fibrous Filters," *Aerosol Sci. Technol.* 1:35-46, 1982.

[7] Lee, K.W., and Liu, B.Y.H., "Theoretical Study of Aerosol Filtration by Fibrous Filters," *Aerosol Sci. Technol*. 1:147-162, 1982.

[8] Monson, D.R., "A Filter Pressure Loss Model for Uniform Fibers," in *Aerosols: Science, Technology and Industrial Applications* (B.Y.H. Liu, D.Y.H. Pui and H.J. Fissan, Ed.), Elsevior, New York, 1984.

F. S. Henry and Teoman Ariman

NUMERICAL MODELLING OF ELECTRICALLY ENHANCED FIBROUS
FILTRATION

REFERENCE: Henry, F. S., and Ariman, T. "Numerical
Modelling of Electrically Enhanced Fibrous Filtration,"
Fluid Filtration: Gas, Volume I, ASTM STP 975, R. R.
Raber, Ed., American Society for Testing and Materials,
Philadelphia, 1986

ABSTRACT: Numerical solutions of the flow through a model
fibrous filter are used to study electrically enhanced
filtration. It is shown that numerical modelling, or
simulation, has advantages over both the analytical and
the experimental approach. Particle trajectories are presented, and filter efficiencies are predicted and compared
to experimental data. The dendritic structure of the particle deposit is also investigated. Differences in
dendrite structure seen in the laboratory between enhanced
and non-enhanced collection are also simulated.

KEYWORDS: fibrous filtration, numerical modelling,
collection efficiency, dendrite growth

The performance of a fibrous filter is characterized by three
parameters: collection efficiency, pressure drop or flow resistance,
and filter life. A filter has a finite useful life because the
resistance to flow eventually increases to an unacceptable level as
the filter's pores become clogged with deposited particles.
Prediction of filter performance is still largely an empirical process
because the mechanics of transient collection in fibrous filters is
still only poorly understood.

It has been shown that placing a fibrous filter in an electric
field can have beneficial effects on both collection efficiency and
pressure drop. Zebel [1] did some of the initial analytical work in
this area, and was able to show that the collection efficiency of both
charged and neutral particles would be increased by the application of
an electric field. Henry and Ariman [2,3] extended Zebel's findings
by considering the effects of neighboring fibers on the flow and
electric fields within a filter. Several researchers have performed
experimental investigations of electrically enhanced fibrous filtra-

Drs. Henry and Ariman are members of the faculty of the
Department of Mechanical Engineering, The University of Tulsa, Tulsa,
OK 74104.

tion. For instance, Bergman et al. [4], performed a series of
measurements on enhanced filters, and reported increased collection
and reduced resistance to flow.

While it is possible to mesure the gross features of filtration
in the laboratory, the structure of fibrous filters makes it difficult
to empirically study the filtration process on a microscopic level.
It is at this level that numerical modelling, or simulation, has an
advantage. The equations governing the motion of small particles in
laminar flow are well understood, and while no general solution exists
for these equations, it is possible, in theory at least, to solve them
to any degree of accuracy using numerical techniques. In this paper
some of the authors' work in this area will be reviewed.

MODEL FORMULATION

Filter Model

A fibrous filter consists of a mass of fibers that are randomly
orientated in a plane that is approximately normal to the mean flow
direction. The random orientation of the fibers led some [5] to consider a stochastic approach to modelling the flow through a fibrous
filter. However, most researchers have opted for a more determanistic
approach by confining their analysis to the flow in the immediate
neighborhood of an individual fiber, and defining a reasonably simple
model of the filter's structure.

Initial theoretical studies of fibrous filtration concentrated on
the flow around a single isolated cylinder. This approach neglects
the influence of the surrounding fibers. The work of Zebel [1] is
typical of this approach. In an attempt to include the effect of the
neighboring fibers, Henry and Ariman [2,3], among others, used the
cell model of Kuwabara [6] to describe the flow around individual
fibers. Kuwabara's cell model is an approximate solution for slow
flow through an array of parallel cylinders.

While the Kuwabara cell model has been shown [6] to be capable of
predicting the pressure drop experienced by a fluid moving slowly
through an array of parallel cylinders, it does not give a complete
description of the flow field within the array. Henry and Ariman [7]
have suggested that this and other deficiencies, limits the utility of
Kuwabara's solution as a filter model.

In an attempt to overcome the limitations of the available analytical flow models, such as Kuwabara's cell model, Henry and Ariman
[8,9,10] solved the flow through an array of parallel cylinders using
a numerical approach. The basic model is shown in Fig. 1. While only
nine cylinders are shown, the array is assumed to be infinite in
extent in all directions. This assumption reduces the problem to the
solution of the flow through the domain shown as a dotted line in Fig.
1. In fact, if all cylinders are assumed to have the same diameter,
the solution domain length can be reduced to one-half of that shown.

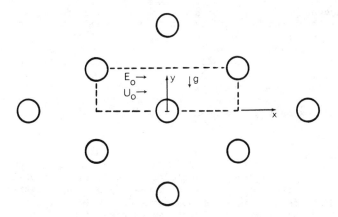

FIG. 1—Array model and solution domain

Flow Field

The equations governing the two-dimensional flow of a Newtonian fluid with constant properties can be given in the following form;

$$\nabla^2 \psi = -\omega \tag{1}$$

and

$$\overline{u} \cdot \nabla \omega = \nu \nabla^2 \omega \tag{2}$$

where ψ is the stream function, ω is the vorticity, u is the velocity, and ν is the kinematic viscosity. The overbar denotes a vector quantity.

The vorticity is the curl of the velocity, and for two-dimensional flow, can be given in Cartesian coordinate as,

$$\omega = \frac{\partial v}{\partial x} - \frac{\partial u}{\partial y} \tag{3}$$

The flow velocity components can be recovered from the stream function using,

$$u = \frac{\partial \psi}{\partial y} \quad , \text{ and } \quad v = -\frac{\partial \psi}{\partial x} \tag{4}$$

The advantage of using the above formulation of the flow equations is that the pressure does not appear, and hence, the number of independent variables is reduced from three to two. However, the pressure drop through the solution domain can be recovered from the stream function and vorticty solutions. It should be noted that Stokes'approximation of vanishing flow inertia has not been invoked in the formulation of the above flow model, and hence, the resulting solutions have more general application than those generated using slow flow approximations, such as Kuwabara' cell model.

Electrical Forces

The presence of a group of dielectric fibers in an otherwise undisturbed electric field causes the lines of constant electric force to become distorted. This is shown schematically in Fig. 2. The electric field polarizes the fibers, and charged particles will be attracted to the side of the fiber that is of opposite polarity to that of the particle. This process in known as electrophoresis. The electrophoretic force, F_e, on a particle of charge q in an electric field of strength E can be given as,

$$\overline{F}_e = q \, \overline{E} \tag{5}$$

If the particles are polarizable they will be attracted to areas of higher electric field density. The areas of highest electric field density will exist at certain locations on the surface of the fibers, and hence, the particles will be attracted to the fibers. This force is known as the dielctrophorectic force, F_d, and can be given as,

$$\overline{F}_d = 2\pi r_p^3 \, \varepsilon_o \frac{\varepsilon_p - 1}{\varepsilon_p + 2} \, \nabla |E|^2 \tag{6}$$

where ε_o is the permittivity of free space, and ε_p is the dielectric constant of the particle.

It can be shown [9] that electrophoresis is more effective at enhancing collection of particles with diameters of .5 microns or less, while dielectrophoreis is more effective for larger particles. While there are other electrical forces that may be important in non-enhanced filters, it can be expected [9] that the above are the domi-nant enhancing mechanisms for filters with applied electric fields.

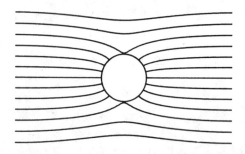

FIG 2--The electric field surrounding a dielectric fiber

NUMERICAL PREDICTIONS

Particle Trajectories

Fuchs [11] has given the governing equations for spherical particles in plane motion. These equations can be written as,

$$St \frac{d}{dt} U_p = U - Up + \beta F_x / U_o \tag{7}$$

and

$$St \frac{d}{dt} V_p = V - V_p + \beta F_y / Uo \tag{8}$$

where the Stokes number, St, is defined as,

$$St = \frac{2 U_o C r_p^2 \rho_p}{9 \mu r_f} \tag{9}$$

and the mechanical mobility of the particle, β, is given as

$$\beta = \frac{C}{6 \pi \mu r_p} \tag{10}$$

U_p and V_p are the x- and y-components of the particle velocity, U and V are the x- and y-components of the flow velocity, F_x and F_y are the components of the sum of all body forces acting on the particle, U_o is the mean flow velocity, C is the Cunningham slip factor, r_p is the particle radius, r_f is the fiber radius, ρ_p is the particle density, and ν is the fluid viscosity. Cunningham's slip factor is included to allow for slip flow at the particle's surface.

In general Eqs. 7 and 8 do not have analytical solutions and numerical techniques have to be used. Henry and Ariman [9] used a predictor-corrector method to solve the particle equations. At each point in the integration the flow velocity components were estimated from the previously calculated flow field. The flow field was calculated by constructing finite-difference versions of Eqs. 1 and 2, and the resulting equations were solved using Gauss-Sidel iteration with point SOR. Further details of the numerical solution of the flow field can be found in Henry and Ariman [10]. The body forces, F_x and F_y, were assumed to consist of only elctrical forces and gravity. The electrical forces were calculated using Eqs. 5, and 6. The electric field was calculated using the method of images.

Typical particle trajectories calculated using the numerical techniques desribed above are shown in Fig. 3. These calculations represent the initial stages of filtration in that the fibers are assumed to be free from previously deposited particles. The particles are assumed to have a density of 2165 kg/m^3 (NaCl). Three separate trajectories are shown in both parts of the figure. The particle shown hitting the fiber is the enhanced case. The trajectory shown by a broken line is for the non-enhanced case. The particle trajectory shown by a dotted line is assumed to have the same mass as the carrier fluid; i.e., this line represents a fluid streamline. The particle diameter is the same for all three trajectories.

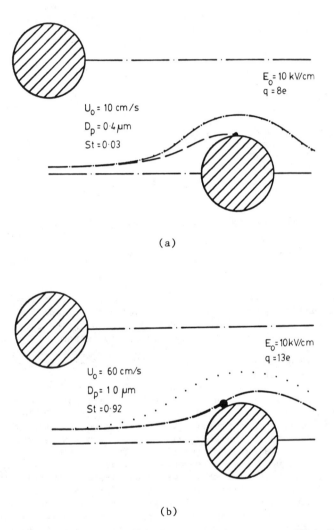

(a)

(b)

FIG. 3--Particle trajectories with and without an applied electric field.

The trajectoires shown in part a of Fig. 3 are for a particle diameter of 0.4 microns, and it can be seen that the application of an electric field greatly enhances collection. The mean flow velocity in

this case is 10 cm/s, and as the non-enhanced particle trajectory closely follows the fluid streamline, it can be assumed that particle
inertia is insignificant in this situation. The trajectories shown in
part b of Fig. 3 are for particles of one micron in diameter, and the
mean flow velocity is 60 cm/s. It can be seen that particle inertia
is more pronounced in this case, and that the effect of the applied
electric field is less significant.

By calculating individual trajectories for various particle
diameters Henry and Ariman [9] were able to estimate initial collection efficiencies for a typical filter. In Fig. 4 the predicted efficiencies for enhanced and non-enhanced collection are compared to the
experimental data of Bergman et al. [4]. The filter was composed of
fibers with mean diameters of 3.75 microns, and was 1 cm thick with a
packing density of 0.008. The mean flow velocity was 66 cm/s. It can
be seen that the numerical predictions match the trends of the data
including giving approximately the same values of most penetrating
particle size.

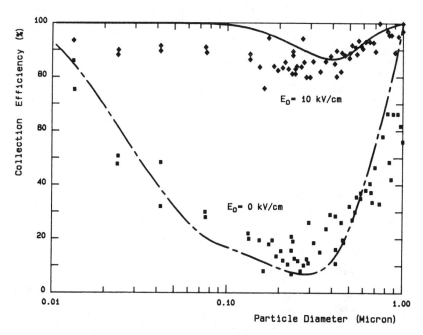

FIG. 4--Predicted filter efficiency compared with the measurements of
        Bergman et al [4].

Pressure Drop in Clean Filters

There are several analytical solutions for the pressure drop
through an array of parallel cylinders. Kuwabara's model, and the
more recent work of Sangani and Acrivos [12] are but two examples.
All such solutions assume that the flow is two dimensional, and that
the Reynolds number of the flow is so small that the flow's inertia

can be ignored. As an array of parallel cylnders is a plausible model
for a fibrous filter these models have been used to predict the
pressure drop experienced by the flow as it passes through a clean
filter.

Henry and Ariman [10] used their numerical model of the flow
though an array of parallel cylinders to check the validity of some of
these analytical solutions and found that they gave reasonable predictions within the limitation of small Reynolds number. Predictions of
the dimensionless drag ($F/\mu U_o$, where F is the drag per unit length of
cylinder) versus distance between cylinders are given in Figs. 5 and 6
It can be seen that for a Reynolds number of one tenth (Fig. 5) the
numerical solution and the analytical solution due to Sangani and
Acrivos [12] are identical. Smilar results would have been found
using Kuwabara's model. However, when the Reynolds number is
increased to unity (Fig. 6), the numerical solution predicts higher
values of drag than does the analytical soluton. This indicates that
inertia is significant even at this low Reynolds number. Also
included in both figures is an analytical solution due to Kaplun [13]
for an isolated cylinder. Kaplun's solution does take the flow's
inertia into account in an approximate manner. It can be seen that a
prediction of the drag, or pressure drop, using an isolated cylinder
model can be expected to be seriously in error when the average
distance between cylinders is small.

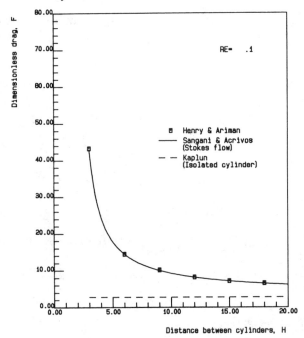

FIG. 5--Drag experienced by an individual cylinder in a square array.
        Re= 0.1.

While it has been demonstrated that various analytical solutions
adequately describe slow flow through an array of parallel cylinders,
it has been shown [8] that none can predict the pressure drop through

a clean fibrous filter to any degree of accuracy. Hence, the validity of using a parallel array of cylinders as a model of a fibrous filter is questionable. The problem would appear to be a result of modelling a complex, three-dimensional flow field as an ordered two-dimensional flow. The actual problem is further complicated by the fact that the actual flow field boundaries are constantly changing as the particles form dendritic structures on the surfaces of the fibers.

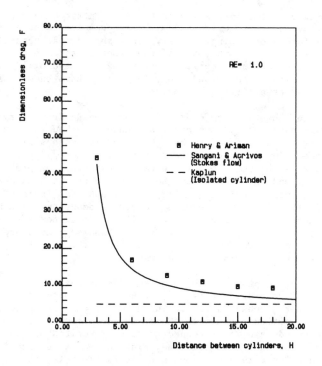

FIG. 6-Drag experienced by an individual cylinder in a square array. Re= 1.0.

Dendrite Growth

The vast majority of filtration studies to date have considered the collection of particles on clean fibers. While this is instructive in determining the magnitude of the various collection mechanisms it only addresses the initial stages of the filtration process. The mechanics of the initial stages of enhanced filtration are now fairly well understood, but the later stages have only recently been considered, and much remains to be resolved. How and why denrites are formed, an how the presence of an electric field affects their formation are areas that require further research.

Billings [14] was apparently the first to exprimentally investigate the formation of dendrites on single glass fibers, and he was able to show that the collection efficiency increased as more par-

ticles were deposited on the fiber. Groups lead by Payatakes and by
Tien have reported several analytical investigations of transient
filtration. See, for example; Payatakes and Gradon [5], and Tien et
al [16]. Analytical investigations of transient filtration have
tended to be concentrated on the formation of dendrites, and have not
progressed to the point where collection efficiency and pressure drop
can be predicted without the aid of empirical correlations.

Most investigations of dendrite growth have concentrated on non-
enhanced filtration, but Nielsen and Hill [17] considered collection
by electrical forces on a single fiber. They showed that dendrites in
enhanced filters tend to cover more of the fiber surface than do the
dendrites in non-enhanced filters. However, Neilsen and Hill [17]
used a two-dimensional model, and Tousi et al [18] have shown that
this produces rather artificial dendrites.

It has been shown experimentally [3] that the dendrite formations
in electrically enhanced filters are less densely packed than in non-
enhanced filters. It is thought that this fact is responsible for the
flow resistance being less in enhanced filters. Practically all
theoretical studies of dendrite growth have considered the problem as
two dimensional, and this restricts the structure of the predicted
dendrite formations. Tousi et al. [18] have shown that it is difficult to tell the difference between individual dendrites in enhanced
filters and those in non-enhanced filters if the dendrites are forced
to grow in a plane. However, Tousi et al [18] showed that there are
noticeable differences in dendrite structure when the dendrites are
allowed to grow in a three-dimensional manner.

Examples of predicted three-dimensional dendrites are shown in
Figs. 7 and 8. Part a of each figure represents the non-enhanced
case, and part b the enhanced case. The dendrite patterns were predicted using a modified version of the array model described above.
The computational domain was extended to allow dendrites to grow
across the lower stagnation streamline. Initial particle locations
were specified using a random number generator. That is, the y- and
z-coordinates of the particle as it entered the collection cell were
randomly determined. Once the particle's z-coordinate was specified,
it was assumed to remain constant. The coordinates of all deposited
particles were stored, and the trajectory of the approaching particle
was constantly checked to ascertain whether it had hit a previously
deposited particle or the fiber. Because of the random nature of the
simulation, several simulatons were performed for a given set of parameters. The dendrite patterns given in Figs. 7 and 8 were picked as
representative of a particular set.

Figs. 7 and 8 clearly indicate the differences in structure between enhanced and non-enhanced dendrites. It is believed that the
enhanced particles cover more of the cylinder's surface because their
trajectories are flatter. The left-hand views of each figure show
that the non-enhanced particles are more densely packed into fewer
dendrites than are the enhanced particles. The non-enhanced dendrites
also protrude further into the flow. These two features could be
responsible for the resistance to flow being greater for non-enhanced
filters.

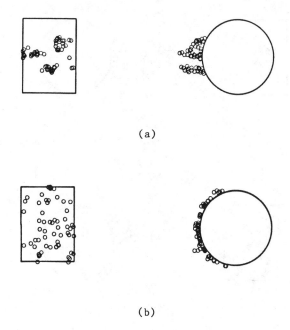

(a)

(b)

FIG. 7—Predictions of dendrite formation for 0.5 micron diameter particles. (a) without enhancement, (b) with enhancement.

Figs. 7 and 8 clearly indicate the differences in structure between enhanced and non-enhanced dendrites. It is believed that the
enhanced particles cover more of the cylinder's surface because their
trajectories are flatter. The left-hand views of each figure show
that the non-enhanced particles are more densely packed into fewer
dendrites than are the enhanced particles. The non-enhanced dendrites
also protrude further into the flow. These two features could be
responsible for the resistance to flow being greater for non-enhanced
filters.

To date, no purely analytical predictions of the time-dependent
nature of the resistance to flow have been made. As it has been shown
that the resistance to flow through clean filters is poorly predicted,
it is to be expected that such predictions will not be forthcoming in
the near future. It is conceivable that the complex structure of a
fibrous filter can be approximated by some fairly simple three-
dimensional model, which will give reasonable predictions of pressure
drop for clean filters, but modelling the changes in the flow field
that would occur in the neighborhood of growing dendrites will probably be beyond the capability of numerical modelling for some time to
come.

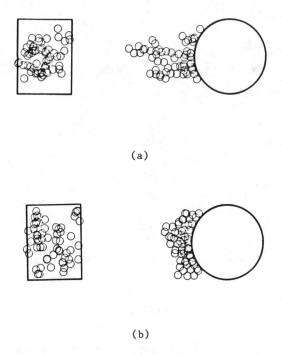

(a)

(b)

FIG. 8--Predictions of dendrite formation for 1.0 micron diameter particles. (a) without enhancement, (b) with enhancement.

CONCLUSION

Some numerical calculation procedures used to investigate
electrically enhanced fibrous filtration have been reviewed. Various
problems associated with modelling the complex structure of a fibrous
filter have been outlined, and their influence on the prediction of
collection efficiency and the resistance to flow of enhanced filters
has been discussed. It was shown that numerical modelling of enhanced
filtration can be used to predict the important features of particle
collection with a fair degree of accuracy. However, the resistance to
flow cannot be accurately predicted with currently available models.
Some recent calculations of dendrite formation have been discussed,
and it was suggested that the three-dimensionality of the filtration
process should not be ignored.

ACKNOWLEDGEMENT

This paper is a partial outcome of the research program supported by the National Science Foundation under Grant No. CPE-8101947 awarded to the University of Tulsa. The authors acknowledge the continued interest and advice of Dr. M. S. Ojalvo, Director of the Particulate and Multiphase Processes program of the National Science Foundation.

REFERENCES

[1] Zebel, G., "Deposition of Aerosol Flowing Past a Cylindrical Fiber in a Uniform Electric Field,'' Journal of Colloidal and Interface Science, Vol. 20, No. 6, 1965, pp. 522-543.

[2] Henry, F.S., and Ariman, T., "Cell Model of Aerosol Collection by Fibrous Filters in an Electrostatic Field," Journal of Aerosol Science, Vol. 12, No. 2, 1981 pp. 91-103.

[3] Henry, F.S., and Ariman, T., "The Effect of Neighboring Fibers on the Electric Field in a Fibrous Filter," Journal of Aerosol Science, Vol. 12, No. 2, 1981, pp. 137-149.

[4] Bergman, W., Hebard, H., Taylor, R., and Lum, G., "Electrostatic Filters Generated by Electric Fields," Proceedings of the Second World Filtration Conference, London, 1979.

[5] Spielman, L., and Goren, S.L., "Model for Predicting Pressure Drop and Filtration Efficiency in Fibrous Filter Media," Environmental Science and Technology, Vol. 2, No. 4, April 1968, pp. 279-287.

[6] Kuwabara, S., "The Forces Experienced by Randomly Distributed Parallel Circular Cylinders or Spheres in Viscous Flow at Small Reynolds Numbers," Journal of the Physics Society of Japan, Vol. 14, No. 4, 1959, pp. 527-532.

[7] Henry, F.S., and Ariman, T., "An Evaluation of the Kuwabara Model," Particlulate Science and Technology, an International Journal, Vol. 1, No. 1, 1983, pp. 1-20.

[8] Henry, F.S., and Ariman, T., "A Staggered Array Model of a Fibrous Filter with Electrical Enhancement," Particlulate Science and Technoloqy, an International Journal, Vol. 1, No. 2, 1983, pp. 139-154.

[9] Henry, F.S., and Ariman, T., "Numerical Calculation of Particle Collection in Electrically Enhanced Fibrous Filters," accepted for publication in Particulate Science and Technology, an International Journal.

[10] Henry, F.S., and Ariman, T., "Numerical Calculation of Flow Through Arrays of Parallel Cylinders," to be submitted to Particulate Science and Technology, an International Journal.

[11] Fuchs, N.A., The Mechanics of Aerosols, Macmillan, New York, 1964.

[12] Sangani, A.S., and Acrivos, A., "Slow Flow Past Periodic Arrays of Cylinders with Application to Heat Transfer," International Journal of Multiphase Flow, Vol. 8 No. 3, 1982, pp. 193-206.

[13] Kaplun, S. "Low Reynolds Number Flow Past a Circular Cylinder," Journal of Mathematics and Mechanics." Vol. 6, No. 5, 1957, pp 595-603.

[14] Billings, C.E., "Effects of Particle Accumulation in Aerosol Filtration," Ph. D. Dissertation, California Institute of Technology, Pasadena, CA, 1966.

[15] Payatekes, A.C., and Gradon, L., "Dendritic Deposition of Aerosol Particles in Fibrous Media by Inertial Impaction and Interception," Chemical Engineering Science, Vol. 35, 1980, pp 1083-1095.

[16] Tien C., Wang, C.S., and Barot, D.T., "Chain-like Formation of Particle Deposits in Fluid-particle Separation," Science, Vol. 196, 1977, pp. 983-985.

[17] Nielsen, K.A., and Hill, J.C., "Particle Chain Formation in Aerosol Filtration with Electrical Forces," AIChE Journal, Vol. 26, 1980, pp. 678-679.

[18] Tousi, S., Henry, F.S., and Ariman, T.,"Simulation of Electrically Enhanced Fibrous Filtration," Proceedings of the First World Congress in Particle Technology Nuremburg, April 16-18, 1986.

Donald R. Monson

KEY PARAMETERS USED IN MODELING PRESSURE LOSS OF FIBROUS
FILTER MEDIA

REFERENCE: Monson, D. R., "Key Parameters used in
Modeling Pressure Loss of Fibrous Filter Media", Fluid
Filtration: Gas, Volume I, ASTM STP 975, R. R. Raber,
Ed., American Society for Testing and Materials,
Philadelphia, 1986

ABSTRACT: Accurate pressure loss modeling of randomly
oriented fibrous filter structures over a broad range
of solidity requires that one accurately account for
the following effects:
1)   Fiber crossings and the resulting effective transverse
boundary proximity and shape parameters and the fiber dynamic
shape factor which all vary with the average solidity.
2)   Slip and transition flow effects on small fibers or at
low gas densities.
3)   Perturbations due to manufacturing processes and multi
modal fiber size distributions which cause fiber curvature,
random orientation and solidity gradients in both the transverse
and axial directions.

Prominent pressure loss models of both the pore and fibrous type
have been found lacking in their ability to predict measured
pressure loss of randomly oriented fibrous filters made of near
uniform size fibers over a broad range of solidity. A mitered
cylinder layered screenlike pressure loss model is described which
is capable of accurately accounting for the effects described.

KEYWORDS: pressure loss, filter media, modeling, slip flow

In his 1976 thesis [1], Schaefer devised a method of fabricating
semi-ideal filter models covering a volume solidity range c from 0.1
to 0.47. The filter structure had essentially uniform diameter fibers
all substantially normal to the flow and were sufficiently well dispersed and of sufficient number of diameters in thickness so that the
structure could be considered to have nearly uniform solidity in the
transverse plane. The fibers, however, were randomly arrayed as
occurs in most manufactured media. Pressure loss measurements of
these samples, reduced to a dimensionless fiber drag F*, were used to
test the validity of prominent pressure loss models. Subsequently,
the data of Lee and Liu [2] and some tests at the Donaldson Company

Mr. Monson works as an applied technologist in the Technology
Group at the Donaldson Co. Inc., P. O. Box 1299, Mpls.,MN 55440

on five glass samples and six stainless steel fiber samples provided a data base that covered a range in solidity of two orders
of magnitude from 0.008 to 0.808.

The evaluation, reported by Schaefer [1] and continued by
Schaefer and Liu [3], included the channel type models of Kozeny-
Carman [4], Langmuir [5] and Davies [6] and a large variety of fiber
drag models. The drag models include the square array of infinite
cylinders perpendicular to the flow studied by Hasimoto [7]; the isolated row model studied by Tamada and Fujikawa [8] and Mijagi [9]; the
staggered infinite cylinder array (cell model) studied by Happel [10]
and Kuwabara [11]; the three-dimensional ordered fiber array of
Iberall [12]; the semi-random fan model array studied by Kirsch and
Fuchs [13]; the screen layer model of Chen [14] with equal interfiber
distance in all three directions; and the three semi-random and fourth
isotropically random fiber arrays of Spielman and Goren [15]. The
results indicated that none of the investigated models fit the data
well over the full range of solidity. For solidities less than 0.2,
the Langmuir model with fiber axes parallel to the flow [5] tends to
pass through the center of the data scatter. For solidities greater
than 0.2, the Kozeny-Carman capillary tube model [4] tends to pass
through the center of the data scatter while the Langmuir theory tends
to take on values approximately 2/3 that of the capillary tube model.

In order to successfully model the uniform fiber semi-idealized
random fiber structure, one must account for the fiber crossings, the
resulting effective transverse interfiber space and it's effective
boundary shape as well as the dynamic shape factor of the resulting
finite length fibers. Since the number of crossings increases with
increasing solidity, these parameters change with solidity. Once
the no-slip characteristics are defined, one then needs to account
for slip and transition flow effects encountered with very small
fibers or at low gas densities.

Once one can model the semi-idealized random structure, then
one needs to account for perturbations encountered in practice.
These perturbations can be attributed to the effects of manufacturing
processes, operating conditions and design choices and include the
following:

1. Fiber curvature.
2. Random fiber orientation.
3. Fiber size distributions.
4. Solidity gradients.
5. Blocking effects of media add-on's.

The above description lists the key parameters that must be
addressed in modeling pressure loss in fibrous filter media. In
reference [16], Monson described briefly a drag type pressure loss
model for uniform fibers that has the capability of accurately
accounting for most of the effect described above. Subsequently,
the use of a more rigorous boundary definition and some further
curve fitting has resulted in equation simplification. Thus, it
may be instructive to review the development of the model in a little
more detail. Parameters 3-5 mentioned above are beyond the scope of
the present paper and will receive minimal or no further discussion.

THEORY

The Mitered Cylinder Model

The Fundamental Relations: Total pressure loss ΔP_t is related to three interrelated dimensionless ratios: volume solidity c, the ratio of filter thickness t to fiber diameter d_f, and dimensionless fiber drag F* by means of the Kirsch-Fuchs relation [13],

$$\Delta P_t = \frac{4\mu U_\infty}{\pi d_f} c \frac{t}{d_f} F^* , \tag{1}$$

where μ is the fluid viscosity and U_∞ is the mean approach velocity. The dimensionless fiber drag is defined as

$$F^* = \frac{\text{total fiber drag/total fiber length}}{\mu U_\infty} = \frac{n D_f}{\mu U_\infty L_f} , \tag{2}$$

where n is the number of fibers, D_f is the drag on an individual fiber, and L_f is the total length of constant diameter fibers in the matrix. The grouping $\pi d_f^2/4$ in Eq 1 represents the average cross sectional area of all the fibers in the matrix so when there is a single mode distribution of fiber sizes, the most representative fiber size is the d-squared mean defined as

$$d_f = \overline{d}_f = \left(\frac{\sum d_{fi}^2 n_i}{\sum n_i} \right)^{1/2} , \tag{3}$$

where n_i is the number of fibers in the ith size interval and d_{fi} is the mid point of the interval.

Pressure loss modeling involves finding a suitable matrix geometry that will yield relations for the unknown quantities n, D_f and L_f of Eq 2. In spite of the fact that the dimensionless drag of an isolated infinitely long cylinder varies with Reynolds number, it has long been known that if t/d_f is sufficiently large, the non-slip dimensionless fiber drag of random fiber filters, denoted by F^*_o, exhibits a Stokes type flow behavior and depends only on solidity, i.e., $F^*_o = F^*_o(c)$; see e.g. Davies [6]. The range of fiber Reynolds number $Re_f \equiv \rho U_\infty d_f/\mu (1-c)$ (where ρ is the fluid density) over which the actual drag does not exceed the Stokes drag by more than 1% varies with solidity. The following ranges are estimated: zero at zero solidity (fibers infinitely long and infinitely far apart), Re_f <0.001 for c=0.125, and for very high solidities, Re_f<21 for c=0.9.[1]

This suggests that the effective length L of the fibers is finite and that on the average, the length varies with solidity. This is consistent with the observed fact that random fibers in a matrix touch

---

[1]The details of this estimate are beyond the scope of the present paper.

or cross at various points and that the number of crossings increases as the fibers are compressed to higher matrix solidities. Unlike an infinitely long fiber whose shear stress and pressure distribution is the same everywhere along its length but varies with Reynolds number, the matrix fibers have secondary flows and modifications to their shear stress and pressure distributions at the crossing points as well as effects from neighboring fibers (boundaries). See Fig. 1.

It is assumed this produces drag characteristics similar to a finite length fiber in a bounded viscous flow. This drag varies with fiber length to diameter ratio $L/d_f = R$ (or aspect ratio) and boundary proximity. It will be shown that if t/d_f is less than some minimum value $(t/d_f)_{min}$ at any solidity, then F^*_o depends on both c and t/d_f, i.e. $F^*_o = F^*_o(c, t/d_f)$. For the remainder of the paper continuum flow will be assumed so we will drop the o subscript.

The Matrix Geometry: The actual fibrous matrix is randomly dis-

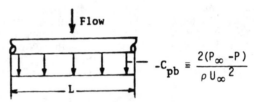

$$-C_{pb} \equiv \frac{2(P_\infty - P)}{\rho U_\infty^2}$$

a. Base pressure distribution on a segment of an infinite cylinder.

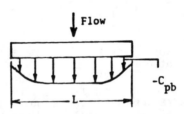

b. Base pressure distribution on an equal length finite cylinder.

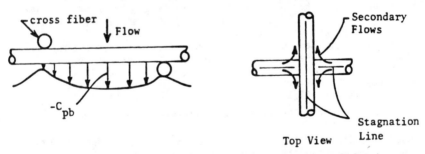

c. Secondary flows set up at a fiber crossing and resulting base pressure distribution.

Fig. 1   Schematic diagram showing the mechanism by which fibers in a real matrix approximate the drag of finite fibers rather than that of infinite fibers.

persed and cannot be treated mathematically except perhaps in some
statistical manner. Thus, the geometry must be idealized in some
manner. Historically, the first drag models used the drag on an
infinite isolated cylinder which included the Reynolds number term,
e.g. Iberall [12]. We see that the Iberall model did not account for
either fiber crossings or the boundary proximity effect of neighboring fibers. The infinite staggered cylinder array studied by
Kuwabara [11] accounts for the boundary effect of neighboring fibers
but not the fiber crossings. Since the fibers are still infinitely
long, but the Reynolds number terms are not present in this solution,
it is specifically valid only for zero fiber Reynolds number. A more
complete solution for this geometry over the full range of solidities
possible was presented by Sangani and Acrivos [17]. Their results
show that the Kuwabara approximation predicts values of F* less than
six percent below the more exact theory for c<0.55. An interesting
aspect of this geometry is that for a given solidity, it provides the
smallest transverse interfiber space of any geometry. As such it
provides an upper bound on F* which cannot be exceeded by any more
random geometry.

The geometry of Chen [14], which modeled the fibers as equivalent to layers of square mesh screen with the same mesh dimension L
in all three directions, seemed to have the capability of modeling
the fiber crossing effect and the transverse boundary proximity
effect. It was found, however, that the large distance L between
screen layers produced too small a transverse interfiber space at a
given solidity. Also, Chen did not detail the geometry of the fiber
intersections which is necessary for accuracy at high solidity. Two
modifications of this geometry which seemed more representative of
real matrices were described in [16].

For the same solidity each geometry will have a different value
of R and thus a different fiber drag. If one imagines the real
matrix to be divided into layers one d_f thick along its thickness t
and the resulting material in each layer is reconstituted into circular fibers arranged into a square screen of mesh length L, it was
found that the mitered cylinder geometry most closely reproduces this
average transverse mesh length. Kanaoka, et al. [18] have shown
experimentally that it is far more important to have the correct
transverse interfiber spacing than the correct longitudinal spacing
at a given solidity. The mitered cylinder geometry, shown in Figures
2a and b, consists of layers of screen with all cylindrical elements

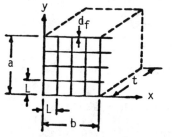

a. Macroscopic sketch of
    mitered cylinder geometry.

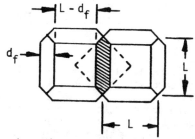

b. Microscopic sketch of
    mitered cylinder geometry.

Fig. 2 The mitered cylinder geometry.

mitered at the juncture. Successive layers are randomly oriented and spaced d_f apart on centers in the flow direction. The pores in this idealized structure close off at a maximum solidity c = 0.9041.

The following relations can be derived from the geometry of Fig. 2: The mean fiber aspect ratio R is related to c by

$$R = \left[\frac{3}{8} \pi - \sqrt{\frac{9\pi^2}{64} - \frac{3}{2} c} \right]^{-1} . \tag{4}$$

The equivalent spherical diameter d_s of the mitered cylinder

$$d_s = d_f \left[\frac{3}{2} R - \frac{2}{\pi} \right]^{1/3} . \tag{5}$$

The projected area solidity S in any layer is related to R by

$$S = \frac{2R-1}{R^2} . \tag{6}$$

A boundary proximity parameter λ defined as the ratio of d_s to the hydraulic diameter D_h of the effective square boundary (shown dotted, Fig. 2b)

$$\lambda \equiv \frac{d_s}{D_h} = \frac{\sqrt{2}}{R} \left[\frac{3}{2} R - \frac{2}{\pi} \right]^{1/3} . \tag{7}$$

The ratio of total length of constant diameter fiber L_f to total length of mitered cylinders nL in the matrix

$$\frac{L_f}{nL} = 1 - \frac{4}{3\pi R} . \tag{8}$$

The Stokes drag D_f on any finite length particle (mitered cylinder) in a bounded fluid can be given by Eq 9, Monson [19],

$$D_f = \frac{3\pi\mu U_\infty d_s}{K_\infty K_{b2}} , \tag{9}$$

in which K_∞ is the shape factor of the mitered cylinder and K_{b2} is a boundary correction factor. Substituting Eq's 5, 8 and 9 into Eq 2 gives F* in terms of parameters relative to the present mitered cylinder model:

$$F_{mc}^* = \frac{3\pi(3/2)^{1/3}}{(1-c)(R - \frac{4}{3\pi})^{2/3} K_\infty K_{b2}} . \tag{10}$$

Dupuit's relation $U_\infty/(1-c)$ for the mean velocity in the matrix has also been used; see, e.g. [4], p. 8.

The shape factor of the mitered cylinder could not be represented by that for a cylinder of the same aspect ratio as the mitered cylinder because the fibers in the real matrix are curved between intersections. The ideal matrix was thus modified as shown schematically in Fig. 3. Curved fibers have less drag than straight fibers of the same aspect ratio. In this model it was assumed that the

shape factor was similar to that for segments of a torus (Fig. 3a)
of the same aspect ratio as the mitered cylinder would have at a
given solidity. The shape factor was given [16] by a linear transition
as a function of R between that for a full torus at c = 0 to that for
a cylinder when R $<\pi$ (c>0.43). Semi-empirical shape factor relations
for a torus, $K_{\infty t}(R)$ and $K_{\infty c}(R)$ for a cylinder were given by Monson
[19]. This transition curve is illustrated in Fig. 4.

This is a first order approximation since there apparently is no
experimental or theoretical information available on the dimensionless
drag of segments of a torus of constant cross sectional diameter.

The difficulties of experimentally measuring such drags and of
the existing theories were discussed briefly in [19] and in more
detail in [20]. The error in assuming a linear decrease in drag at a
given aspect ratio as a cylindrical segment is curved progressively
toward a complete torus certainly cannot exceed half the maximum
fractional drag change between a full torus and a cylinder. This is
seen in Fig. 3 of [19] to be $(K_{\infty t}-K_{\infty c})/2K_{\infty c}=0.075$ at R=11.1 or
$\pm$ 7.5%. More realistically, the maximum error is probably less than
$\pm$ 3%.

At the upper end of the transition curve (R=π) where the ends of
the closed torus segment contribute considerably to the total drag of
the segment, the transition curve is heavily weighted toward the
straight cylinder so errors in the toroidal segment drag estimate
contribute little to the overall error. By the same reasoning, we
see that at R=1000, Fig. 3 of [19], the error in the toroidal segment
drag estimate is probably less than $\pm$ 2%. This corresponds to a
solidity of 0.0016, a value lower than what can be achieved in most
practical filters. Finally, as solidity continues to decrease toward
zero, we see that the transition curve becomes heavily weighted toward
the torus drag curve but that in the limit, the aspect ratio of the
various shapes becomes infinite resulting in identical drag for all
geometries. Thus, as solidity decreases, the drag of the ends of the
toroidal segment gradually becomes negligible relative to the drag of
the segment between the ends.

a. Low solidity   b. High solidity

Fig. 3 Schematic diagram of idealized curved fiber matrix.

A simplified relation for this transition curve has been developed which matches the more precise relation of [16] within $\pm$ 0.5% for the range of interest, $0.001 < c < 0.4325$:

$$K_\infty = \left[\frac{0.4154}{c^{0.543}} + 0.7661 \ c^{0.340} \right]^{-1} . \tag{11}$$

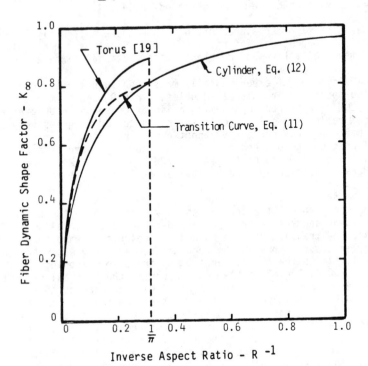

Fig. 4   Shape factor of tori, cylinders, and the assumed transitional geometry.

When c is greater than 0.4325, there is negligible curvature between the more numerous fiber crossings so the shape factor for straight cylinders [19] can be used:

$$K_\infty = \frac{3}{8} R_e^{1/3} \left[\frac{(2R_e^2 - 3) \ \mathrm{arccosh} \ R_e}{(R_e^2 - 1)^{3/2}} + \frac{R_e}{R_e^2 - 1} \right] , \tag{12a}$$

wherein the aspect ratio R_e of a prolate spheroid of equal shape factor is given by

$$R_e = \sigma_c R , \tag{12b}$$

with σ_c given by

$$\sigma_c = 0.835 \exp (0.527 \ R^{-0.69}) . \tag{12c}$$

Physical justification for this choice is evident in photographs of the DACRON® polyester media tested by Schaefer [1], Fig. 5. Fig. 5a shows substantially more curvature between fiber crossings at a solidity of 0.11 than is seen between the more numerous crossings at a solidity of 0.43, Fig. 5b.

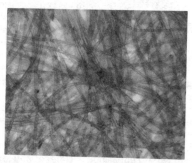

a. Schaefer's 13.4 m DACRON® media (c = 0.0119); 56X.

b. Schaefer's 13.4 m DACRON® media (c = 0.4341); 56X.

Fig. 5   The effect of solidity on the amount of curvature between fiber crossings.

When the surrounding effective boundary is porous (other fibers) rather than solid, the boundary correction factor K_{b2} is assumed to also be a function of porosity (1-c) in addition to the parameters K_∞, λ and k, the boundary shape parameter, in Brenner's general first order boundary theory [21]. Furthermore, K_{b2} was assumed to be separable into two parts, i.e.

$$K_{b2} = \phi_1(1-c)\ \phi_2(K_\infty,\lambda,k) \quad . \tag{13}$$

The porosity effect was modeled after the form given by Maude and Whitmore's general sedimentation theory [22], i.e., $\phi_1 = (1-c)^m$. The exponent m was empirically found to be 2.0 ± 0.25 for best fit to the data. The second part was assumed to take the form of Brenner's first order boundary theory $K_{b1}(K_\infty, \lambda, k)$ except the boundary shape constant k was replaced by an effective value which varies with solidity via λ, i.e., $k_e(\lambda)$. There are several reasons for doing this which may not be readily apparent.

First, Brenner's first-order boundary theory applies for an arbitrarily shaped particle moving past a solid boundary, or conversely for the fluid and a solid boundary moving past an arbitrarily shaped particle fixed in space. In the present case we have flow past an arbitrarily shaped particle surrounded by other such particles which act as a porous boundary of some sort. Lack of a detailed theory has forced two a priori assumptions to be made. One is that which lead to Eq 13 and the other is that the form of Brenner's first-order equation is still valid except that the boundary shape constant would be different from that of a solid boundary and that it would probably vary with solidity, especially at high values where there are greater differences between the actual and ideal structures. This also allows empirical correction for higher order terms.

Secondly, the proximity of one layer of fibers with the next requires consideration of a longitudinal boundary correction to the drag in addition to the transverse boundary effect which has been shown to have the largest effect [18]. The mitered cylinder screen model assumes the longitudinal spacing to be substantially independent of solidity. The real structure is formed from a network of fibers crossing each other, some of which may come from adjacent layers and thus the longitudinal boundary correction may be a weak function of solidity.

Rather than treat the above mentioned effects separately, one might envision the effective boundary around each fiber to be analogous to a porous oblate spheroid with the minor axis parallel to the flow direction whose axis ratios change with solidity. All the above mentioned effects are believed to be adequately accounted for by use of the effective boundary shape parameter $k_e(\lambda)$ which is established as follows. The experimental values F^*_{exp} [3] and the capillary tube model (where applicable) were used in Eq 10 to obtain an empirical curve fit for k_e, Fig. 6. Thus,

$$\phi_2(K_\infty, \lambda, k_e) \equiv K_{b1} = 1 - \frac{k_e}{K_\infty} \lambda \qquad (14)$$

and

$$k_e^{-1} = 0.4998 + 0.8414 \lambda - 0.0462 \lambda^2 + 0.1444 \lambda^3 . \qquad (15)$$

This curve fit relaxes the boundary shape parameter at high solidity so that at c = 0.9041, F* does not become infinite.

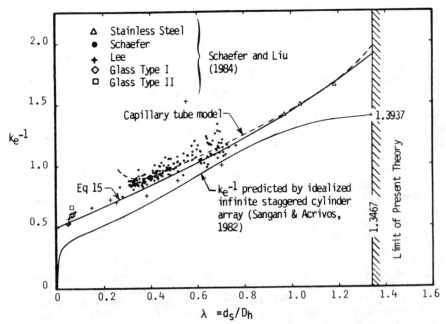

Fig. 6. Variation of effective boundry shape parameter k_e with boundary proximity parameter λ.

The final relation is compared in Fig. 7 with the data of Schaefer and Liu [3]. Also shown are the highest values of F* which can be achieved with a <u>regular</u> infinite fiber array (Sangani and Acrivos [17] and the predicted effect on F* if we have a transverse log-normal distribution in solidity throughout the media depth given by values of geometric standard deviation in solidity $\sigma_g(c) > 1.0$. These curves are solved numerically. If we had assumed straight fibers, the dashed curve would have resulted.

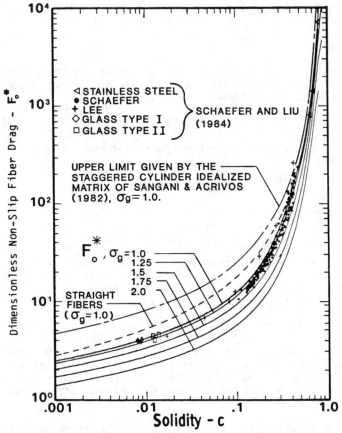

Fig. 7   Dimensionless non-slip drag F^*_0 as a function of solidity and geometric standard deviation of solidity; comparison with experimental values of F^*_{exp}

Thickness Effects

The Swiss Cheese Analogy: In Swiss cheese, the maximum size of the gas bubbles is controlled by the manufacturing process. If we have a sufficiently thick block of cheese, none of the holes go all the way through and the local solidity at various transverse locations tends to be nearly uniform. As the cheese is sliced progressively thinner, however, more and a broader range of hole sizes appear causing an increasingly broader transverse distribution of solidity. Likewise in filter media, there exist bubbles of low solidity media whose maximum size is determined by the manufacturing process. Analogous to the Swiss cheese then there should be some minimum thickness to fiber diameter ratio at any average solidity value above which one can say that the transverse solidity distribution is uniform, i.e. $\sigma_g(c)=1.0$. This minimum thickness ratio is called a unity layer designated by t_u and is quantified as follows.

Unity Layer: Utilizing the mitered cylinder geometry, the probability of a pore being covered by a single layer would be

$$P_c = S, \qquad (16)$$

where P_c is the probability of a pore being covered and the projected area solidity S is related to fiber aspect ratio R and thus volume solidity c by Eq 6. For r layers

$$P_c = 1-(1-S)^r . \qquad (17)$$

Since, in this model, each layer is one fiber diameter thick the number of layers is

$$r=t/d_f . \qquad (18)$$

Combining this with Eq 17 and solving for t/d_f we obtain

$$(t/d_f)_{min} = \frac{\ln (1-P_c)}{\ln (1-S)} . \qquad (19)$$

The actual random fiber array has a probability less than one that S=1.0 as predicted by Eq 6 as R→1 at large solidity (c→0.904). Let P_s be the probability that the random geometry has an area solidity S in any given layer as defined by Eq 6. Then Eq 19 becomes

$$t_u \equiv (t/d_f)_{min} = \frac{\ln (1-P_c)}{\ln (1-P_s S)} . \qquad (20)$$

P_s must vary with solidity. It should have a value close to unity for low to medium solidities and decrease as higher solidities are attained where the actual matrix deviates progressively more from the assumed screen layer geometry which requires a mitered screen mesh geometry in each layer.

Comparison of this function with the carefully chosen data used in this report suggests the following tentative values:

$$P_c=0.999 \qquad (21)$$

$$P_s=e^{-2c} . \qquad (22)$$

Fig. 8 shows a plot of minimum thickness to diameter ratio for an ideal structure as given by Eq 19 using Eq 21 for P_c. This is indicated by the dashed line. The solid line gives t_u as a function of solidity for the non-ideal structure as estimated by Eqs 20-22. Some of the data from [3] and other Donaldson Co. tests are shown on this graph. Darkened symbols indicate tests which resulted in F* values which were 10% or more below the theory curve for $\sigma_g(c)=1.0$. A limited amount of data have suggested some tentative values of $\sigma_g(c)>1$ for media which have thicknesses less than t_u as a function of solidity.

Piekaar and Clarenburg [23] have shown that the random laying down of fibers in a plane forms a log-normal pore size distribution having a geometric standard deviation of 1.9 independent of solidity. One can infer the same for the geometric standard deviation of the transverse solidity distribution, i.e., $\sigma_g(c)=1.9$ independent of solidity. They have incorrectly assumed this to be true for all media regardless of thickness. One can see in Fig. 8 that the information of Piekaar and Clarenburg and the tentative line for $\sigma_g(c)=2.0$ map out a zone of media thicknesses where one would expect $\sigma_g(c)$ to be approximately 2.0 for well dispersed random fiber structures. It is possible to have such poor fiber dispersion that $\sigma_g(c)>2.0$ but these structures make such poor use of the fine fiber component that they are not considered a practical filter.

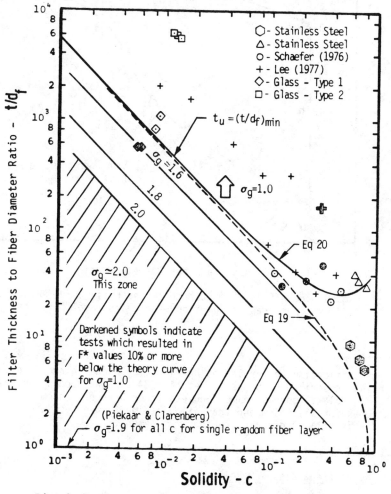

Fig. 8 Regime of applicability for uniform and nonuniform solidity models.

The increase of F* and the decrease of $\sigma_g(c)$ approaching unity with increasing t/d_f is further illustrated in Fig. 9. Here one through ten layers of CEREX® were stacked up and the resulting values of F* are plotted against t/d_f. A single layer of CEREX had the equivalent of $\sigma_g(c)=1.43$ according to the mitered cylinder model, Fig. 7, for c = 0.189. For thickness ratios greater than approximately 35, F* was within 10% of the predicted value for uniform solidity ($\sigma_g(c)=1.0$). An unavoidable compression of the filter stack in the test rig increased the final solidity of the stack to an estimated value of 0.20 with a corresponding increase in the uniform solidity dimensionless fiber drag from 25.3 to 27.5.

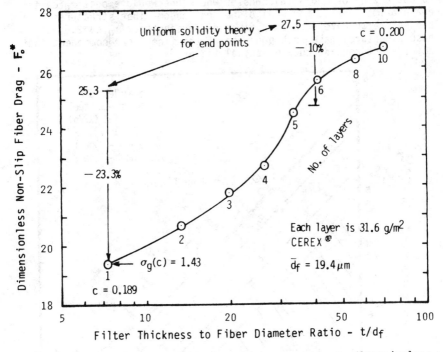

Fig. 9   The effect of filter thickness ratio on dimensionless fiber drag; comparison with uniform solidity theory.

Slip Effects

A modification to a relation developed by Fuchs, Kirsch and Stechkina, described by Schaefer, et al. [24], shows good potential for being able to correct dimensionless fiber drag of random fiber structures for slip flow and transition regime flow effects. This relation is still undergoing further refinement.

DISCUSSION

Apparent Discrepancies - Resolved

The data for glass - Type I at a solidity of 0.008, Fig. 7,

yielded an experimental value of F^*_{exp} = 4.17, 4.3% higher than the
theoretical value for curved fibers, F^*_{mc}=3.997. This media was a
furnace type filter having a greater fraction of rather stiff, nearly
straight, 45.7 m diameter fibers. If all the fibers were straight
and normal to the flow, the theory for straight fibers would predict
F^*_n = 4.835. At low solidity, however, the fibers are not all normal
to the flow but are isotropically randomly dispersed. Spielman and
Goren [15] studied one case where the fibers could take on all angles
from normal to parallel to the flow. In many cases, however,
including the present case, the length of fibers in the matrix L'
exceeds the thickness of the media t. In this case then, we would
have a limited angle isotropic random fiber dispersion. If we take
θ to be zero in the transverse plane and $\pi/2$ for fibers parallel to
the flow, the maximum angle of the dispersion cannot exceed

$$\theta_{max} = \sin^{-1} t/L' .$$ (23)

Going one step beyond Spielman and Goren, we normalize their frequency function and incur the maximum angle θ_{max} to the isotropic
random dispersion. Then the ratio of dimensionless fiber drag for
the limited angle random isotropic case F^*_i to that for straight
fibers normal to the flow F^*_n is given by

$$\frac{F^*_i}{F^*_n} = \frac{1}{3}\left[\cos^2\theta_{max} + 2 + \frac{D_p}{D_n}\sin^2\theta_{max}\right] ,$$ (24)

where D_p/D_n is the ratio of the drags on a fiber of mean aspect
ratio R between fiber crossings for the case of the fiber major axis
parallel and normal to the flow respectively. The value for R at
any solidity is determined using Eq 4. The ratio D_n/D_p can be estimated with sufficient accuracy using the ellipsoid approximation
which was given by Langmuir [5], Table V.

For the present example, fiber lengths were found to range
from 1.27 cm to 4.45 cm for an average of L' $\simeq$ 2.87 cm. The mean
filter thickness was t=1.89 cm so that Eq 23 predicts θ_{max}=41.5°.
For c = 0.008, R=196, so $D_n/D_p \simeq$ 1.7. Then Eq 24 predicts
F^*_i= 0.94F^*_n =4.545. We would expect the actual structure to have
a value for F^* somewhere between F^*_i=4.545 (all fibers straight,
isotropically dispersed and with a maximum angle θ_{max}=41.5°) and
F^*_{mc}=3.997 (all fibers curved and normal to the flow). A reasonable
estimate of 50% of each of the two cases would predict F^*=4.26. This
is within 2% of the measured value. It is perhaps fortuitous that
setting θ_{max}=$\pi/2$ in Eq 24 results in the prediction that F^*_i=4.17,
the measured value. The main point of this exercise is to show that
the curved fiber mitered cylinder model will also closely predict the
performance of structures having a substantial fraction of straight
fibers that are isotropically randomly dispersed. At high solidities
D_n/D_p approaches unity and we see from Eq 24, that the ratio F^*_i/F^*_n
also approaches unity.

We see a tendency for the highest solidity polyester fiber data
taken by Schaefer, Fig. 7, to fall approximately 18% higher than the
mitered cylinder model predicts and yet at the higher solidities of
the stainless steel data, the agreement between experiment and theory
is within 9%. It has been found that the polyester media forms with

a linear longitudinal gradient in solidity. This gradient causes larger deviations from uniform solidity theory as solidity increases due to the nonlinear increase of the dimensionless fiber drag.

As an example, Schaefer's data point at c = 0.434 yielded an experimental value $F*_{exp}$=202.9 whereas the mitered cylinder model predicts $F*_{mc}$=171.5. By measuring the fiber length per unit area for the top two layers on the two faces and in the center of the media, local solidities can be inferred using photo-micrographs. The absolute level needed some adjustment but the slope was identified with the result that the fractional change in solidity $\Delta c/c$=0.318 for the polyester samples. A numerical integration through the depth of the media resulted in a prediction for the axial solidity gradient case of $F*_c$=187. This is within 9% of the measured value and thus accounts for half of the discrepancy.

Comparison with Single Layer Screen Models

Comparison of the mitered cylinder model to models for viscous flow over plain square weave screen represents a test of the mitered cylinder model for a limiting geometry. One would expect the results for woven screen two fiber diameters thick to be very sensitive to the precision of the mesh spacing and that $\sigma_g(c)$ should be greater than 1.0 for most manufactured samples. This geometry has been tested and modeled three different ways [25-27]. All were expressed in terms of inertial type loss coefficients. These have been converted to dimensionless fiber drag as follows. MacDougall [25], 0.13<c<0.50:

$$ F* = \frac{8.4823\ R}{\sqrt{R^2-1}} \left(\frac{R}{R-1} \right)^{3.74} . $$
(25)

Benardi et al. [26], 0.16<c<0.47:

$$ F* = \frac{299.5\ R^2}{\sqrt{R^2+1}}\ \exp[-7.01\ (1-1/R)^2] . $$
(26)

Armour and Cannon [27], 0.24<c<0.65:

$$ F* = \frac{2.641\ \sqrt{R^2+1}}{\left[R - \frac{\pi}{4}\ \sqrt{R^2+1} \right]^2} . $$
(27)

In all cases if we approximate the weave with straight line segments, the fiber aspect ratio R is related to volume solidity by

$$ c = \frac{\pi}{4}\ \frac{\sqrt{R^2+1}}{R^2} , $$
(28)

and to projected area solidity by Eq 6. The results are shown in Fig. 10 relative to the curved fiber mitered cylinder model. We should actually be comparing these to the straight fiber mitered cylinder model. The MacDougall curve seems to be paralleling the straight fiber model at some $\sigma_g(c)$>1 but overall we can see that these models are represented well by the curved fiber mitered cylinder model for $\sigma_g(c)$ between 1.25 to 1.5. The variation is no doubt due to the variance in the mesh spacings as delivered by the manufacturer, which in turn affects $\sigma_g(c)$.

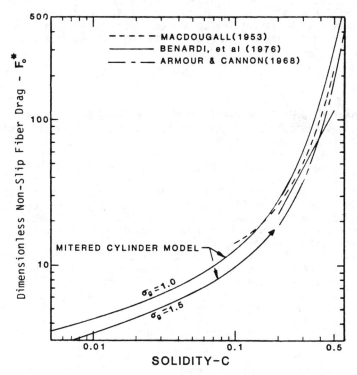

Fig. 10   Previous modeling of viscous flow
over single layer woven screen

CONCLUSIONS

The key parameters encountered in modeling pressure loss through fibrous filters have been identified. A mitered cylinder pressure loss model has been described which can accurately account for the effect of most of these parameters over at least two orders of magnitude in solidity. It has been shown that making bi-layer structures more regular, or random structures sufficiently thick, significantly reduces $\sigma_g(c)$ to values approaching unity. The model has been found useful in predicting initial pressure loss in both homogeneous, non-homogeneous and layered filter structures of a single fiber size and can be extended to multimodal fiber sizes. It can be used to infer an effective mean aerodynamic fiber size but this can be done only if the sample has a sufficient t/d_f so that $\sigma_g(c)$ 1.0. It has the potential of extending the solidity range over which number frequency distributions can be estimated from pore sizing data and it has been found useful in predicting specific flow resistance of sound absorbing materials.

ACKNOWLEDGEMENTS

The author would like to acknowledge with appreciation many fruitful discussions on various aspects of the pressure loss model held with Professor B. Y. H. Liu of the University of Minnesota and Mr. James W. Schaefer of the Donaldson Company. Appreciation is also extended to Mary Ann Johnson for typing this manuscript.

REFERENCES

[1]   Schaefer. J.W. "An Investigation of Pressure Drop Across Fibrous Media," M.S. Thesis, Mech. Eng. Dept., University of Minnesota, Minneapolis, MN., 1976.

[2]   Lee, K.W. and Liu, B.Y.H., "Experimental Study of Aerosol Filtration by Fibrous Filters," Aerosol Science and Technology, Vol. 1, 1982, pp. 35-46.

[3]   Schaefer, J.W. and Liu, B.Y.H., "An Investigation of Pressure Loss Across Fibrous Filter Media," in Aerosols, Elsevier Science Publishing Company, New York, 1984, pp. 555-558.

[4]   Carman, P.C., Flow of Fluids in Porous Media, Butterworths Scientific Publications, 1956, pp. 1-26.

[5]   Langmuir, I. "Report on Smokes and Filters," Part IV of Report on "Filtration of Aerosols and the Development of Filter Materials," CWS-15:NL-B 34, by Langmuir, and LaMer, 1942, pp. 4-57.

[6]   Davies, C.N., Air Filtration, Academic Press, New York, 1973.

[7]   Hasimoto, H., "On the Periodic Fundamental Solutions of the Stokes Equations and their Application to Viscous Flow Past a Cubic Array of Spheres," J.Fluid Mech., Vol.5, 1959, pp.317-328.

[8]   Tamada, K. and Fujikawa, H., "The Steady Two-dimensional Flow of a Viscous Fluid at Low Reynolds Numbers Passing Through an Infinite Row of Equal Parallel Cylinders," Quart. J. Mech. Appl. Math., Vol. 10, 1957, pp 425-432.

[9]   Mijagi, T., "Viscous Flow at Low Reynolds Numbers Past an Infinite Row of Equal Circular Cylinders," J. Phys. Soc. Japan, Vol. 13, 1958, pp. 493-496.

[10]   Happel, J., "Viscous Flow Relative to Arrays of Cylinders," A.I.Ch.E.J., Vol. 5, 1959, pp. 174-177.

[11]   Kuwabara, S., "The Forces Experienced by Randomly Distributed Parallel Circular Cylinders or Spheres in a Viscous Flow at Small Reynolds Numbers," Journal of the Physical Society of Japan, Vol. 14, 1959, pp. 527-533.

[12]   Iberall, A.S., "Permeability of Glass Wool and Other Highly Porous Media," Journal of Research of the National Bureau of Standards, Research Paper 2150, Vol. 45, 1950, pp. 398-406.

[13]   Fuchs, N.A. and Kirsch, A.A., "Studies on Fibrous Aerosol Filters - II. Pressure Drops in Systems of Parallel Cylinders," Ann. Occup. Hyg. Vol. 10, 1967, pp.23-30.

[14]   Chen, C.Y., "Filtration of Aerosols by Fibrous Media," Chemical Review, Vol. 55, 1955, pp. 595-623.

[15]   Spielman, L. and Goren, S.L., "Model for Predicting Pressure Drop and Filtration Efficiency in Fibrous Media," Environmental Science and Technology, Vol. 2, No. 4, 1968, 279-287.

[16]   Monson, D.R., "A Filter Pressure Loss Model for Uniform
        Fibers," in Aerosols, Elsevier Science Publishing Company,
        Inc., New York, 1984, pp. 551-554.
[17]   Sangani, A. S. and Acrivos, A., "Slow Flow Past Periodic
        Arrays of Cylinders with Application to Heat Transfer,"
        Int. J. Multiphase Flow, Vol. 8, No. 3, 1982, 193-206.
[18]   Kanaoka, C., Emi, H. and Degushi, A., "Effect of Inter-Fiber
        Distance on Collection Efficiency of a Single Fiber in a
        Model Filter Composed of Parallel Fibers in a Row," in
        Aerosols, Elsevier Science Publishing Company, Inc., New York,
        1984, pp. 563-566.
[19]   Monson, D.R., "The Effect of Transverse Curvature on the Drag
        and Vortex Shedding of Elongated Bluff Bodies at Low Reynolds
        Number," J. Fluids Eng., Vol. 105, 1983, pp. 308-322.
[20]   Monson, D.R., "The Effect of Transverse Curvature on the Drag
        and Vortex Shedding of Elongated Bluff Bodies at Low Reynolds
        Number," ASME Paper 81/WA/FE-4, 1981.
[21]   Brenner, H., "Effect of Finite Boundaries on the Stokes
        Resistance of an Arbitrary Particle," J. Fluid Mech., Vol. 12,
        1962, pp. 35-48.
[22]   Maude, A.D. and Whitmore, R.L., "A Generalized Theory of
        Sedimentation," British Journal of Applied Physics, Vol. 9,
        1957, pp. 477-482.
[23]   Piekaar, H.W. and Clarenburg, L.A., "Aerosol Filters-Pore
        Size Distribution in Fibrous Filters," Chemical Engineering
        Science, Vol. 22, 1967, pp. 1399-1408.
[24]   Schaefer, J.W., Barris, M.A., and Liu, B.Y.H., "Filter Media
        Design for High Purity Air Application," in Proceedings, 32nd
        Annual Technical Meeting of the IES, May, 1986.
[25]   MacDougall, D.A., "Pressure Drop Through Screens," M.S. Thesis,
        Ohio State University, Columbus, Ohio, 1953.
[26]   Benardi, R.T., Linehan, J.H. and Hamilton, L.H., "Low Reynolds
        Number Loss Coefficient for Fine Mesh Screens," J. Fluids Eng.,
        December, 1976, pp. 762-764.
[27]   Armour, J.C. and Cannon, J.H., "Fluid Flow Through Woven
        Screens," AICh E Journal, Vol. 14, No. 3, 1968, pp. 415-420.

Ken W. Lee

PARTICLE COLLECTION MECHANISMS PERTINENT TO GRANULAR BED
FILTRATION

REFERENCE: Lee, K. W., "Particle Collection Mechanisms
Pertinent to Granular Bed Filtration", Fluid Filtration:
Gas, Volume I, ASTM STP 975, R. R. Raber, Ed., American
Society for Testing and Materials, Philadelphia, 1986.

ABSTRACT: Flow fields and particle collection mechanisms
in the maximum penetration regime are reviewed. Simple
equations are introduced for predicting the size of the
aerosol particles that most effectively penetrate a
granular bed, and the corresponding minimum collection
efficiency. Influences of filter operating velocity
and granule size are also discussed. The results show
that the most penetrating particle size for a granular
bed filter exists as a result of either the Brownian
diffusion mechanism combined with gravitation or Brownian
diffusion with the interceptional mechanism.

KEYWORDS: granular beds, filtration, dust collection,
most penetrating particle size

The utilization of of granular beds for removing aerosol
particles from gas streams is an old concept; however, recent
interest in the removal of chemically reactive aerosols and in the
filtration of particulate matter from high temperature gases has
caused granular beds to be studied more thoroughly. As in other
particulate separation techniques, the pressure drop and collection
efficiency of granular beds are the critical considerations in their
design and operation.

Dr. Lee is Associate Manager of the Environmental Physics and
Chemistry Section of Battelle's Columbus Division, Columbus, Ohio
43201.

FLOW FIELDS

As a result of various measurements and theoretical studies concerning fluid flows through porous media, it is now well known that for low values of Reynolds number, the pressure drop results primarily from viscous forces and is linearly proportional to the flow velocity. As the flow rate is increased, the inertia of the fluid and turbulence become more important and the pressure drop eventually depends on the square of flow velocity. Among the available mathematical models describing the flow field in a system of many spheres, the cell model, adopted first by Happel and subsequently by Kuwabara [1,2], may be used in predicting the pressure drop through granular beds at low Reynolds number.

The flow fields of Happel and Kuwabara were obtained from solutions of the creeping flow equations for a system of stationary spheres arranged in a staggered manner. By imagining a concentric boundary around each sphere, the packing density of the media or bed is accounted for in the flow solutions. Happel imposed the condition of no tangential stress on the outer boundary surface in addition to the usual boundary conditions of zero radial and circumferential velocities at the inner sphere. The Kuwabara flow field is based on the same configuration except with a different condition on the outer boundary. He argued that the vorticity should vanish there.

If the pressure drop across a packed bed, ΔP, is assumed to be a simple addition of drag forces, F, around each sphere, then the pressure drop can be written as:

$$\Delta P = \frac{3F\alpha L}{4\pi a^3} ,$$

(1)

where F = drag force, α = packing density of the bed, L = bed depth and a the sphere radius. Using Equation (1), the Happel flow field gives the following expression for the pressure drop across the bed:

$$\Delta P = \frac{9\alpha\mu u_0 (1 + \frac{2}{3}\alpha^{5/3})L}{2(1 - \frac{3}{2}\alpha^{1/3} + \frac{3}{2}\alpha^{5/3} - \alpha^2)a^2} ,$$

(2)

where μ = gas viscosity and u_0 = approaching flow velocity at the center of the outer boundary. Utilization of the Kuwabara flow conditions yields the following expression for the pressure drop:

$$\Delta P = \frac{9\alpha\mu u_0 L}{2(1 - \frac{9}{5}\alpha^{1/3} + \alpha - \frac{1}{5}\alpha^2)a^2} .$$

(3)

Although the appropriateness of either the Happel or Kuwabara flow fields must be validated with experimental data covering a wide range of the variables, the applicability of these solutions to flow through packed beds is discussed here by comparing only predicted pressure drops with measured values. In packed beds operating at low flow rates, the spheres comprising the beds remain stationary and the packing density is usually high compared to those of moving beds or fluidized beds. If all of the spheres are assumed uniform in size, packing densities may range from 0.524 for cubic packing to 0.740 for rhombohedral packing. However, the packing density of most stationary packed beds in practical use ranges from 0.57 to 0.62.

Gebhart, et al [3], and Yung, et al [4], have measured the pressure drop across the packed beds. Figs. 1 and 2 compare the pressure drop predicted by Equations (2) and (3) and these experimental data. The data appearing in Fig. 1 correspond to the results of Gebhart, et al, [3] with sphere radii ranging from 0.0925 to 2 mm and with face velocities ranging from 0.44 to 10.6 cm/sec. The volume fractions of the four packed beds were all measured to be 0.615. For the velocity, u_0 appearing in Equations (2) and (3), the superficial velocity at the face of the packed beds has been used. It can be seen that prediction from the Kuwabara flow fields is in better agreement with the data although it still underpredicts the pressure drop somewhat.

Fig. 2 illustrates the comparison with the pressure drop measurements of Yung, et al, [4]. A bed material of uniform spheres 520 μm in diameter was used and the packing density was measured to be between 0.57 and 0.59. The agreement between both predictions and the data are seen to be about equal. The Reynolds number for the data of 50 cm/sec face velocity is about 17, which seems to be rather large to be predicted by the viscous flow equations. However, the measured pressure drop in dimensionless form remains roughly constant in spite of the high Reynolds number indicating that the cell model may still be valid. When compared with the data of Gebhart, et al, in Fig. 2, the data appearing in Fig. 2 illustrate that the pressure drop across the bed is strongly dependent on packing density. Namely, the average value of measured dimensionless pressure drop is seen to vary from 215 for $\alpha = 0.58$ to 350 for 0.615. This strong dependence is predicted by both models.

Despite the fact that the actual shape of the outer boundary in a closely packed bed may be far different from that of concentric spheres as assumed in the cell models, the preceding comparisons lead to the conclusion that the Happel-Kuwabara flow fields can be applied to flow in packed beds with reasonable accuracy. Further, it is shown that the Kuwabara solution yields a somewhat better estimate of the pressure drop across packed beds than the Happel flow field.

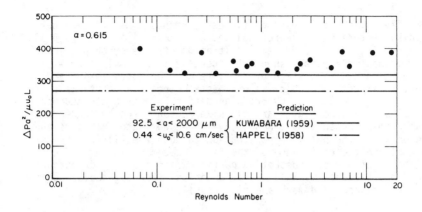

FIG. 1--Comparison of the pressure drop of Gebhart, et al,
with the cell models

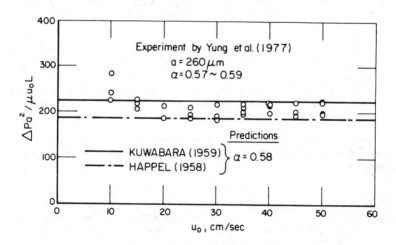

FIG. 2--Comparison of pressure drop data of Yung, et al,
with the cell models

PARTICLE FILTRATION MECHANISMS

Filtration of aerosol particles by granular bed filters has been subject to many theoretical and experimental studies [3,5-13]. As a result, the dependence of filtration efficiency on particle size is now well established. An increase in particle size will cause increased filtration by the interception, gravitation and inertial impaction mechanisms whereas a decrease in particle size will enhance collection by Brownian diffusion. As a consequence, there is an intermediate particle size region where two or more mechanisms operate simultaneously yet none dominate. This is the region where the particle penetration through the filter becomes maximal and the efficiency of the filter becomes minimum. The existence of a most penetrating particle size and the corresponding minimum efficiency is well realized not only for a granular bed but also for fibrous and membrane filters [14,15].

For theoretical analyses of aerosol collection by granular beds, collection efficiency is commonly represented using the concept of single sphere efficiency. A single sphere efficiency is defined as the ratio of the cross-sectional circular area surrounded by limiting streamlines of the flow approaching the collection sphere to the projected area of the sphere. The limiting streamlines are such that all particles passing within them will be collected by the sphere while all the particles outside the streamlines will escape it. With this representation, the concentration ratio of the particles collected inside the granular bed to those approaching it or the overall efficiency of particle collection by the bed can be shown to relate to the single sphere efficiency as follows:

$$E = 1 - \exp\left[\frac{-3\alpha \eta L}{2(1-\alpha)\, d_g}\right] \qquad (4)$$

where E = overall collection efficiency of granular bed, η = single sphere efficiency, α = solidity or solid volume fractions of the bed, and d_g = diameter of the sphere or the packed bed media.

In Equation (4), E increases monotonically with increasing η. Therefore, the particle size that gives the minimum single sphere efficiency also results in the highest penetration through the filter.

Several different aerosol collection mechanisms are operative approximately to an equal extent in the immediate neighborhood of maximum aerosol penetration as previously mentioned although the absolute value of each contribution may be small.

The single sphere efficiency due to Brownian diffusion can be written in the following generalized form from previous studies [7,10].

$$\eta_D = c\, \beta^{1/3}\, Pe^{-2/3} \qquad (5)$$

where η_D = single sphere efficiency due to Brownian diffusion, Pe =
Peclet number (= ud_g/D), u = flow velocity inside the bed, d_p =
particle diameter, D = diffusion coefficient of particle ($=kTC/3\pi\mu d_p$),
k = Boltzmann constant, T = absolute temperature, C = Cunningham
slip correction factor, and c = constant. According to Reference
6, the value of a is 3.5 and according to Reference 10, it is about
4.0. The difference originates from the method of boundary layer
analysis used in deriving the diffusional efficiency. β = function
of solid volume fraction, α, of granular bed and depending upon the
flow fields used in the analysis the following expressions can be
used [1,2]:

$$\beta = (1-\alpha)/\left(1 - \frac{9}{5}\alpha^{1/3} + \alpha - \frac{1}{5}\alpha^2\right): \text{ Kuwabara flow field} \qquad (6)$$

$$\beta = (1-\alpha^{5/3})/\left(1 - \frac{3}{2}\alpha^{1/3} + \frac{3}{2}\alpha^{5/3} - \alpha^2\right): \text{ Happel flow field.} \qquad (7)$$

The equivalent expression for the Neale and Nader flow field [16]
is rather complicated and not included here.

The interceptional collection efficiency is written as [17]

$$\eta_R = 1.5 \beta \frac{R^2}{(1+R)^\gamma} \qquad (8)$$

where η_R = single sphere efficiency due to interception, R = interception parameter (= d_p/d_g), and γ = $(1+2\alpha)/(3-3\alpha)$.

In aerosol collection by a granular bed, the bed media size is
usually large and the gravitational collection mechanism can become
as important as the interceptional collection mechanism.

The gravitational collection efficiency is given as [6,11,13]

$$\eta_G = \frac{G}{1+G} \qquad (9)$$

where η_G = single sphere efficiency due to gravitation, G = sedimentation parameter ($=\rho gd_p^2 C/18\mu u$), ρ = particle density, and g = gravitational constant.

Inertial impaction is an important filtration mechanism for
large particles at high velocity filtrations. However, collection
efficiency due to this mechanism generally decreases very sharply
as particle size is reduced to that of maximum penetration and, in
general, its contribution to the minimum efficiency is rather weak.
Numerical calculations by Gutfinger and Tardos [5] show that in the
vicinity of the maximum penetration regime, contribution of inertial
impaction mechanism is insignificant for Stokes numbers below 0.1.
The Stokes numbers that correspond to the maximum penetration

obtained by Gebhart, et al [3] and Gutfinger and Tardos [5] are all calculated to be much less than the above value. For this reason, contribution of the inertial impaction mechanism is considered negligible and will not be considered here.

In combining several individual efficiencies, it is customary in aerosol filtration studies to assume that to a first approximation, they are independent and additive. Thus, the combined efficiency, η, may be written as

$$\eta = \eta_D + \eta_R + \eta_G. \tag{10}$$

MOST PENETRATING PARTICLE SIZE

The most penetrating particle size is obtainable by differentiating Equation (10) with respect to particle size and settling the resulting value to be equal to zero. However, due to the nonlinear dependence of each term on the particle size, it is necessary to approximate Equations (5), (8), and (9) with a simpler form. The first simplification is to approximate $(1+R)$ with unity that appears on the righthand side of Equation (8). Since $R(= d_p/d_g)$ is usually much smaller than unity for most granular bed applications, this simplification can be easily justified. Since G is also negligible compared to unity, we similarly approximate Equation (9) with G. Finally, the Cunningham slip correction, C, contained in the Peclet number, Pe, of Equation (5) and in the sedimentation parameter, G, of Equation (9) is not to be accounted for. The last simplification would not deteriorate excessively the accuracy of the analysis. As a consequence of the above approximations, we rewrite

$$\eta = c \, \beta^{1/3} \left(\frac{3\pi\mu u d_g d_p}{kT} \right)^{-2/3} + 1.5\beta \left(\frac{d_p}{d_g} \right)^2 + \frac{\rho g d_p^2}{18\mu u} \tag{11}$$

Noting that the combined collection efficiency given by Equation (11) is now written as a linear function of particle size, one finds the intended differentiation becomes straightforward and can be carried out to obtain the following particle size:

$$d_{p,min} = A \, \beta^{1/8} \left(\frac{kT}{\mu u d_g} \right)^{1/4} \Big/ \left[\frac{\rho g}{18\mu u} + \frac{3}{2d_g^2} \beta \right]^{3/8} \tag{12}$$

where $d_{p,min}$ = most penetrating particle size and A = constant whose value is 0.61 or 0.63 depending upon whether 3.5 or 4.0 is used, respectively, for c appearing in Equation (11).

Equation (12) is graphically shown in Fig. 3 as a function of media size, air velocity, and solid volume fraction using $\rho = 1$ g/cm^3 and $\mu = 1.84 \times 10^{-4}$ dyne-sec/cm^2. Equation (6) was used

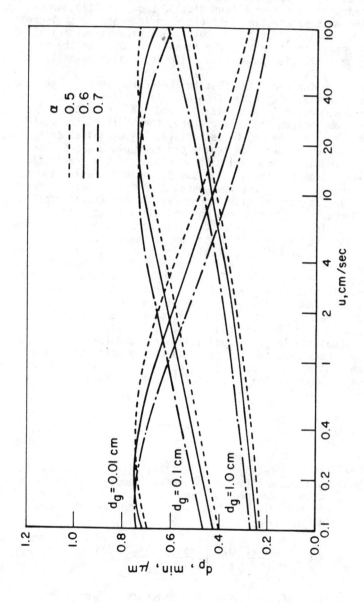

FIG. 3--Most penetrating particle size as a function of α, d_g and u

for β and 0.61 was used for A. There are several interesting points
to be noted from the figure and in Equation (12) with regard to the
air velocity and the media size dependencies of the most penetrating
particle size. For a bed of large granular size, the second term
of the denominator on the righthand side of Equation (12) becomes
small and the most penetrating particle size increases with increasing velocity. However, for a bed whose media size is sufficiently
small, the second term overrides the first term to cause the most
penetrating particle size to decrease with increasing velocity.

It should be further noted that the numerator on the righthand
side of Equation (12) originates from Brownian diffusion, the first
term of the denominator from gravitation and the second term of the
denominator from interception. Therefore, the shift of the most
penetrating particle size may be mechanistically explained as follows.
The maximum aerosol penetration regime can take place where, as
controlling mechanisms, Brownian diffusion plus either gravitation
or interception are operative. When the Brownian diffusion and the
gravitation are operative, the most penetrating particle size shifts
to a larger size as air velocity increases. If the Brownian diffusion and interceptional mechanisms control air velocity and with
decreasing granule size. In order to validate Equation (12) through
comparison with available experimental data, the filtration efficiency
measurement results of Gebhart, et al, [3] as shown in Fig. 4 which
covered a wide range of variables have been used. In determining
the most penetrating particle size, the penetration data plotted
against particle size were graphically interpolated to locate the
size that yields a maximal penetration. In calculating the theoretical curves, the average flow velocity inside the bed was used for
the velocity, u, appearing in Equation (12). In the theoretical
calculation, Equation (6), 0.61 and 1.46 were used for β, A and B,
respectively. Both experimental data and theory demonstrate that
the most penetrating particle size can be increased or decreased
depending on the combination of controlling mechanisms.

MINIMUM EFFICIENCY

Once the most penetrating particle size, $d_{p,min}$, is calculated,
the corresponding minimum efficiency becomes readily obtainable by
substituting Equation (12) into Equation (10) based on Equations (5),
(8), and (9). For a simpler expression, Equation (11) has been
used instead. Thus we have

$$\eta_{min} = B \; \beta^{1/4} \left(\frac{kT}{\mu u d_g} \right)^{1/2} \left[\frac{\rho g}{18\mu u} + \frac{3}{2d_g^2} \beta \right]^{1/4} \tag{13}$$

where η_{min} = minimum efficiency and B = constant 1.46 or 1.59
depending on the used value for c.

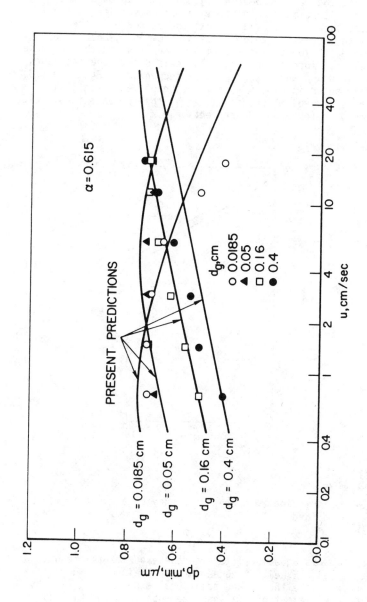

FIG. 4--Comparison of predicted most penetrating particle size with the experimental data measured by Gebhart, et al [3]

Unlike in the expression for the most penetrating particle size, Equation (13) dictates that the minimum efficiency always decreases either with increasing air velocity or increasing granule size. Relating the terms appearing in Equation (13) with the corresponding filtration mechanisms in a manner similar to the previous discussion, it is evident that the decrease in the efficiency with increasing velocity is a result of decreases of both Brownian diffusion and gravitation mechanisms and the decrease in efficiency with increasing granule size is due to the decreases of Brownian diffusion and interceptional mechanisms. Equation (13) is graphically shown in Fig. 5 as a function of media size, air velocity, and solid volume fraction. Again the Kuwabara flow field and 1.46 were used for β and B, respectively.

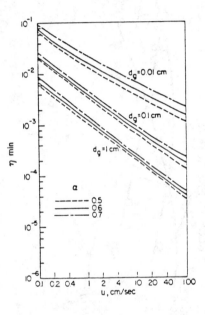

FIG. 5--Minimum single sphere efficiency as a function of α, d_g and u

Fig. 6 compares Equations (13) and the experimental results of Gebhart, et al [3]. Excellent agreement is obtained. Fig. 7 is a comparison between the theory and additional experimental results. A correlation coefficient of 0.75 has been obtained for Fig. 7.

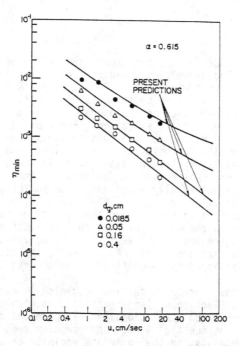

FIG. 6--Comparison of predicted minimum efficiency with the experimental data measured by Gebhart, et al [3]

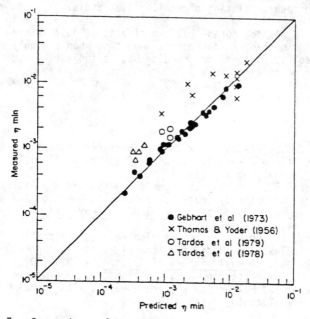

FIG. 7-- Comparison of predicted η_{min} with measured η_{min}

CONCLUSIONS

As a result of the present study, the following conclusions can be drawn. The Happel and Kuwabara flow fields describe the flow in granular beds with reasonable accuracy. A most penetrating particle size exists for which the collection efficiency becomes minimal and its dependency on the operating flow and the granule size can be mechanistically explained. For a bed of large media operating at a low velocity, important mechanisms in the regime are found to be Brownian diffusion and gravitational sedimentation. For a bed of small granules operating at a high velocity, Brownian diffusion and interception become dominant. Otherwise, all three mechanisms are about equally important. Due to these different regimes of controlling mechanisms, the size of a most penetrating particle shifts to a larger size in the diffusion-sedimentation controlling regime and to a smaller size in the diffusion-interception controlling regime either as the operating velocity increases or as the media size decreases. As a result, the most penetrating particle size is found to be confined below a certain upper limit.

Although the inertial impaction mechanism is expected to become important for collection of large particles at high velocities [18], good agreement between predictions and experimental data without accounting for the inertial impaction mechanism suggests that the role of this mechanism in the immediate vicinity of the maximum penetration point is considered minor. Regardless of the controlling collection mechanisms, it is found that the minimum efficiency can be increased by operating the bed at a low velocity, or by employing the bed consisting of smaller size media.

It should also be noted that the effects of dust loading on the filtration mechanisms can become important when a highly co ent-trated aerosol is filtered or granular bed is subject to use for an extended period of time [19].

REFERENCES

[1]   Happel, J., "Viscous Flow in Multiparticle Systems: Slow Motion of Fluid Relative to Beds of Spherical Particles", AIChE Journal, 4, 197 (1958).

[2]   Kuwabara, S., "The Forces Experienced by Randomly Distributed Parallel Circular Cylinders or Spheres in Viscous Flow at Small Reynolds Numbers", Journal of the Physical Society of Japan, 14, 527 (1959).

[3]   Gebhart, J., Roth, C., and Stahlhofen, W., "Filtration Properties of Glass Bead Media for Aerosol Particles in the 0.1-2 μm Size Range", Journal of Aerosol Science, 4, 355 (1973).

[4]   Yung, S. C., Patterson, R. G., Calvert, S., and Drehmel, D. C., "Granular Bed Filter Study", presented at 70th Annual Meeting of APCA at Toronto, Ontario, Canada.

[5]   Gutfinger, C. and Tardos, G. I., Atomspheric Environment, 13,
      853 (1979).
[6]   Lee, K. W., "Maximum Penetration of Aerosol Particles in
      Granular Bed Filters", Journal of Aerosol Science, 12, 79
      (1981).
[7]   Lee, K. W. and Gieseke, J. A., "Collection of Aerosol Parti-
      cles by Packed Beds", Environmental Science & Technology, 13,
      466 (1979).
[8]   Lee, K. W., Reed, L. D., and Gieseke, J. A., "Pressure Drop
      Across Packed Beds in the Low Knudsen Number Regime", Journal
      of Aerosol Science, 9, 557 (1978).
[9]   Schmidt, E. W., Gieseke, J. A., Gelfand, P., and Lugar, T.
      W., "Filtration Theory for Granular Beds", Journal of the Air
      Pollution Control Association, 28, 143 (1978).
[10] Tardos, G. I., Abuaf, N., and Gutfinger, C., "Dust Deposition
      in Granular Bed Filters: Theories and Experiments", Journal
      of the Air Pollution Control Association, 28, 354 (1978).
[11] Tardos, G. I., Yu, E., Pfeffer, R., and Squires, A. M.,
      Journal of Colloid Interface Science, 71, 616 (1979).
[12] Thomas, J. W. and Yoder, R. E., "Aerosol Size for Maximum
      Penetration Through Fiberglass and Sand Filters", AMA Arch.
      Industrial Health, 13, 545 (1956).
[13] Yoshioka, N., Emi, H., Kanaoka, C., and Yasunami, M., Kagaku
      Kogaku (Chem. Eng. Japan), 36, 313 (1972).
[14] Lee, K. W. and Liu, B.Y.H., "On the Minimum Efficiency and
      the Most Penetrating Particle Size for Fibrous Filters",
      Journal of the Air Pollution Control Association, 30, 377
      (1980).
[15] Liu, B.Y.H. and Lee, K. W., "Efficiency of Membrane and
      Nuclepore Filters for Submicrometer Aerosols", Environmental
      Science & Technology, 10, 345 (1976).
[16] Neale, H. N. and Nader, W. K., AIChE Journal, 20, 530 (1974).
[17] Lee, K. W. and Gieseke, J. A., "Note on the Approximation of
      Interceptional Collection Efficiencies", Journal of Aerosol
      Science, 11, 335 (1980).
[18] D'Ottavio, T. O. and Goren, S. L., "Aerosol Captive in Granu-
      lar Beds in the Impaction Dominated Regime", Aerosol Science
      and Technology, 2, 91 (1983).
[19] Tien, C., Turian, R. M., and Pendse, H., AIChE Journal, 25,
      385 (1979).

Chi Tien

EFFECTS OF PARTICLE DEPOSITION ON THE PERFORMANCE OF GRANULAR FILTERS

REFERENCE: Tien, C., "Effects of Particle Deposition on the
Performance of Granular Filters," Fluid Filtration: Gas,
Volume I, ASTM STP 975, R. R. Raber, Ed., American Society
for Testing and Materials, Philadelphia, 1986.

ABSTRACT: A review is presented of the effect of particle
deposition on the performance of granular filters. The
extent to which this effect can be incorporated into the
design calculations and the major problems to be solved are
discussed.

KEYWORDS: Granular Filters, Filtration, Particle Deposition,
Loading Effect, Aerosol Filtration

Granular filtration operated in the fixed-bed mode is inherently
an unsteady-state process because of the continuous accumulation of
deposited matter within filters and the changes in the filter media
structure. As a result, both the effluent concentration and the pressure drop necessary to maintain a given flow rate change with time.
A typical set of experimental data exhibiting this kind of time-dependent behavior is given in Figure 1.

The effect of particle deposition on filter performance, also
referred to as the loading effect, is well known to practitioners in
filtration. However, as a subject of investigation, it has been
largely ignored until recently. During the past decade, some serious
attempts have been made to examine the deposition effect on a more
fundamental basis. The following pages will review some of the highlights of these studies.

DESCRIPTION OF THE DEPOSITION EFFECT

The practical motivation for studying the loading phenomenon is
its significant effect on filter performance. It is therefore
necessary to relate explicitly and quantitatively the increase in

Chi Tien is Professor of Chemical Engineering, Department of
Chemical Engineering and Materials Science, Syracuse University,
Syracuse, N.Y. 13244-1240.

FIG. 1 CONTINUED

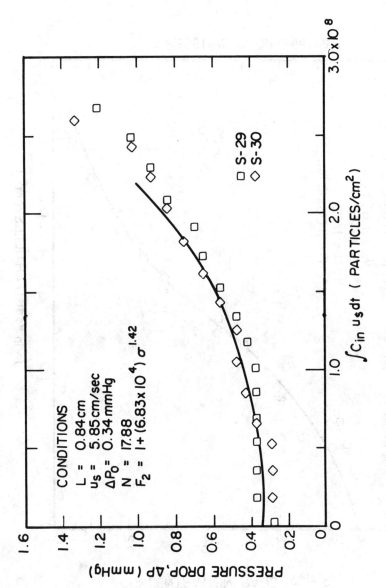

CONDITIONS

$L = 0.84 \, cm$
$u_s = 5.85 \, cm/sec$
$\Delta P_0 = 0.34 \, mmHg$
$N = 17.88$
$F_2 = 1 + (6.83 \times 10^4) \, \sigma^{1.42}$

□ S-29
◇ S-30

$\int C_{in} \, u_s \, dt$ (PARTICLES/cm^2)

PRESSURE DROP, ΔP (mmHg)

FIG. 1 EXPERIMENTAL DATA ON EFFLUENT CONCEN-
TRATION AND PRESSURE DROP OBTAINED FROM GRANU-
LAR FILTERS

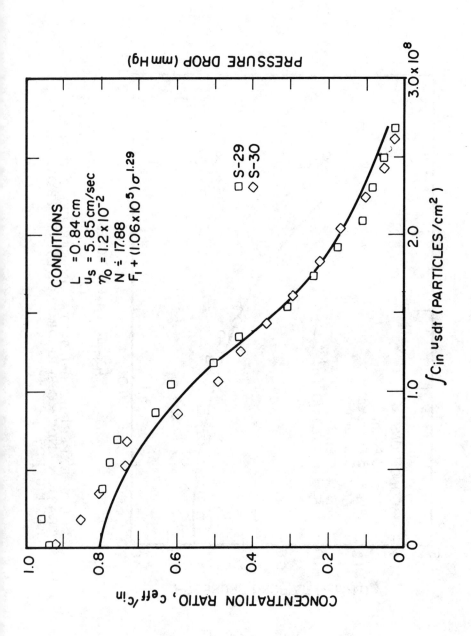

deposition within a filter with the performance of the filter. The performance of a filter, of course, is measured by the effluent quality of the filter as well as the power necessary to maintain its operation (that is, the pressure drop across the filter). The problem at hand, therefore, is predicting the dynamic behavior of granular filtration while taking into account the effect of depositions.

The macroscopic equations of granular filtration are given by Tien and Payatakes [1] to be

$$u \frac{\partial c}{\partial z} + \frac{\partial \sigma}{\partial \theta} = 0 \tag{1}$$

$$\frac{\partial \sigma}{\partial \theta} = u \cdot \lambda \cdot c \tag{2}$$

$$\frac{\lambda}{\lambda_o} = \frac{\eta}{\eta_o} = F_1(\underline{\alpha}, \sigma) \tag{3}$$

$$\Delta P = (\frac{\partial P}{\partial z})_o \int_0^L F_2(\underline{\beta}, \sigma) \cdot dz \tag{4}$$

$$\frac{(\partial P/\partial z)}{(\partial P/\partial z)_o} = F_2(\underline{\beta}, \sigma) \tag{5}$$

where c and σ denote the particle concentration of the fluid and the amount of deposited particles per unit volume of filter bed (vol/vol), respectively. σ, known as the specific deposit, can therefore be considered the particle concentration in the stationary phase.   The independent variables are the axial distance, z, and the corrected time θ. defined as $t-z \, \varepsilon/u$. The parameter λ, known as the filter coefficient, characterizes the rate of filtration and can be directly related to the single collector efficiency, η [3]. The effect of the particle deposition is expressed in terms of its effect on λ and the pressure gradient ($\partial P/\partial z$) at a given flow rate. The ratio of λ/λ_o, where λ_o is the value of λ initially (or the bed is clean), is an indication of the change in the rate of filtration due to deposition. Similarly, $(\frac{\partial P}{\partial z})/(\frac{\partial P}{\partial z})_o$ gives the effect of deposition on pressure drop. The effect of deposition on filter performance can be quantitatively assessed if F_1 or F_2 are known.

THEORETICAL INVESTIGATIONS

The primary reason an increase in the extent of deposition causes changes in filter performance is that the filter media structure changes. Accordingly, with a postulated relationship between particle deposition and resultant changes in filter media, one can obtain the necessary relationship describing the effect of deposition on filter performance (that is, F_1 or F_2).

An example of applying this kind of approach is given below. One may argue that the deposition of particles outside filter grains leads to the presence of a deposited layer outside filter grains. This layer may be assumed to be approximately uniform in thickness. The results of deposition are an increase in the effective grain diameter and a decrease in the filter porosity. The filter diameter (d_g), the filter porosity (ε), and the specific deposit (σ) and their respective initial values (denoted by subscript o) are related to one another by the following equations:

$$\frac{d_g}{d_{g_o}} = (\frac{1-\varepsilon}{1-\varepsilon_o})^{1/3} \tag{6}$$

$$\varepsilon_o - \varepsilon = \sigma/(1-\varepsilon_d) \tag{7}$$

where ε_d is the porosity of the deposited layer.

If one assumes that the pressure-drop/flow-rate relationship of filter beds is given by the Carman-Kozeny equation, the ratio of the pressure gradient to that of the clean filter is found by Tien et al. [3] to be

$$\frac{(\partial P/\partial z)}{(\partial P/\partial z)_o} = [1 + \frac{\sigma}{(1-\varepsilon_d)(1-\varepsilon_o)}]^{4/3} [1 - \frac{\sigma}{\varepsilon_o(1-\varepsilon_d)}]^{-3} \tag{8}$$

Experimentally, it is known that at $\sigma \cong 10^{-4}$ the pressure gradient ratio is of the order of two or higher. This condition cannot, however, be met by Equation (8) unless $1-\varepsilon_d$ assumes absurdly small values. Similar conclusions can be obtained regarding the change in λ (or η).

The fact that Equation (8) failed to correctly predict the effect of deposition on filter performance clearly called into question the validity of the assumption used in assessing the deposition effect-- that is, that deposited particles form smooth layers outside filter grains.

An important consideration in aerosol filtration that deserves attention is that the suspensions to be treated are inevitably extreme- ly dilute. The distribution of particles throughout a suspension is uniform only in a global sense. On the other hand, the manner in which deposited matter is formed is stochastic. Accordingly, even if a relatively smooth deposit layer is formed outside filter grains ulti- mately the smooth-layer morphology does not apply throughout the en- tire period of filtration. This argument is substantiated by a number of experimental observatins in the deposition of aerosols on single collectors and model filters. At least in the earlier stage, deposi- tion can better be characterized as that of dendritic growth as shown in Figure 2. Smooth layers of deposited particles on filter cakes are formed only in the later stages, when neighboring particle dendrites are merged and become indistinguishable.

Payatakes and Tien [4] developed a formal theory describing dendritic growth by a series of ordinary differential equations. The

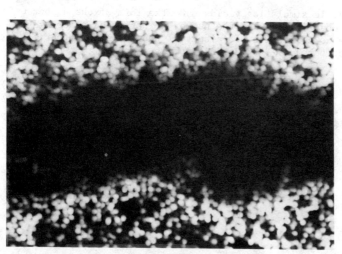

FIG. 2 MICROGRAPHS SHOWING DEPOSIT MORPHOLOGY
IN PARALLEL-FIBER MODEL FILTERS (Ref. 15)

approach was further extended by Payatakes and coworkers [5,6,7] and
applied to the study of the loading effect in fibrous filtration.
These studies have yielded explicit expressions relating the increase
in λ (or η) with relevant operating variables. In obtaining these
expressions, however, it was found necessary to introduce extraneous
assumptions concerning the spatial distribution of particle dendrites.

Tien et al. [8,9] proposed a more direct, but perhaps more tedious,
way to predict the buildup of deposited particles based on stochastic
simulation. In essence, this method called for a re-creation of
particle deposition through computer experiments. The suspensions
to be treated in filtration are inevitably dilute. The specific
positions occupied by aerosol particles as they approach the collector
can, therefore, be assumed to be randomly distributed over the entire
domain of interest. Whether a given aerosol particle will be deposited depends upon (a) whether the trajectory in the particle will
intercept the collector or any one of the previous deposited particles,
and (b) the nature and magnitude of the surface-interaction forces
between the particle and the collector upon the particle's impact.
Once a particle is deposited on a collector, that part of the collector surface where the deposition takes place and its immediate sur-
roundings are no longer available for deposition. Instead, the deposited particle acts as an additional collector, leading to the
presence and growth of particle dendrites, as shown in Figure 2.

The work of Pendse and Tien [10] represents a specific example
of applying the stochastic simulation to the study of the deposition
effect in granular filtration. To characterize filter beds, these
investigators employed the constricted-tube porous-media model, which
likens the deposition problem occuring in a filter bed to the deposition
of particles from suspensions flowing through a constricted tube (see
Figure 3). The positions occupied by the various entering particles
at the inlet are assumed to be randomly distributed over the inlet
cross-section of the tube. Once the inlet position is specified, the
particle trajectory can be determined; the deposition of the particles
can be then ascertained from the particle trajectory. An illustration
of the particle deposits formed within a constricted-tube collector
is shown in Figure 4.

The stochastic simulation yields take the form of the number of
particles collected (m_c) versus the number of particles entering (m_{in}).
The instantaneous collection efficiency of the collector (that is,
the constricted tube), η, by definition, is given as

$$\eta = \frac{dm_c}{dm_{in}} \tag{9}$$

If m_c can be fitted as a polynomial of m_{in},

$$m_c = \sum_i a_i m_{in}^i \tag{10}$$

then, from the results of the simulation, the collection efficiency,
η, can be given as

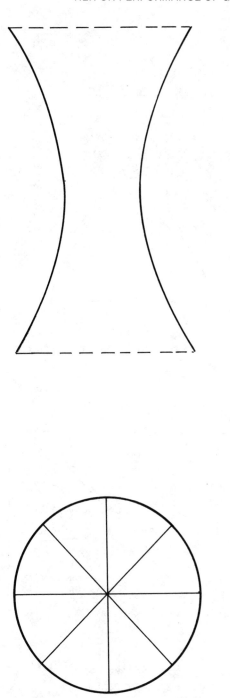

FIG. 3 FRAMEWORK USED IN STOCHASTIC SIMULATION

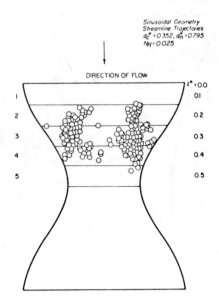

FIG. 4 DEPOSIT PATTERN OBTAINED FROM STOCHASTIC SIMULATION (Ref. 10)

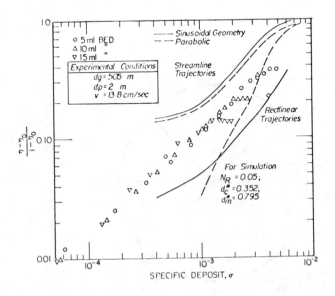

FIG. 5 COLLECTION EFFICIENCY INCREASE AS A FUNC-TION OF   (Ref. 10)

$$\eta = \sum_i ia_i m_{in}^{i-1} \tag{11}$$

Furthermore, the initial efficiency of the constricted-tube collector (in the clean-collector efficiency) is the limit of the values of at $t \to 0$:

$$\eta_o = \lim_{m_{in} \to 0} \frac{dm_c}{dm_i} \tag{12}$$

The extent of the deposition (expressed as the specific deposit, σ) can be calculated as

$$\sigma = \frac{(\frac{4}{3} \pi a_P^3)m_c}{\ell/N} \tag{13}$$

where N is the number of constricted tubes per unit cross-sectional area of the filter and ℓ is the axial length of the unit bed element defined as $[\frac{\pi}{6(1-\epsilon)}]^{1/3} <d_g>$ where $<d_g>$ is the average filter grain diameter.

From the simulation results and the relationships given above, one can obtain, for a specified set of operating conditions, the changes in collection efficiency as a function of σ. An example is given in Figure 5, in which the change in η is expressed as $(\eta-\eta_o)/(1-\eta_o)$, that is, the ratio of the increase in η and the ultimate increase. As stated earlier, this stochastic simulation requires knowledge of the trajectories of aerosol particles as they flow through the constricted tube. Determining these trajectories, in turn, requires knowledge of the flow field through the tube, which varies with the extent of deposition and, in general, is not known. For this reason, Pendse and Tien carried out the simulation limiting particle trajectories to two conditions: particles possess high inertia such that the trajectories are rectilinear; and particles possess little inertia such that their trajectories follow fluid streams based on clean collector conditions. Comparisons of the results of these two limiting conditions and experimental data are shown in Figure 5.

Generally speaking, the development of the theoretical analysis of the loading effect has not reached the stage of yielding results for practical uses.

EXPERIMENTAL DETERMINATION OF F_1 AND F_2

At present, experimentation remains the only practical way to obtain useful information on the effect of deposition on filter performance, specifically the determination of the functional relationships, F_1 and F_2.

The effect of the deposition, generally speaking, is a local phenomenon, since the extent of deposition is not uniform throughout a filter bed. On the other hand, such quantities as effluent concentration and pressure drop, which can be readily measured from experimental filters, are macroscopic in nature. For example, effluent concentration is determined by the particle-collecting capabilities of a number of collectors (that is, filter grains) connected in a series, each of which is under a different deposition effect. The complication can be ameliorated to a degree by carrying out measurements using shallow filters. However, even with the use of a bed 0.5 cm high and filter grains with a diameter of 500 μm, the bed represents, roughly, ten collectors connected in a series. Furthermore, using shallow filters for aerosol filtration experiments poses considerable difficulties. The consistency and accuracy of the results obtained are often uncertain.

In a more recent study, Walata et al. [11] systematically examined three different methods for obtaining information about F_1 (or F_2) from effluent concentration and pressure-drop measurements. The following method was found to be the best. Filtration experiments can be conducted by using filters of differing heights under otherwise identical conditions. For measurements obtained from a given filter, the filter coefficient and the average specific deposit at different times can be evaluated from the effluent-concentration data by using the following equations:

$$\lambda = (\frac{1}{L})/\ln \frac{c_{in}}{c_{eff}} \qquad (14)$$

$$\bar{\sigma} = \frac{u}{L} \int_{0}^{L} (c_i - c_{eff}) \cdot d\theta \qquad (15)$$

An empirical correlation of the following form is then sought:

$$\frac{\lambda}{\lambda_o} = 1 + \bar{\alpha}_1 \bar{\sigma}^{\bar{\alpha}_2} \qquad (16)$$

By applying the same procedure to data obtained from filters of differing heights, different $\bar{\alpha}_1$ (and $\bar{\alpha}_2$) values at different L can be obtained. The limiting values of $\bar{\alpha}_1$ and $\bar{\alpha}_2$ as L→0 are then taken as the correct value describing the effects of deposition on λ. In other words, F_1 is given as

$$F_1 = 1 + \alpha_1 \sigma^{\alpha_2} \qquad (17)$$

where α_1 and α_2 are the limiting values of $\bar{\alpha}_1$ and $\bar{\alpha}_2$ as L→0.

Similar procedures can be followed to obtain F_2. The average pressure gradient $(\overline{dP/dz})$ is calculated from the overall pressure drop to be

$$(\frac{\partial P}{\partial z}) = \frac{\Delta P}{L} \qquad (18)$$

An expression similar to Equation (16) is then sought:

$$\frac{\overline{(\partial \overline{P}/\partial z)}}{(\partial P/\partial z)_o} = 1 + \overline{\beta}_1 \sigma^{\overline{\alpha}_2} \tag{19}$$

The correct expression of F_2 is assumed to be

$$F_2 = 1 + \beta_1 \sigma^{\beta_2} \tag{20}$$

where β_1 and β_2 are the limiting values of $\overline{\beta}_1$ and $\overline{\beta}_2$ as $L \to 0$.

This method was subsequently taken up by Takahashi et al. (12). Based on data collected from experiments using mono-dispersed aerosol suspensions (diameter 2.02 µm) and filter grains of the glass spheres of different sizes, F_1 and F_2 are found to be given by Equations (19) and (20), with constants α_1, α_2, β_1, and β_2 as given by the following expressions:

$$\alpha_1 = [3.42 \times 10^{-5} + 0.0292 \ N_R^{1.5}] \ N_{St}^{-3.8} \tag{21}$$

$$\alpha_2 = 0.26 \ \frac{1}{N_{St}} - 0.23 \tag{22}$$

$$\beta_1 = [1.84 \times 10^{-5} + 4.32 \times 10^{-2} \ N_R^{1.5}] N_{St} \tag{23}$$

$$\beta_2 = 0.52 + 0.14 \ ha \ \frac{1}{N_{St}} \tag{24}$$

These results should not be regarded as definitive, since they are based on a limited amount of data. It is also important to note that in practical applications suspensions to be treated by granular filtration are not mono-dispersed but contain particles of different sizes. It is not clear what kind of average values of α_1, α_2, β_1, and β_2 are appropriate for such cases.

AEROSOL DEPOSITION IN MODEL FILTERS

It is obvious that, because of the complex nature of the deposition phenomenon, the study of the deposition effect and the determination of F_1 and F_2 would be greatly facilitated if better insights into the depositic processes were available. Such insights can be obtained by microscopically observing the deposition processes on a nonintensive and continuous basis.

Tien and coworkers [13,14] have recently explored the use of two-dimensional model filters to obtain direct and microscopic information about aerosol deposition. These model filters were fabricated by photoetching plate glass and consisted of many relatively small (approximately 680 μm in diameter) cylinders arranged in three different configurations. These configurations were found to share certain characteristics with granular filters. By using these model filters, it was possible to observe the local deposition phenomenon (around each cylinder) over a relatively large spatial domain simultaneously. One of the important findings is that the extent of deposition around each collector at the same filter height varies significantly and depends upon the size of the pore space immediately above the cylinder. Another observation of possible significance is that, for each individual collector, significant deposition occurs at the front stagnation point of the collector. One might therefore argue that the aerosol deposition, on a local basis, may be described as a two-dimensional jet of aerosols impinging on a collector. However, the impinging jet flow field has never been considered by theoretical analysts. Yoshida and Tien [14] also obtained an expression for F_2 based on model filter experimental data, which agreed qualitatively with the data of Takahashi et al. [12].

The results of the model filter studies prove that the model filters provide a powerful tool for obtaining insights into the deposition process. However, to obtain quantitative information using model filters in the study of the effect of deposition on filter performance, it is necessary to establish an equivalence criterion between model filters and granular filters. Such a study is currently under way in the author's laboratory.

ACKNOWLEDGMENT

This study was performed under Contract DE-AC02-79ER10386, Department of Energy, Office of Basic Energy Science.

REFERENCES

[1] Tien, C. and Payatakes, A. C., "Advances in Deep Bed Filtration," AIChE Journal, Vol. 25, No. 5, Sept. 1979, pp. 737-759.

[2] Payatakes, A. C., Tien, C. and Turian, R. M., "Trajectory Calculation of Particle Deposition on Deep Bed Filtration: Part I Model Formulation," AIChE Journal, Vol. 20, No. 5, Sept. 1974, pp. 889-900.

[3] Tien, C., Turian, R. M. and Pendse, H., "Simulation of the Dynamic Behavior of Deep Bed Filtration," AIChE Journal, Vol. 25, No. 3, May 1979, pp. 385-395.

[4] Payatakes, A. C. and Tien, C., "Particle Deposition in Fibrous Media Wall Dendrite-Like Pattern: A Preliminary Model," Journal of Aerosol Science, Vol. 7, No. 2, 1976, pp. 85-100.

[5] Payatakes, A. C., "Model of Transient Aerosol Particle Deposition in Fibrous Media with Dendrite Pattern," AIChE Journal, Vol. 23, No. 2, March 1977, pp. 192-203.

[6] Payatakes, A. C. and Graydon, L., "Dendritic Deposition of Aerosol Particles on Fibrous Media by Inertial Impaction and Interception," Chem. Eng. Sci., Vol. 35, No. 5, 1980, pp. 1083-1095.

[7]   Payatakes, A. C., and Okuyama, K., "Effects of Aerosol
      Particle Deposition on the Dynamic Behavior of Uniform
      and Multilayer Fibrous Filters," Proceedings of International
      Symposium on Powder Technology '81, The Society of Powder
      Technology, Kyoto, 1982, pp. 501-508.

[8]   Tien, C., Wang, C. S. and Barot, D. T., "Chainlike Formation
      of Partial Deposits in Fluid Particle Separation," Science,
      Vol. 196, No. 4298, May 1977, pp. 983-985.

[9]   Wang, C. S., Beizaie, M. and Tien, C., "Deposition of Solid
      Particles on a Collector: Formulation of a New Theory,"
      AIChE Journal, Vol. 23, No. 6, Nov. 1977, pp. 879-889.

[10] Pendse, H. and Tien, C., "A Simulation Model of Aerosol
      Collection in Granular Media," Journal of Colloid and
      Interfacial Science, Vol. 87, No. 1, May 1982, pp. 225-241.

[11] Walata, S. A., Takahashi, T. and Tien, C., "Effect of Deposition
      on Granular Aerosol Filtration," Aerosol Science and Technology,
      Vol. 5, No. 1, 1986, pp. 23-37.

[12] Takahashi, T., Walata, S. A. and Tien, C., "Transient Behavior
      of Granular Filtration of Aerosols," AIChE Journal, Vol. 32,
      No. 4, April 1986, pp. 164-190.

[13] Ushiki, K. and Tien, C., "In-Situ Observation of Aersol
      Filtration on a Two-Dimensional Mode Filter," AIChE Sym.
      Ser., No. 241, Vol. 80, 1985, pp. 137-148.

[14] Yoshida, H. and Tien, C., "Dynamic Behavior of Aerosol Filtra-
      tion on a Two-Dimensional Model Filter," Aerosol Science in
      Technology, Vol. 4, No. 4, 1985, pp. 365-381.

[15] Tsiang, R. C., Wang, C. S. and Tien, C., "Dynamics of Particle
      Deposition on Model Fiber Filters," Chem. Eng. Sci., Vol. 37,
      No. 11, 1982, pp. 1661-1673.

Kenneth L. Rubow and Benjamin Y. H. Liu

CHARACTERISTICS OF MEMBRANE FILTERS FOR PARTICLE COLLECTION

REFERENCE: Rubow, K. L. and Liu, B. Y. H., "Characteristics of Membrane Filters for Particle Collection," Fluid Filtration: Gas, Volume I, ASTM STP 975, R. R. Raber, Ed., American Society for Testing and Materials, Philidelphia, 1986

ABSTRACT: A summary is given of the aerosol filtration models for membrane filter media. While the capillary tube model has been shown to accurately predict the particle collection characteristics of Nuclepore membrane filters, the fibrous filter model gives more accurate prediction for the conventional solvent-cast membranes. Excellent agreement has been found between the effective fiber diameter used in the model and the diameter of the fiber-like structures in the conventional membrane. Results are presented showing the comparisons of the measured filter penetration with the theoretically predicted value for various particle sizes and filtration velocities. Excellent agreement has been obtained both in the degree of particle penetration through the filter and the location of the most penetrating particle size. For the conventional membranes the most penetrating particle size is typically in the range of 0.05 to 0.2 um diameter.

KEYWORDS: Gas filtration, membrane filters, aerosol filtration, filtration theories, filter modeling

Membrane filters are thin, porous structures used to separate particles from the suspending gas. They are widely used for the collection of particulate matter for analysis or for gas purification. Due to their small pore-like structure, they can achieve high filtration efficiency. Once collected, the particles can be analyzed to determine their size distribution, concentration and/or chemical composition.

Membrane filters consist of an integral porous structure of

Drs. Rubow and Liu are respectively Research Associate and Manager, and Professor and Director, Particle Technology Laboratory, Mechanical Engineering Department, University of Minnesota, Minneapolis, MN 55455.

either plastic or metal with a substantial void volume. The membrane is usually quite thin with a thickness in the range of 10 to 150 um. As a result of their chemical composition and the techniques of manufacturing, membrane filters generally attain a fairly uniform structure. This structure consists of a myriad of interconnected void spaces (pores). The size of the pores can be controlled such that membranes can be produced with effective pore diameters ranging from about 0.01 to 40 um. Due to this uniform pore structure, the resulting membrane has absolute particle retention characteristics for all particles larger than the rated pore size of the membrane.

Membrane filters can be classified generically according to their internal pore structure, chemical composition and/or method of manufacture. Materials of construction include cellulose esters, polyvinyl chloride, nylon, polypropylene, polyvinylidene fluoride (PVDF), polytetrafluoroethylene, polycarbonate, sintered metal and others. The manufacturing process can be casting, stretching, additive leach, sintering, track-etching, etc. As a result of these various manufacturing processes, the void spaces within the structure usually attain a complex form which results in a tortuous flow path. The primary exception is the Nuclepore filter which has a straight capillary-tube like pore structure.

Micrographs of several different types of membrane filters are shown in Figure 1. The upper surface of the Nuclepore filter shows the openings of the cylindrical pores. The other filters are examples of the conventional membrane filters. Micrographs of the Millipore Type SC and Durapore membranes are shown. These membranes, which are solvent-cast, are composed of cellulose-ester and PVDF respectively. The SC membrane has a rated pore size of 8.0 um and the Durapore, 0.2 um. A cross-sectional view of the SC filter is shown in Figure 1d. This view, together with that of the upper surface shown in Figure 1c, show the interconnected void space of this membrane. Also evident is the fiber-like nature of the connecting links in the filter structure.

As a result of the different types of membrane structure, two different filter models have been developed to describe the particle collection characteristics of each type. These models are the capillary tube model and the fibrous filter model. The capillary tube model can be used for filters such as the Nuclepore. The pores in this membrane are cylindrical and oriented perpendicular to the filter face. The fibrous filter model more closely resembles the structure of the conventional membrane in that the membrane consists of a series of interconnected fiber links between the adjacent open void spaces.

FILTER MODELS

The goal of modeling a porous media is to mathematically describe the fluid flow and particle retention characteristics of the porous media. Ideally, the model would be based on an exact geometric representation of the real structure. However, this is not possible

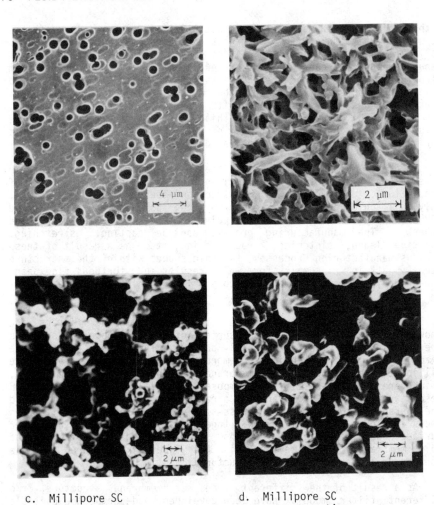

c. Millipore SC
    -front surface

d. Millipore SC
    -cross section

FIG. 1--Micrographs of various membrane filters.

in practice due to the complex and random nature of the void spaces within porous materials. Thus, simplified models are developed to describe the porous material.

In the case of aerosol filtration by porous media, two filtration models have been developed. These models are the capillary tube and the fibrous models.

Capillary Tube Model

The capillary tube model is one of the simplest and most widely used model for porous media. This model has been used for several decades as the basis for predicting the fluid pressure drop across porous media [1,2].   However, it was not until the 1960's that particle deposition theories based on the capillary tube model were developed and applied to membrane filters. The initial work was done by Spurny and Pich [3-5] and Megaw and Wiffen [6].   During the past fifteen years this particular model has been further refined and additional particle deposition theories have been developed by other investigators.   Numerous reviews of filtration theories have been made [7-14].

Geometric Description: In the capillary tube model the filter is represented by an assembly of straight capillary tubes of cylindrical cross section.   The tubes are parallel to each other and perpendicular to the filter face.   The diameter of the tubes is assumed to be equal to the diameter of pores in the real filter and is usually taken to be constant for all tubes.

The porosity P of the filter is the fraction of void volume in the filter. It is equal to $1 - \alpha$, where α is the solid volume fraction. For the capillary tube model, the porosity is

$$P = N_p A_p \tag{1}$$

where N_p is the number of pores per unit filter surface area and A_p is the cross-sectional area of each capillary tube.   If one assumes a hexagonal lattice arrangement for the pore openings, as shown in Fig. 2, the relation between the distance between pore centers d_p', pore diameter d_p, and porosity is

$$P = \frac{\pi}{2\sqrt{3}} \left(\frac{d_p}{d_p'}\right)^2 \tag{2}$$

The average velocity U of the fluid through each pore, based on the Dupuit relation [1], is related to the face velocity U_o by

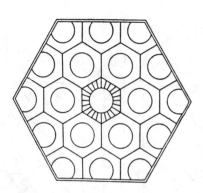

FIG. 2--Schematic diagram of the filter surface in the capillary tube model showing the pores in a hexogonal lattice.

$$U = \frac{U_o}{A_p N_p} = \frac{U_o}{P} \quad . \tag{3}$$

Particle Deposition Theories: The overall filtration theories are formulated in terms of individual deposition mechanisms. The mechanisms include inertial impaction, diffusion, interception, gravitational settling and electrostatic deposition. In most cases theories are formulated to deal with each mechanism individually. However, in a few cases, the theories are based on the simultaneous operation of two mechanisms, for example, inertial impaction and direct interception or direct interception and diffusion.

Table 1 summarizes, by deposition mechanism, the available particle deposition theories for the capillary tube model. Note that the theories all deal with incompressible fluids and particle deposition on a "clean" filter. Additional assumptions common to all theories include:

1. Fluid velocity and particle size distributions upstream of filter are uniform.
2. Pores are parallel to the direction of fluid flow.
3. Pores are uniformly distributed over the entire filter surface.
4. Fluid flow into pores is steady and axisymmetric.
5. All filter surfaces are smooth.
6. No particle bounce or blowoff from the filter surfaces.

The primary particle deposition mechanisms are diffusion, direct interception and inertial impaction. Other deposition mechanisms may become significant depending upon the fluid flow rate, pore size, particle size and state of the electrical charge of the filter-particle system. The most important additional mechanisms are

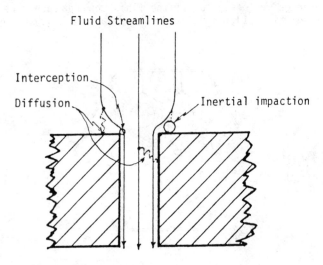

FIG. 3--Schematic diagram of a capillary tube showing the fluid streamlines and the mechanisms of particle collection.

TABLE 1

Summary of Capillary Tube Particle Deposition Mechanism Theories

Particle Deposition Mechanism	Reference	Solution Method	Remarks
Impaction	Pich (1964) [16]	Analytical	Approximated shape of converging streamline as parabolic
Impaction and Interception	Smutek and Pich (1974)[21]	Analytical	Approximated fluid field flow
	Smith and Phillips (1975) [18]	Numerical	Solution in Graphic form
	Manton (1978) [22]	Numerical	Obtained approximation expression from numerical results
	Kanaoka et al. (1979) [23]	Numerical	
Interception	Spruny et al. (1969) [7]	Analytical	Assumed uniform velocity profile at pore entrance
	Smith et al. (1976) [17]	Analytical	Assumed parabolic velocity profile at pore entrance
	John et al. (1978) [19]	Analytical	Assumed non-uniform velocity profile at pore entrance
Pore Wall Diffusion	Gormley and Kennedy (1949) [26]	Analytical	Fully developed velocity profile
	Twomey (1962) [25]	Numerical	Fully developed velocity profile
	Smith et al. (1976) [17]	Empirical	For 3.0 μm Nuclepore
Face Diffusion	Manton (1979) [28]	Numerical	Obtained approximation expression from numerical results
Pore Wall and Face Diffusion	Smith et al. (1976) [17]	Empirical	For 3.0 μm Nuclepore
Pore Wall Diffusion and Interception	Smutek (1972) [27]	Numerical	Fully developed slip flow velocity profile

electrical deposition [15] and gravitational sedimentation [3,9,10].

A schematic diagram of a capillary tube-like pore in a membrane filter is shown in Fig. 3. The fluid streamlines and primary mechanims of particle particle collection are also illustrated.

Inertial Impaction: Particles are deposited on the filter face by the mechanism of inertial impaction due to the finite mass and monentum of the particle and the change in direction as fluid converges into the pore openings. The first theory [3,16] involves the assumption that the flow field near the pore entrance can be approximated by converging streamlines which are assumed to be parabolas. The resulting particle collection efficiency η_I for inertial impaction is

$$\eta_I = \frac{2 E_i}{1 + \beta} - \frac{E_i^2}{(1 + \beta)^2} \tag{4}$$

where $E = 2 St\sqrt{\beta} + 2 St^2 \beta \exp\left(-\frac{1}{St\sqrt{\beta}}\right) - 2 St^2 \rho_p$

and $\beta = \dfrac{\sqrt{P}}{1 - \sqrt{P}}$

The Stokes number St is defined as

$$St = \frac{C D_p^2 \rho_p U_o}{9 \mu d_p} \tag{5}$$

where D_p is the particle diameter, ρ_p the particle density, μ the fluid viscosity and C the Cunningham slip correction.

Direct Interception: Particles are also deposited on the filter face by the mechanism of direct interception. Interception occurs when the streamline carrying a particle passes within a distance equal to or less than the particle's radius from the edge of the pore opening, thus, allowing the particle to be "intercepted." For particles greater than the diameter of the pore openings, all particles are intercepted.

As shown in Table 1, three different theories have been developed. Spurny et al. [7] developed an expression assuming uniform fluid velocity and particle concentration profiles at the pore entrance. The resulting collection efficiency η_R and penetration P_R expressions were shown to be

$$\eta_R = 2 R - R^2$$
$$P_R = (1 - R)^2 \tag{6}$$

where the interception parameter R is the ratio of the particle to pore diameter, D_p/d_p. Smith et al. [17], assuming a fully developed, parabolic velocity profile at the pore entrance (based on the numerical analysis of the flow field [18]) developed the penetration expression

$$P_R = (1 - R)^2 (2 - (1 - R)^2) \qquad (7)$$

John et al. [19] derived the expression

$$n_R = (2 R - R^2)^{3/2} \qquad (8)$$

Experimental work [7,20] has showed that Eqs 7 and 8 better predict the performance of Nuclepore filters than does Eq 6.

Impaction and Interception: The interception of particles at the entrance to the pore opening is enhanced by inertial effects. Because of the finite particle mass and momentum, the trajectory of particles would depart from the trajectory of the fluid as the latter approaches the pore opening. This results in additional collection.

Smutek and Pich [21] derived an expression for particle deposition on the filter surface due to the combined mechanisms of inertial impaction and direct interception. They used the same assumptions as Pich [16] and obtained the following equation for the collection efficiency

$$n_{IR} = \frac{(3+k)(3+2k) - P(5+3k) + PR(2-R)(1+k)(k^2+4k+5)}{(1+k)(3+k)^2}$$

$$+ \frac{2(1-R)(P(1+k)(2+k)(k(3+k) + 2P - PR(2-R)(1+k)(2+k)))^{1/2}}{(1+k)(3+k)^2} \qquad (9)$$

where k is a dimensionless impaction parameter defined as:

$$k = \frac{1}{Re\ St} \qquad (10)$$

Because 1/k is generally quite small, Eq 9 can be approximated by expanding it into a Taylor series and dropping the higher order terms. The resulting expression is

$$n_{IR} = 2(1-\sqrt{P}) \frac{1}{k} + 2PR + (8\sqrt{P}-3P-5) \frac{1}{k^2} - PR^2 + 2\sqrt{P} (1+P-2\sqrt{P}) \frac{R}{k} \qquad (11)$$

The theory is limited to the case where filter porosities are near one since the fluid velocity in the pore is assumed to be the same as to the face velocity.

Several authors [18,22-24] have used numerical methods to determine particle deposition due to the combined mechanisms of impaction and interception. They solved the fluid flow equations numerically and determined the resulting particle trajectories by numerical integration of the particle equation of motion. The results from three of the studies [18,23,24] are presented only in graphical form. In addition to graphically presenting his results, Manton [22] derived an approximate equation for the collection efficiency. A Reynolds number of zero and fully developed velocity profile at the pore entrance were assumed. The approximate expression for filters with porosities in the range of 0.04 to 0.36 is as follows

$$\eta_{IR} = (R' (2-R'))^{2/(1 + a_i R' + 2 b_i R')} \tag{12}$$

where $a_i = (a_1 I^2 - a_2 I)/(I + a_3)$

$b_i = b_1 I + b_2 I^{1/2} + b_3 I^{1/4}$

$a_1 = 0.688500 - 1.473959 P + 0.286458 P^2$

$a_2 = 7.754343 - 46.535261 P + 72.737760 P^2$

$a_3 = 1.296234 - 4.525388 P + 6.5544101 P^2$

$b_1 = 2.720000 - 33.125000 P + 65.625000 P^2$

$b_2 = 16.117649 - 36.715870 P + 27.829932 P^2$

$b_3 = -10.611610 + 28.831488 P - 31.098401 P^2$

and $I = \dfrac{St}{R'^2}$; $R' = \dfrac{D_p}{2P^{1/2}}$

Diffusion: As aerosol flows through a filter, particle deposition due to Brownian diffusion occurs both on the filter face and on the pore walls. Diffusion to the pore wall is the dominant mechanism.   The effects of face diffusion has been completely neglected until recently.

The original capillary tube wall diffusion theory as applied to membrane filters was proposed by Spurny and associates in the early 1960's.   They used a diffusion theory developed by Twomey [25] to describe particle deposition from a fully developed laminar flow field in a cylindrical tube. The expression used to predict particle penetration P_D through a capillary tube is

$P_D = 0.81904 \exp (-3.6568 N_D) - 0.9752 \exp (-22.3045 N_D)$

$-0.03248 \exp (-56.95 N_D - 0.0157 \exp (-107.6 N_D) \tag{13}$

where N_D is the dimensionless parameter defined by

$$N_D = \frac{4 L D}{d_p^2 U} \tag{14}$$

and D is the particle diffusion coefficient and L the filter thickness.   Spurny et al. [7] used this expression for $N_D > 0.001$. but for smaller N_D, they used the following expression obtained by Gormley and Kennedy [26],

$$\eta_D = 2.56 N_D^{2/3} - 1.2 N_D - 0.177 N_D^{2/3} \tag{15}$$

Lastly, expressions have been developed to predict diffusional deposition to pore walls for the case of slip flow in the pore [27] and particle diffusion to the filter face [17,28].

Overall Filter Efficiency: The total filter efficiency can be obtained by combining the expressions for the individual particle collection mechanisms. It is usually assumed that there is no flow interaction between neighboring pores and that each deposition mechanism operates independently.

One method to obtain an expression for the overall penetration P_T is to assume that P_T can be expressed as the product of the penetrations for the individual mechanisms. This assumes that the mechanisms are mutually independent. The overall penetration and total efficiency E_T, can be written as

and
$$P_T = P_I \, P_R \, P_D \tag{16}$$

$$E_T = \eta_I + \eta_R + \eta_D - \eta_I \, \eta_R - \eta_R \, \eta_D - \eta_I \, \eta_D + \eta_I \, \eta_R \, \eta_D \tag{17}$$

Other variations on the above expressions have also been used. Spurny et al. [7] developed the expression

$$E_T = \eta_I + \eta_D + \Delta \eta_R - \eta_I \, \eta_D - \Delta \, \eta_I \, \eta_R \tag{18}$$

where Δ is an empirically determined coefficient equal to 0.15 for Nuclepore filters. This coefficient is needed to correct for the overprediction of the total efficiency by Eq 17 in the interception regime. Spurny and Madelaine [20] further modified the overall efficiency equation to correct the underestimation of the interception effects which sometimes occurs with Eq. 18. Their expression is

$$E_T = \eta_I + (1 - \eta_I)(\eta_D + \eta_R^{\omega \, (1-R)}) \tag{19}$$

where $\omega = 0.63$ is an empirically determined constant for Nuclepore filters. The interception efficiency expression used with Eqs 18 and 19 is Eq 6.

Fibrous Model

Numerous theories have been developed to describe the pressure drop and particle penetration through fibrous filters. A review of all of these in detail is beyond the scope of this paper. Therefore, only the salient points required in the application of this approach to the modeling of membrane filters will be described. Additional details of the fibrous filter model are given by Liu and Rubow in another section of this book.

In the fibrous model, the filter is assumed to consist of an assemblage of straight, cylindrical elements (fibers). The fibers all lie parallel to each other and perpendicular to the direction of fluid flow. Each fiber has a circular cross section of the same diameter. All fibers are distributed uniformly throughout the filter.

The flow field around each fiber can be obtained by one of two basic approaches. In one approach, the fiber is treated as an isolated cylinder; whereas in the second approach the filter is treated as a system of cylinders. This latter approach, called a cell model, accounts for the interference caused by adjacent cylinders.

The most widely used fluid flow field for the cell model is that derived by Kuwabara [29]. This flow field is a solution of the Navier-Stokes equation for the case of viscous flow. The Kuwabara flow field originally was derived for the continuum flow regime. Pich [30] extended the theory to include the slip flow.

The Knudsen number Kn can be used to determine the approximate flow regime as the fluid flows around a cylinder [9,10]. The Knudsen number is defined as

$$Kn = \frac{2\lambda}{D_f} \tag{20}$$

where λ is the mean free path of gas molecules and D_f the fiber diameter. The continuum flow regime exists for Kn <0.001 while the slip flow regime extends from Kn values of 0.001 to 0.25. Kn equals 0.13 for a fiber diameter of 1 um in air at standard conditions of 20^0 C and 1 atm pressure.

The mathematical approach generally taken to determine the efficiency of a filter is to consider filtration by the individual mechanisms of diffusion, interception and inertial impaction. These individual mechanisms are then combined to give the overall efficiency.

In the region of the maximum penetration, the dominant mechanisms of particle collection are diffusion and direct interception. Assuming that these two mechanisms act independently, the single fiber efficiency can be written as

$$\eta_s = \eta_D + \eta_R \tag{21}$$

where η_D and η_R are the single fiber efficiency due to particle diffusion and interception respectively.

The relationship between single fiber efficiency η_s and the overall filter collection efficiency E_T and penetration P_T is

$$P_T = 1 - E_T = \exp\left(-\frac{4 \alpha L}{\pi D_f} \frac{\eta_s}{\epsilon}\right) \tag{22}$$

where L is the filter thickness and ϵ is the inhomogeity factor. The relationship accounts for the exponential decrease in particle concentration with penetration distance into the filter.

The inhomogeneity factor ϵ is an empirically determined factor correlating the pressure drop and single fiber efficiency η_{sf} of a real filter to a given model prediction [31,32]. This factor accounts for the random fiber orientation and non-uniform separation distance between fibers not accounted for in the model but present in real filters. The inhomogeneity factor can be defined in terms of pressure drop

$$\varepsilon = \frac{\Delta P_{theoretical}}{\Delta P_{experimental}} \tag{23}$$

or single fiber efficiency

$$\varepsilon = \frac{\eta_s}{\eta_{sf}} \tag{24}$$

where η_s is the theoretical single fiber efficiency and η_{sf} is the actual filter single fiber efficiency. Based on single fiber eficiency, the inhomogeity factor ε is approximately 1.67 for the polyester filter filters studied by Yeh and Liu [33] and Lee and Liu [34].

Deposition via diffusion results when particles collide with the fiber due to their random Brownian motion. This motion becomes more pronounced as the particle diameter becomes smaller. The dimensionless characteristics parameter used to describe the degree of deposition is the Peclet number, Pe, defined as:

$$Pe = \frac{U \, D_f}{D} \tag{25}$$

where D is the particle diffusion coefficient and U is the mean fluid velocity within the filter. This velocity is equal to $U_0/(1-\alpha)$. The magnitude of deposition increases as Pe decreases.

A particle is deposited via the interception mechanism if a particle of finite size is brought within one particle radius from the fiber surface and is caught by the fiber. The dimensionless interception parameter R characterizing this deposition mechanism is defined as

$$R = \frac{D_p}{D_f} \tag{26}$$

Particle deposition generally increases with increasing values of R.

The selection of the appropriate fibrous filtration theories for application to membrane filters is based on consideration of two filter parameters, namely, the filter solid volume fraction α and fiber diameter D_f. The solid volume fraction, also referred to as packing density or solidity, is the ratio of the volume of filter material per unit filter volume. For the conventional membrane filter, the solid volume fraction is of the order of 0.2 as compared to less than 0.1 for most fibrous filters. The "effective" fiber diameter of the membrane filter is usually less than 1 um. This means that at atmospheric pressure conditions, where the gas mean free path of air is 0.0653 um, the flow must be regarded as in the slip flow regime. Consequently, the criteria used to select the theories for review were:
1. The theory be valid in the slip flow regime.

2.   It has been derived to include terms on the order of α.
Most theories are eliminated by these criteria because they do not
appear to be completely relevant to the case of membrane filtration.

Rubow [35] derived expressions for the single fiber efficiency
due to particle diffusion and interception which are valid for both
the slip flow regime and higher values of the solid volume fraction.
The theory is based on the Kuwabara flow field.   The derivation of
the expression is based on the boundary layer approach.   The approach
used is similar to that used by Lee and Liu [36] for the continuum
regime, but treats the effect of slip in a manner similar to that
used by Pich [30,37].   The resulting expressions for single fiber
efficiency due to particle deposition via diffusion and interception,
η_D and η_R respectively are

$$\eta_D = 2.86 \frac{(1-\alpha)^{1/3}}{K^{1/3} \, Pe^{2/3}} \, C_D \qquad (27)$$

where

$$C_D = 1 + 0.389 \, \xi' \left(\frac{(1-\alpha) \, Pe}{K} \right)^{1/3} \qquad (28)$$

and

$$\eta_R = \left(\frac{1-\alpha}{K} \right) \frac{R^2}{1 + R} C_R \qquad (29)$$

where

$$C_R = 1 + \frac{2 \, \xi'}{R} \, . \qquad (30)$$

The hydrodynamic factor K [37] is given by the expression

$$K = -3/4 - 1/2 \, \ln\alpha + \alpha - 1/4 \, \alpha^2 + \xi' \, (-1/2 - \ln\alpha + 1/2 \, \alpha^2) \qquad (31)$$

A value of 1.15 Kn was assumed for the dimensionless coefficient of
slip ξ' [38,39].

COMPARISONS OF THEORIES WITH EXPERIMENTAL DATA

Nuclepore Filter

Comparisons of the theoretically predicted collection efficiency
to the experimental values obtained by Spurny et al. [7] for
Nuclepore filters are shown in Fig. 4 and 5.   The theoretical
efficiency is based on Eqs 4, 6, 13, 15, and 18.   Results in Fig. 4
show the effect of particle size on the efficiency.   The effect of
face velocity can be seen in Fig. 5 for particle collection in the
diffusion regime.   These data were obtained for Nuclepore filters
with pore sizes of 2 and 5 um diameter.

Researchers [14,20,23,17] have made additional comparisons
between experiment data and the theoretical values predicted by the

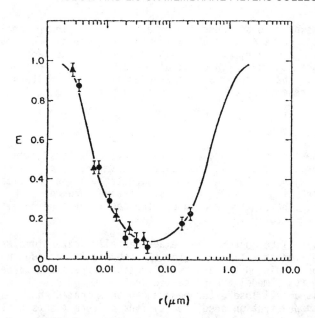

FIG. 4--Comparison of the collection efficiency as theoretically predicted and experimentally determined for a 5 um pore diameter Nuclepore filter at a face velocity of 5 cm/sec [7].

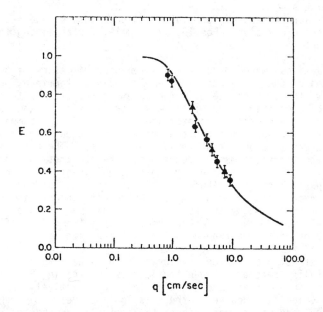

FIG. 5--Comparison of the theoretically predicted and experimentally determined collection efficiency for a 2.0 um diameter Nuclepore filter with 0.024 um diameter test aerosol [7].

various expressions described earlier. Their analysis show differences between theoretical and experimental results in some instances. However, while additional refinements can be made in the theory, the current theoretical expressions are adequate in describing the essential characteristics of filters with pore structures resembling capillary tubes.

Conventional Membranes

For the case of the Nuclepore and fibrous filters, the choice of an appropriate model is obvious. However, for the conventional membranes with a tortuous pore structure, the choice of an appropriate model is not as straightforward. Historically, investigators have attempted to use the capillary tube model but with little success [8,20].

Capillary Tube Model: The experimentally determined penetration data [35] for the Millipore Type SC membrane filter together with the penetration predicted by various filtration expressions based on the capillary tube model are presented in Fig. 6 and 7. The Millipore SC membrane is a cellulose ester membrane with a rated pore size of 8 um. The penetration predicted by four different sets of theoretical expressions are shown in Fig. 6. The sets involved the following combinations of theoretical expressions:
    Set 1: Eqs 4, 6, 13, 15, 16.
    Set 2: Eqs 4, 7, 13, 15, 16.
    Set 3: Eqs 4, 6, 13, 15, 18.
    Set 4: Eqs 4, 6, 13, 15, 19.
Two expressions were used for interception, Eqs 6 and 7, together with three expressions, Eqs 16, 17, and 19, for obtaining the overall penetration.

Calculations were made for assumed pore diameters of 2 and 8 um. Comparisons of the predicted to empirically determined penetration data show that, while the differences among the various theoretically determined penetrations are small, there is a significant difference between the theoretical and experimental values. The comparison also shows that the pore diameter has a greater affect on penetration than the differences among the various theories. The results indicate that as the pore diameter decreases, the theoretical penetration decreases. This decrease is most significant for the smaller diameter particles, i.e. those particles in the diffusion regime. However, regardless of which pore size or theory is used, the penetration predicted by theory is, in general, much higher than the actual values. The difference is more than two orders of magnitude in the region of the most penetrating particle size.

The effect of face velocity on the theoretically and experimentally determined penetration is shown in Fig 7. The comparison shows that the theory only qualitatively predicts the effect of velocity. However, there is no face velocity range where the theory adequately describes the actual penetration. For most particle sizes, the theory overpredicts the degree of particle penetration. Furthermore, the theory underpredicts the effect of velocity on penetration. The experimentally determined maximum

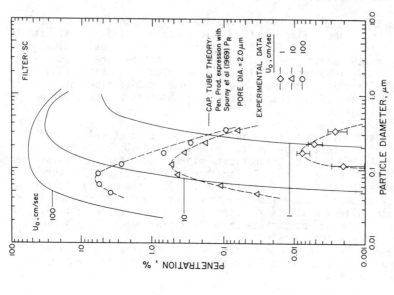

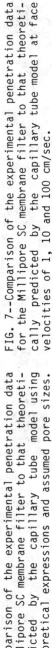

FIG. 6—Comparison of the experimental penetration data for the Millipore SC membrane filter to that theoretically predicted by the capillary tube model using various theoretical expressions and assumed pore sizes.

FIG. 7—Comparison of the experimental penetration data for the Millipore SC membrane filter to that theoretically predicted by the capillary tube model at face velocities of 1, 10 and 100 cm/sec.

penetration increases by 3 orders of magnitude as the velocity changes from 1 to 100 cm/sec while the theory only predicts less than one order of magnitude change.

Fibrous Filter Model: In the application of the fibrous filter model to membrane filters, the filter is treated as if it is equivalent to a fibrous filter. Specifically, the actual values for the filter thickness and solid volume fraction are used in the modeling. A inhomogeneity factor equivalent to that found for fibrous filters is also used. Lastly, an effective fiber diameter is determined to obtain the best fit of the theoretical penetration to the experimental data.

Fig. 8 and 9 show a comparison of the theoretically predicted penetration to that experimentally determined based on the fibrous filter model [35]. The theoretical expressions for diffusion and interception, Eqs 27 and 29, together with Eqs 21 and 22, were used to determine the theoretical penetration. An inhomogeity factor of 1.67 was assumed. Comparison are shown for the Millipore Type SC and AA membrane filters which have rated pore sizes of 8 and 0.8 um respectively. The comparison shows that the theory closely predicts the degree of particle penetration as well as the location of the most penetrating particle size. The differences between the theoretically and experimentally determined most penetrating particle sizes were less 10%.

Analysis of the experimental data show that these conventional membrane filters attain a low particle penetration (high particle collection) at a most penetrating particle size which is essentially independent of filter pore size. The most penetrating particle size was found to decrease from 0.18 to 0.05 um as the face velocity increased from 1 to 100 cm/sec. The maximum penetration for the 8 um pore size filter ranged from 0.008 to 8.0 % over the face velocity range of 1 to 100 cm/sec, respectively, while for the 0.8 um pore size, the penetration ranged from less than 0.001 to about 0.008 % over the same velocity range.

The comparison between the theoretical and experimental penetration curves presented in Fig. 8 and 9 shows excellent agreement in the shape of the penetration curves as a function of particle size. The theory, however, tends to over-predict the effect of velocity. The relative magnitude of this over-prediction increases as the face velocity increases or decreases from midrange values. The maximum theoretical over- or under- prediction is at most a factor of 2 and 10 for face velocities of 100 and 1.0 cm/sec respectively. For face velocities of 100 cm/sec and greater, this over-prediction can be partly contributed to the added increase in particle deposition due to impaction as the theory only assumes particle deposition via the mechanisms of diffusion and interception.

Excellent agreement was found between the effective fiber diameter used in the fibrous filter model and the diameter of the fiber-like structures in the membrane filter. The effective fiber diameters of the SC and AA were found to 0.87 and 0.555 um compared to the average diameter, measured from photomicrographs, of the fiber-like structures in the membrane of 0.8 and 0.5 um respectively.

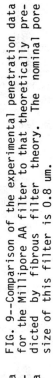

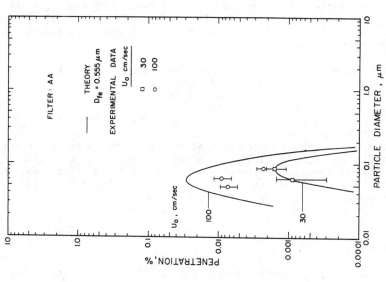

FIG. 9—Comparison of the experimental penetration data for the Millipore AA filter to that theoretically predicted by fibrous filter theory. The nominal pore size of this filter is 0.8 um.

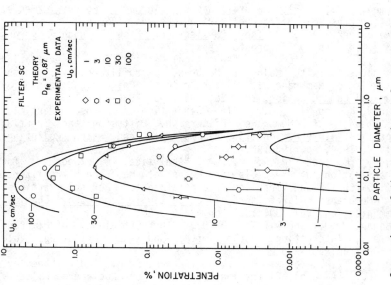

FIG. 8—Comparison of the experimental penetration data for the Millipore SC filter to that theoretically predicted by fibrous filter theory. This filter has a nominal pore size of 8.0 um.

SUMMARY

Two filter models have been used to describe the particle collection characteristics of membrane filters. The capillary tube model has been shown to apply to the Nuclepore filter while the fibrous model more closely predicts particle deposition in conventional membranes.

In the case of the conventional membrane filter, the fibrous model allows certain essential attributes of the filter to be understood on a theoretical basis. These include:
1. High particle collection efficiency.
2. Most penetrating particle size in the range of of 0.05 to 0.2 um for typical membrane filters.
Furthermore, the theory indicates that diffusion and interception are the dominant particle collection mechanisms in the conventional membrane filter and that the efficiency of the filter is determined largely by the size of the physical elements in the filter rather than by the nominal pore size.

ACKNOWLEDGEMENTS

This work is supported under the U. S. Amry Contract No. DAAA15-85-C-0046. The support by the Army is gratefully acknowledged.

REFERENCES

[1] Carmen, P., Flow of Gases Through Porous Media, Butterworths Scientific Publications, London, 1956.
[2] Scheidegger, A. E., The Physics of Flow Through Porous Media, University of Toronto Press, 1974.
[3] Spurny, K. and Pich, J., "Analytical Methods for Determination of Aerosols with Help of Membrane Ultrafilters: VI: On the Mechanism of Membrane Ultrafilter Action," Collection Czechoslov Chemical Communication, Vol. 28, 1963, pp. 2886-2893.
[4] Spurny, K. and Pich, J., "The Separation of Aerosol Particles by Means of Membrane Filters by Diffusion and Inertial Impaction," International Journal of Air and Water Pollution, Vol. 8, 1964, pp. 193-196.
[5] Spurny, K. and Pich, J., "Analytical Methods for Determination of Aerosols by Means of Membrane Ultrafilters: VII: Diffusion and Impaction Precipitation of Aerosol Particles by Membrane Ultrafilters," Collection Czechoslov Chemical Communication, Vol. 30, 1965, pp. 2276-2286.
[6] Megaw, W. J. and Wiffen, R. D., "The Efficiency of Membrane Filters," International Journal of Air and Water Pollution, Vol. 7, 1963, pp. 501-509.
[7] Spurny, K. R., Lodge, J. P., Jr., Frank, E. R. and Sheesley, D. C., "Aerosol Filtration by Means of Nuclepore Filters: Structural and Filtration Properties," Environmental Science and Technology, Vol. 3, 1969, pp.453-464.

[8] Spurny, K., "Aerosol Filtration by Means of Analytical Pore Filters", in Assessment of Airborne Particles, Charles C. Thomas, Springfield, Illinios, 1972, pp. 54-80.

[9] Pich, J., "Theory of Aerosol Filtration by Fibrous and Membrane Filters," in Aerosol Science (C. N. Davies, Ed.), Academic Press, London, 1966, pp. 223-285.

[10] Pich, J., "Gas Filtration Theory," in Filtration-Principles and Practices (C.Orr, ed.), Marcel Dekker, Inc, New York, 1977, Chap. 1.

[11] Cadle, R., The Measurement of Airborne Particles, John Wiley and Sons, New York, 1975.

[12] Davies, C.N., Air Filtration, Academic Press, London, 1973.

[13] Heidam, N. Z., "Review: Aerosol Fractionation by Sequential Filtration with Nuclepore Filters," Atmospheric Environment, Vol. 15, No. 6, 1981, pp 891-904.

[14] Gentry, J. W., Spurny, K. R. and Schoermann, J., " Diffusional Deposition of Ultrafine Aerosols on Nuclepore Filters," Atmospheric Environment, Vol. 16, No. 1, 1982, pp. 25-40.

[15] Zebel, G., "A Simple Model for the Calculation of Particle Trajectories Approaching Nuclepore Filter Pores with Allowance for Electrical Forces," Journal of Aerosol Science, Vol 5, 1974, pp. 473-482.

[16] Pich, J., "Impaction of Aerosol Particles in the Neighbourhood of a Circular Hole," Collection Czechoslov Chemical Communication, Vol. 29, 1964, pp. 2223-2227.

[17] Smith, T. N., Phillips, C. R. and Melo, O. T., "Diffusive Collection of Aerosol Particles on Nuclepore Membrane Filter," Environment Science and Technology, Vol. 10, 1976, pp. 274-277.

[18] Smith, T. N. and Phillips, C. R., "Inertial Collection of Aerosol Particles at Circular Aperture," Environmental Science and Technology, Vol. 9, 1975, pp. 564-568.

[19] John, W., Reischl, Goren, S., and Plotkin, D., "Anomalous Filtration of Solid Particles by Nuclepore Filters," Atmospheric Environment, Vol. 12, 1978, pp. 1555-1557.

[20] Spurny, K. and Madelaine, G., "Analytical Methods for Determination of Aerosols by Means of Nuclepore Filter by Means of Latex Aerosols," Collection Czechoslov Chemical Communication, Vol. 36, 1971, pp. 2857-2866.

[21] Smutek, M. and Pich, J., "Impaction of Particles on the Surface of Membrane Filters," Journal of Aerosol Science, Vol. 5, 1974, pp. 17-24.

[22] Manton, M. J., "The Impaction of Aerosols on a Nuclepore Filter," Atmospheric Environment, Vol. 12, 1978, pp. 1669-1675.

[23] Kanaoka, C., Emi, H. and Aikura, T., "Collection Efficiency of Aerosols by Micro-Perforated Plates," Journal of Aerosol Science, Vol. 10, 1979, pp. 29-41.

[24] Parker, R. D. and Buzzard, G. H., "A Filtration Model for Large Pore Nuclepore Filters," Journal of Aerosol Science, Vol. 9, 1978, pp. 7-16.

[25] Twomey, S., "Equations for the Decay of Diffusion of Particles in an Aerosol Flowing through Circular and Rectangular Channels," Bull. Obs. Puy de Dome, Vol 10, 1962, pp. 173-180.

[26] Gormley, P. G. and Kennedy, M., "Diffusion from a Stream Flowing through a Cylindrical Tube," Proc. Roy. Irish Acad., Vol 52A, 1949, pp. 163-169.

[27] Smutek, M., "On the Separation of Air-Borne Particles by Diffusion," _Journal of Aerosol Science_, Vol. 3, 1972, pp 337-343.

[28] Manton, M. J., "Brownian Diffusion of Aerosol to the Face of a Nuclepore Filter," _Atmospheric Environment_, Vol. 13, 1979, pp. 525-531.

[29] Kuwabara, S., "The Forces Experienced by Randomly Distributed Parallel Circular Cylinders or Spheres in Viscous Flow at Small Reynolds Numbers," _Journal of Physics Society of Japan_, Vol 14, 1959, pp. 527-532.

[30] Pich, J., "The Filtration Theory of Highly Dispersed Aerosols," _Staub_, Vol. 5, May, 1965, pp 16-23, English Translation.

[31] Kirsch, A. A. and Fuchs, N. A., "Studies on Fibrous Aerosol Filters-III: Diffusional Deposition of Aerosols in Fibrous Filters," _Annals of Occupational Hygiene_, Vol. 11, 1968, pp. 299-304.

[32] Yeh, H. C. and Liu, B. Y. H., "Aerosol Filtration by Fibrous Filters-I. Theoretical," _Journal of Aerosol Science_, Vol. 5, 1974, pp. 191-204.

[33] Yeh, H. C. and Liu, B. Y. H., "Aerosol Filtration by Fibrous Filters-II. Experimental," _Journal of Aerosol Science_, Vol. 5, 1974, pp 205-217.

[34] Lee, K. and Liu, B. Y. H., "Experimental Study of Aerosol Filtration by Fibrous Filters," _Aerosol Science and Technology_, Vol. 1, 1982, pp. 35-46.

[35] Rubow, K. L., "Submicron Aerosol Filtration Characteristics of Membrane Filters," _Ph.D. Thesis_, University of Minnesota, Mechanical Engineering Department, Minneapolis, Minnesota, 1981.

[36] Lee, K. W. and Liu, B. Y. H., "Theoretical Study of Aerosol Filtration by Fibrous Filters," _Aerosol Science and Technology_, Vol. 1, 1982, pp. 147-162.

[37] Pich, J., "Wirkungsgrad des Sperreffektes in Faserfiltren bei Kleinen Knudsenschen Zahlen," _Staub_, Vol 26, 1966, pp. 267-270.

[38] Albertoni, S., Cercignani, C. and Gotusso, L., "Numerical Evaluation of the Slip Coefficient," _Phys. Fluids_, Vol. 6, 1963, pp. 993-996.

[39] Kirsch, A. A., Stechkina, I. B. and Fuchs, N. A., "Effect of Gas Slip on the Pressure Drop in a System of Parallel Cylinders at Small Reynolds Numbers," _Journal of Colloid and Interface Science_, Vol. 37, 1971, pp. 458-461.

Applications and Testing: Flat Media Testing

Bernard Miller, Ilya Tyomkin, and John A. Wehner

QUANTIFYING THE POROUS STRUCTURE OF FABRICS
FOR FILTRATION APPLICATIONS

REFERENCE: Miller, B., Tyomkin, I., and Wehner, J. A.,
"Quantifying the Porous Structure of Fabrics for Filtration
Applications," Fluid Filtration: Gas, Volume I, ASTM STP 975,
R. R. Raber, Ed., American Society for Testing and Materials,
Philadelphia, 1986

ABSTRACT: Quantification of the porous nature of a flat
filter medium can be accomplished in a variety of ways.
However, some of the commonly used methods are either inadequate or irrelevant. For example, incremental porosimetry does not produce pore size distributions that can be
related to filter performance. This paper presents a method
for determining types of pore constriction frequency distributions that are more directly applicable to filtration and
flow-through processes. A recommended procedure for determining air permeability is shown to be inadequate both
for predicting the resistance of a fabric to air flow and
for comparing different fabrics. An extended air permeability test methodology is described that can be used to
give a better quantification of flow-through resistance.

KEYWORDS: pore size, pore size distribution, minimum
bubble pressure, air permeability, filters

INTRODUCTION

It is general practice to classify or rate flat filter media in
terms of two measurable quantities: (1) the largest open passage through
it (by way of a minimum bubble pressure test), and (2) air permeability
under a single pressure gradient. For a more detailed description, a
pore size distribution is obtained, usually by means of mercury intrusion porosimetry. This paper discusses the inadequacies of such

Drs. Miller and Tyomkin are Associate Director of Research and Staff
Scientist, respectively, at Textile Research Institute, P. O. Box 625,
Princeton, NJ 08542; Mr. Wehner is a Textile Research Institute Fellow
and a candidate for the Ph.D. in the Department of Chemical Engineering
at Princeton University.

practices and describes alternate procedures that can quantify the porous nature of such materials in terms that are more relevant to filtration and flow-through phenomena.

PORE VOLUME DISTRIBUTIONS

In most laboratories, pore size distributions are obtained using mercury intrusion porosimetry. Unfortunately, this technique can produce misleading results with fibrous materials and does not give distributions that are pertinent to flow-through and filtration processes. The procedure requires covering the degassed sample with mercury, slowly increasing the external pressure, and monitoring the volume increments of liquid that penetrate the porous structure as a function of the pressure gradient. The critical pore radius, R_c, that will allow mercury to enter at a given gradient, ΔP, is

$$R_c = \frac{-2 \, \gamma \, \cos \theta_a}{\Delta P} \qquad (1)$$

where γ = surface tension of mercury, and
 θ_a = advancing contact angle.

The surface tension of mercury is very high (~460 mN/m), so that considerable pressure is needed to force it into small pores. In consequence, during the measurement the structure of a fibrous network can be altered and even the individual fiber components distorted. The apparent pore sizes would therefore not be the same as those present in the uncompressed material. There are several other serious uncertainties involved in mercury porosimetry concerning the correct value of the advancing contact angle and the significant back pressure that can be generated if even a small amount of air remains inside the pores. These problems have been discussed in detail by Good [1] and Winslow [2]. If one wishes to obtain the same kind of data that is produced by mercury porosimetry without these uncertainties, it is recommended that the analogous technique of liquid extrusion [3] be employed. This method is based on the same general principle expressed by Eq 1, but starts with the pores filled with liquid and thus involves the receding contact angle, θ_r. The pressure gradient across the sample is then increased in stages and the imbibed liquid drains out of the pores in a sequence depending on their radii, the largest ones emptying first. A wide choice of liquids is possible in all cases, and the problems of excessive pressure and uncertainty in the applicable contact angle are eliminated.

However, both of the methods described are not appropriate for evaluating filters because they do not give the right kind of distribution. The integral curve in Figure 1 (solid line) represents the kind of data obtained by either method; the derivative plot (dashed) calculated from this curve describes the distribution, that is, $\Delta V/\Delta R$ as a function of R. It is important to note that this is *not* a frequency distribution; instead it shows the contribution that each size pore (ranked according to its effective radius) makes in terms of available volume. Thus, a peak does not represent the most prevalent pore size, but rather that size pore that provides more of the internal free volume than any other size. For filtration and other flow-through processes, internal free volume is not likely to be a relevant dimensional property, since it includes dead end pores, crevices between fibers, and other interstices that are not part of any continuous path through the medium. A more appropriate quantification would be the smallest cross section (throat) of each pore and how many of each size are present. As will be shown, a pore throat distribution can be quite different from one based on pore volumes.

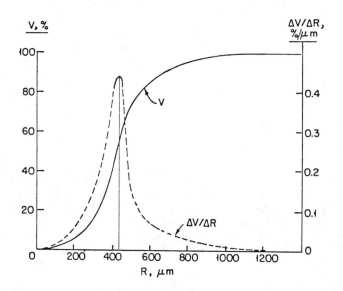

FIG. 1--Volume of liquid in sample (as percent of maximum volume held) vs. critical pore radius (solid line); derivative curve dashed.

PORE THROAT DISTRIBUTIONS

A simple method has been developed to produce frequency distributions that describe those pores that are responsible for the passage of fluid through the plane of a material. It is based on the well known "minimum bubble pressure" principle operative when gas pressure is applied to one side of a wetted fabric while the other side is in contact with liquid. At a critical pressure, the first air bubble(s) comes through the largest pore(s) available within the wetted area (Figure 2). The effective radius of this pore is obtained using Eq 1 with $\cos \theta_r = 1$:

$$R_{max} = \frac{2\gamma}{\Delta P} \tag{2}$$

where ΔP is the pressure gradient across the liquid/gas interface. This relationship can be used even if θ_r is not close to $0°$, as long as it is less than $90°$.

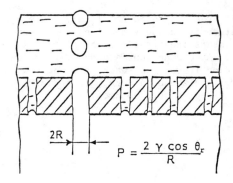

FIG. 2--Principle of the minimum bubble pressure test.

A single bubble pressure experiment identifies only the largest pore within a scanned area. In order to establish the distribution of pore sizes, we assume that on a macroscopic scale this distribution is uniform throughout the sample. If we were to test an entire specimen, such as that shown in Figure 3, only one largest pore, ⑤, would be identified. If the specimen were cut into two parts (e.g., at A) and each part tested separately, we would count one ⑤ and one ④ pore. Another cut to make four parts (e.g., at B) would produce one ⑤, one ④, one ③, and one ②. (The pair of ②s in the lower left quadrant would probably be counted as one pore.) Subdividing the specimen into sixteen portions (by cuts at C, for example) would lead to the following count:

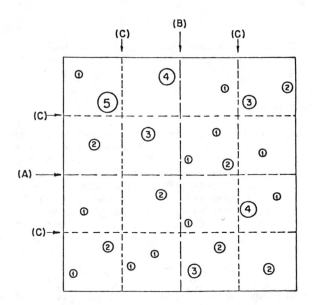

FIG. 3--Hypothetical distribution of pores in a fabric.

Pore size	No. counted	No. actually present
5	1	1
4	2	2
3	3	3
2	5	7
1	5	11

The observed collection differs from the actual because any pore that is not the largest in its subsector is not counted. Thus small pores have little chance of being detected until the scanned areas become small enough.

A reasonably accurate number distribution of pore sizes could be obtained, of course, by dividing a specimen up into many small, uniform regions and making individual bubble pressure measurements over each one. Since this would be impractical, it is necessary to rely on a statistical analysis of relatively few measurements. As previously pointed out, the chances of a small pore being detected increase as the test area is reduced; contrariwise, the probability of finding a large pore goes up with increasing scan area. To obtain a reasonable representation of the distribution, a range of scanning areas must be used.

To carry out bubble pressure measurements over different wetted areas, the multiport chamber shown in Figure 4 has been constructed. Five sets of four holes with cross-sectional areas of 8.04, 2.01, 0.50, and 0.13 mm^2, respectively, were drilled vertically through the lower section so that they connect to the air inlet below. The same size holes were drilled in the upper plate coincident with their counterparts in the lower one. The 15 smaller upper holes were partially drilled out from the top so that their openings from above would all be the same as the largest one. A strip of fabric to be tested is placed between the two plates and clamped in place.

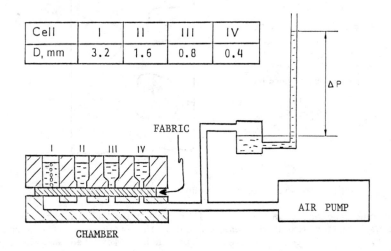

Cell	I	II	III	IV
D, mm	3.2	1.6	0.8	0.4

FIG. 4--Apparatus for multiple scan bubble pressure measurements.

The fabric cell is then placed in enough liquid so that the fabric becomes wet and liquid is present above it in each hole. With all holes open, the air pressure is increased by pumping until the first bubbles appear in one of the holes. After this critical pressure is recorded (giving the effective pore radius through Eq 2), that hole is closed with a plug and additional pressure applied until bubbles appear at another hole. The latter is then plugged also, and the process continued with the remaining holes. Then the fabric strip is moved so that the holes are located over another portion of the specimen, and the bubble pressures are determined in the same manner. The process is repeated until a sufficiently large area of the material has been scanned (usually 5 to 10 locations on the fabric, i.e., 100-200 pressure readings).

The accumulated results reveal the number of pores of each size detected and the scanned area in which they were found. Figure 5 illustrates the logic of the analysis. The first (largest) pore detected is found over a scanned area equal to the total of all the hole cross sections. Each successive pore is found over an area that is reduced from the total by the cumulative areas of all previously tested locations.

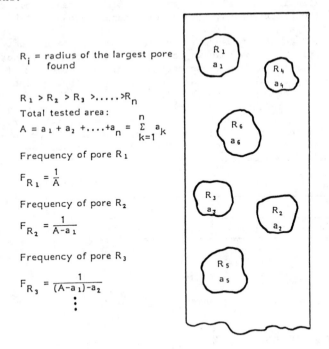

R_i = radius of the largest pore found

$R_1 > R_2 > R_3 > \ldots > R_n$

Total tested area:

$$A = a_1 + a_2 + \ldots + a_n = \sum_{k=1}^{n} a_k$$

Frequency of pore R_1

$$F_{R_1} = \frac{1}{A}$$

Frequency of pore R_2

$$F_{R_2} = \frac{1}{A - a_1}$$

Frequency of pore R_3

$$F_{R_3} = \frac{1}{(A - a_1) - a_2}$$
$$\vdots$$

FIG. 5--Analysis for frequency distribution.

As one proceeds down to the smaller pores, the accuracy of the number density calculation decreases because the fraction of the specimen actually scanned becomes smaller. This can be alleviated only by a considerable increase in the number of smaller holes, which would then require a large number of additional bubble pressure measurements.

Since the main use of number distribution data would be to predict flow-through, particulate capture, or some kind of failure phenomenon, the distribution of the larger pores would be most important. It is un-likely to be worth the considerable additional effort to extend the measurement technique to cover the smaller pores in the distribution.

Figures 6 and 7 show the contrasting distributions obtained for a woven fabric by the liquid extrusion technique and by pore throat distribution analysis. The former shows the typical bimodal form that is found with all woven materials; the large pores are those between yarns and the small ones between the fibers. The pore throat frequency distribution (points in Figure 7) covers only the interyarn pores, that is, the ones involved in transporting fluid through the fabric. There is no peak; instead, the frequency, F, of the pores increases exponentially as their size decreases (note the logarithmic scale). For flow-through processes one might consider that throat cross-sectional areas are more important than throat diameters. An area distribution ($F \times \pi R^2$ vs. R) is included in Figure 7, which is somewhat different from the simple F vs. R plot. If the fluid used has a significant viscosity, a more appropriate distribution would be based on the Poiseuille re-lationship which states that flow is proportional to the fourth power of the radius. A distribution plot of $F \times R^4$ vs. R is also shown in Figure 7; when expressed in these terms, the contribution of each size pore to flow-through is no longer a monotonic function.

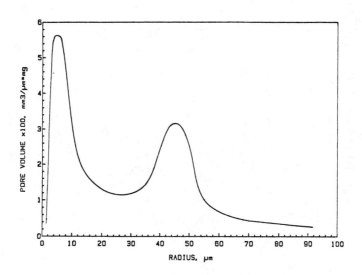

FIG. 6--Pore volume distribution for a woven fabric.

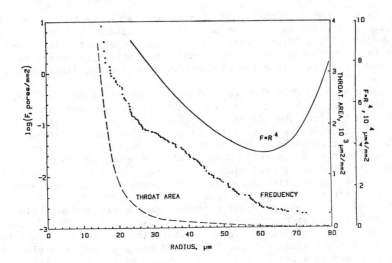

FIG. 7--Pore throat size distribution for the same woven fabric as in
Figure 6 (data points). Calculated area and Poiseuille
distributions: $F \times \pi R^2$ vs. R (---) and $F \times R^4$ vs. R (——).

Another example of the usefulness of pore throat distribution
analysis is shown in Figure 8. In this case data were obtained for
two samples of a 100% cotton woven fabric, one untreated and the other

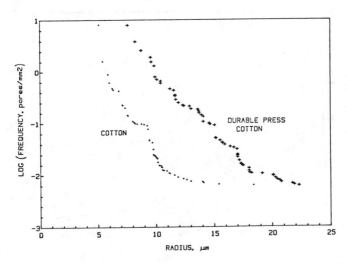

FIG. 8--Pore throat size distributions for a cotton fabric (·) and for
the same fabric treated for durable press (+), both laundered.

treated with a conventional durable press (DP) resin finish. The
original purpose of this comparison was to see if the addition of the
resin reduces the pore dimensions. However, the results show that the
pores are actually somewhat larger in the DP cotton. This is a direct
consequence of the fact that both materials were washed using conventional home laundry equipment before being analyzed. The untreated
cotton shrank more than the DP version, and the relative change in
dimensions is what produced the differences shown in Figure 8.

Figure 9 compares a pore throat analysis with filtration efficiency
data for a nonwoven filter fabric. Although penetration was limited to
particles smaller than 0.2 μm, there were a considerable number of pore
openings present in the 1.5-3.0 μm range. These data indicate that
particle capture by this medium was not at all dependent on a sieving
process, but was due to impaction on the fiber surfaces.

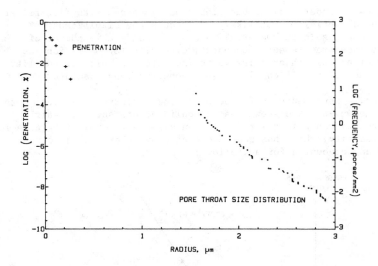

FIG. 9--Comparison of pore throat size distribution for a nonwoven
        filter (·) with particle penetration data (+).

AIR PERMEABILITY MEASUREMENTS

The ASTM has presented two standard methods for determining the
air permeability of fabrics used in textile [4] and filtration [5]
applications. The former specifies that the air flow rate through
a test fabric should be measured at one pressure drop (usually 0.5
inches of water). A single measurement such as this may be adequate
for quality control purposes, but in the more general case where the
fabric may be required to perform under different conditions, either a
range of pressure drops should be experimentally investigated, or some
mathematical relationship must be established to extrapolate from the
result of the single test value. This is necessary because, in many
cases, air velocity and pressure drop are not linearly related.

The more comprehensive method of determining the air permeability of filtration materials [5] recognizes the shortcomings of measuring the air flow through a sample at only one pressure drop and describes the possible causes of nonlinearity. There are three reasons why the pressure drop-velocity curve might not be linear: (1) at high enough Reynolds numbers there could be an inertial contribution to the pressure drop; (2) increased air flow rates could lead to physical changes in the specimen, such as compression; or (3) the compressibility of air could cause air density variation across the specimen thickness if the pressure drop is considerable.

Considering the last of these possibilities first, a pressure difference of 20 inches of water would be necessary to cause a 5% change in air density. For a typical fabric such a pressure drop would require an air velocity of about 100 m/min. For the great majority of fabric applications, therefore, the air density may be assumed to be constant.

Compression under typical gas flow conditions and with flat filter media such as fabrics is not likely to produce any significant structural changes in the material unless a considerable number of free fiber ends are present. Comparative results for woven fabrics made with either smooth or hairy yarns indicate that air permeability is not influenced significantly by such constructional factors.

However, measurements performed on a wide variety of fabrics, both woven and nonwoven, have shown that a nonlinear pressure drop-air velocity relationship is quite common; an example is shown in Figure 10. This nonlinearity, which has been documented by others [6,7], can be explained and accounted for as follows:

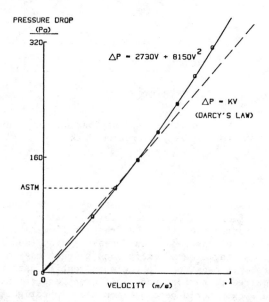

FIG. 10--Pressure drop vs. air velocity data for a polyester fabric (o).
Solid curve represents best fit to the equation $\Delta P/V = a + bV$.
Dotted line is based on ASTM single point test.

The pressure drop, ΔP, of a fluid flowing through a porous medium such as a fabric at a velocity, V, can be described by

$$\Delta P/L = a\,V^2 + b\,V \qquad (3)$$

where L is the fabric thickness, and a and b are quantities that depend on both fluid properties and fabric structure. The first term on the right side of Eq 3 represents the contribution of inertial forces to the measured pressure drop, while the second term represents the influence of viscous forces. We have found that the inertial term may not be insignificant, even at relatively moderate air velocities.

As recommended in the ASTM filtration standard [5], the experimental data can be handled better if all the terms in Eq 3 are divided by V:

$$\Delta P/LV = a\,V + b \qquad (4)$$

Thus, a plot of $\Delta P/V$ versus V yields a straight line, as shown in Figure 11. If there were no inertial effect (if Darcy's law held), the slope of this line would be zero. The coefficients a and b can be obtained from the linear plot of Figure 11; for this particular set of data they have been used to get the solid line in Figure 10, which coincides very well with the experimental points. The broken line in Figure 10 is the Darcy relationship using the single value obtained at the 0.5 in. pressure gradient to establish the Darcy coefficient K. This leads to some underestimation of permeability at lower pressures and increasing overestimation as pressure is increased.

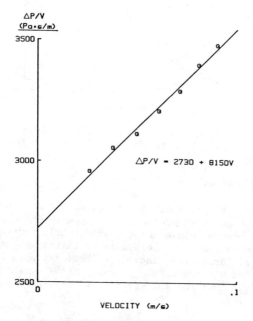

FIG. 11--Best fit line when the experimental pressure drop vs. air velocity data of Figure 10 are plotted according to Eq 4.

One way to rank fabrics would be in terms of the intercept value
from plots such as that in Figure 11. This intercept value equals the
limiting value of ΔP/V as the air velocity approaches zero, and is
equal to the initial slope of the ΔP versus V graph. As V approaches
zero, fabric compression would be minimized, if it were a factor.
Therefore, the intercept value could serve to compare the openness
of structures under a zero pressure gradient. However, such a ranking
may be misleading because it is possible that fabrics can reverse
their relative permeability rankings as pressure drop increases.
Figure 12 shows an example of such reversal. At low pressures, fabric
B exhibits a greater flow rate; at high pressures, fabric A is more
permeable. The crossover velocity, 1.29 m/s, is well within the range
that many filter fabrics would experience.

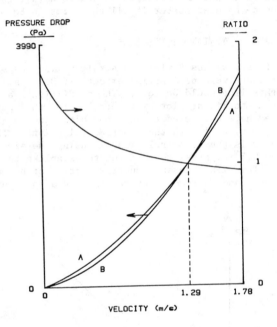

FIG. 12--Pressure drop vs. air velocity relationships for two fabrics,
and the ratio of their pressure drops as a function of air
velocity.

SUMMARY AND CONCLUSIONS

A useful quantification of the porous nature of a fabric or paper
requires that the measured properties be relevant to the application
of the material. Pore volume measurements, while pertinent to liquid
absorption processes, are not relatable to filtration or flow-through
operations. Pore throat frequency distributions, obtained by an
elaboration of the minimum bubble pressure technique, can be used to
describe materials in terms that predict flow-through and sieving
filtration behavior.

Air flow measurements at a single pressure gradient are not adequate to define the permeability of fabrics. As recommended by the ASTM Standard [5] measurements taken at a few different gradients can be used to establish a mathematical relationship that describes the permeability of a material over a wide range of conditions. To best compare different materials, one must be sure that their permeabilities are known at a pressure drop that is close to that which they will experience in use.

ACKNOWLEDGEMENTS

The contribution of Gerard McDonald in obtaining the results shown in this paper is gratefully acknowledged. This work is one aspect of the project "Surface Controlled Interactions of Fluids with Fibrous Materials," supported by a group of Corporate TRI Participants.

REFERENCES

[1] Good, R. J., "The Contact Angle of Mercury on the Internal Surfaces of Porous Bodies," in Surface and Colloid Science (E. Matijevic and R. J. Good, Eds.), Plenum Press, Vol. 13, 1984, pp. 283-286.

[2] Winslow, D. N., "Advances in Experimental Techniques for Mercury Intrusion Porosimetry," in Surface and Colloid Science (E. Matijevic and R. J. Good, Eds.), Plenum Press, Vol. 13, 1984, pp. 259-282.

[3] Miller, B., and Tyomkin, I., "An Extended Range Liquid Extrusion Method for Determining Pore Size Distributions," Textile Research Journal, Vol. 56, No. 1, January 1986, pp. 35-40.

[4] ASTM Standard D-737-75.

[5] ASTM Standard F-778-82.

[6] Rainard, L. W., "Air Permeability of Fabrics, II," Textile Research Journal, Vol. 17, No. 3, March 1947, pp. 167-170.

[7] Goodings, A. C. "Air Flow Through Textile Fabrics," Textile Research Journal, Vol. 34, No. 8, August 1964, pp. 713-724.

James E. Moulton III, James C. Wilson, Sr.

THE DUSTRON: AN AUTOMATED FLAT SAMPLE FILTRATION PERFORMANCE TESTING SYSTEM

REFERENCE: Moulton, J.E., Wilson, J. C., "The Dustron: An Automated Flat Sample Filtration Performance Testing System", Fluid Filtration: Gas, Volume I, ASTM STP 975, R. R. Raber, Ed., American Society for Testing and Materials, Philadelphia, PA, 1986.

ABSTRACT: A microprocessor controlled flat sample filtration performance tester called HV DUSTRON has been developed to be a flexible and versatile laboratory or quality control system. A five year record of performance with a standard filter medium has resulted in a coefficient of variation of 6 - 7%. Evidence is presented that the DUSTRON can distinguish one type of filtration surface from another and one type of filter medium from another.

KEYWORDS: filtration test, flat sheet, penetration, contaminant, face velocity, terminal pressure.

INTRODUCTION

The manufacturer of filtration media is constantly faced with a costly process of product evaluation. This evaluation may involve long periods of time and necessitate several material modifications to meet certain field requirements. Once the "in house" laboratories have decided upon the product, another extensive process begins -- field evaluation. Reduction in time and material costs involved in this evaluation can enhance the cost-effectiveness of this process.

J.E. Moulton and J.C. Wilson are Senior Development Engineers at Hollingsworth & Vose Co., Research and Development Laboratory, Rt. 113, Greenwich NY 12834.

The flat sample filtration performance test allows for the early performance evaluation of any potential filter material development. Initial laboratory samples may be tested to efficiently screen potential candidates. Machine produced variations of the laboratory screened candidates can be performance tested, evaluated and selected for further consideration. As a result the filter design engineer will receive a material which has a higher degree of performance predictability.

The flat sample filtration performance test has another key feature. Once the design engineer fabricates the material into some geometric shape and inserts this filter design into an assembly, other influences on filtration performance become a factor in the total product evaluation process. The flat sample performance test evaluates only media, and eliminates the hazard of choosing a "standard test assembly" to test all product developments.

The DUSTRON challenges a flat sample with a solid aerosol contaminant in a controlled upflow air stream. At a predetermined test end point, the system's microprocessor concludes the evaluation and displays pertinent data.

DESIGN PHILOSOPHY

The system was developed to operate as a sample in/data out device to minimize the operator's influence on test results and to maximize testing thruput. The unit was designed compact and mobile enough to be moved from one type of environment to another. The services required by the unit was limited to 110 VAC at 20 Amperes.

The overall design was to be versatile enough to accommodate a variety of test procedures and sufficiently flexible to readily allow modifications to meet future industry filtration performance testing needs.

Pneumatic Considerations

The primary interest in our air filtration testing is in the low velocity regime (under 50 feet per minute) where settling effects become significant. To promote an even deposition of contaminant across the test sample, the sample should be oriented horizontally and the contaminated flow stream should be vertical.

As shown in Figure 1, most of the previous "flat sheet testers" utilize a down flow arrangement and an evacuated system.

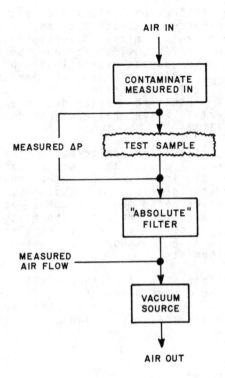

FIGURE I -- PHYSICAL ORIENTATION OF
TYPICAL DOWN AIR FLOW
TESTER

As shown in Figure 2, the Dustron has a pressurized system which allows a smaller and quieter pump than an evacuated system with the same flow. Also, leaks are easily detected in a pressurized system by applying a soap solution to a suspect joint and looking for the appearance of bubbles. The arrows in Figure 2 indicate the flow paths during standard tests as described in this communication.

An upflow arrangement is used in the Dustron to meet the objectives of versatility and flexibility. The upflow of contaminant laden air presents a natural classification condition (Stokes' Law [1]) which is significant at the air velocities of interest. Media and associated assemblies under field conditions can produce similar classification conditions.

Versatility was designed into this equipment by incorporating the unique "Fountain Area". A chamber with specially slotted walls, it allows a portion of the

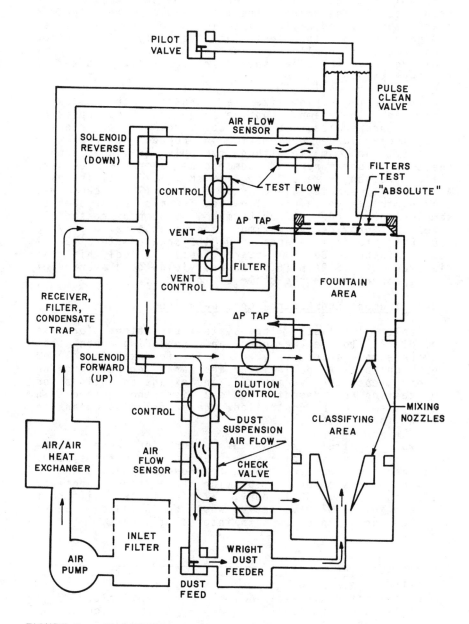

FIGURE 2 -- H/ DUSTRON PNEUMATIC DIAGRAM

contaminant laden air stream to be uniformly removed prior to impinging onto the test sample. The effect is that the aerodynamic particle size can be varied independent of the test face velocity. The upflow arrangement also allows evaluation of cleanable media by applying reverse flow and/or reverse flow impulses to the test sample.

The test sample size was chosen by the following criterion: the sample must be large enough to represent average properties, be not so large as to require excessively large equipment to produce desired face velocity, and smaller than commonly available sample sheets. Laboratory and commercial samples from the field frequently are 22 cm x 28 cm (8-1/2 X 11 inches). An overall sample diameter of 17.8 cm (7.0 inches) was chosen to leave ample allowance for mis-cut or damaged sample sheet edges. Furthermore, a pump of reasonable size can provide a maximum face velocity of 0.1 M/sec (20 FPM) to a test area of 0.0182 square meters (28.27 square inches, 0.1963 square feet). We feel this test diameter of 15.2 cm (6.0 inches) yields good average of media properties in the scale of interest.

Instrument and Control Considerations

Table 1 lists the system parameters (process variables) to be measured and depending upon test requirements, selectively used as feedback for closed loop control. Also included are parameter setpoints or targets (operator variables), and expansion options for new testing requirements. Table 2 lists the equipment in the system to be controlled, again including expansion options for flexibility.

Two parallel modes of operation were to be implemented; a manual system and an automatic system. This approach was taken both to provide a backup and to provide means for calibration and troubleshooting. Following this philosophy, at least two different displays of system variables would be included; a panel meter and an X-Y plotter with internal timebase.

TABLE 1--Instrumentation Required

PARAMETER	TYPE
Filtered Air Flow	Analog
Contaminated Air Flow	Analog
Small Range Sample ΔP	Analog
Large Range Sample ΔP	Analog
Time Base	Analog
For Future Expansion (3)	Analog
Test Type	Numeric
Test Parameter Setpoints	Numeric
Test Initiation	Momentary
Emergency Abort	Momentary

TABLE 2--Control Elements Required

FUNCTION	TYPE
Flow Direction Solenoids (2)	On/Off
Dust Feeder Air Solenoid	On/Off
Reverse Flow Pulse Solenoid	On/Off
Compressor Motor	On/Off
Dust Feeder Motor	On/Off
X-Y Plotter - Start Trace	Momentary
X-Y Plotter - Pen Reset	Momentary
For Future Use (4)	On/Off
Analog Signal Switch - Panel Meter	1 of 8
Analog Signal Switch - X channel	1 of 8
Analog Signal Switch - Y channel	1 of 8
Dust Feeder Air Throttle Valve	Analog
Filtered Air Throttle Valve	Analog
Dilution Air Throttle Valve	Analog
Vent Air Throttle Valve	Analog
For Future Expansion (2)	Analog

Among the control system technology available, we considered electromechanical, discrete logic, and microcomputer based controllers. An electromechanical implementation of the control system would require four analog closed loop controllers and a switching network to connect each input to each throttling valve controller. A massive array of switches would also be required for a particular test. To set up a new test, the unit would need additional wiring installed. Mechanical devices also wear. Because of these reasons, the choice of an electromechanical control system was discarded.

Discrete solid state logic does not suffer from wear like mechanical devices. With thoughtful design, the flexibility requirements could be met. However, the digital circuitry would be extensive, and much analog circuitry would need designing. This choice was discarded due to excessive development cost.

A microprocessor based system was the best choice. We decided to buy off-the-shelf assemblies which would provide most of the functions required, and custom design the interfacing. The small industrial programmable controllers and computers available at the time required a separate development system to program and maintain them. These devices were inappropriate for a one-off project. However, the recently introduced personal microcomputers contained all the necessary programming tools and included a Cathode Ray Tube (CRT) display. A suitable model was chosen for the following reasons; a large number of interface ports, extensive documentation and a powerful programming language.

Electrical/Electronic Considerations

Flexibility requirements also dictated buss oriented electronics, where every signal of a certain type is available to many devices. Two main busses were needed; the sensor output buss (high level analog) and the system control buss (digital). The X-Y plotter, panel meter, and the analog to digital converter need access to the sensor output buss. The X and Y plotter channels and the panel meter may be connected to the output of any sensor via electronic switches. Similarly, either the computer or the front panel can set any final control element to a desired state.

A system this complex could present time consuming difficulties in locating a defective part. Using a modular layout and having a single function on each circuit board allows replacement of suspect units in a minimum of time. Proper selection of purchased components would keep the number of different circuit boards minimal.

Software Considerations

The program to operate the system would also use modular techniques. Any test procedure could be constructed with five main modules; operator prompt and input, self test, parameter conversion and set-up, control and interrupt servicing, termination and reset. The main modules would consist of smaller modules. Using this approach, a new test procedure would use appropriate existing modules where possible, and only the totally different actions would need new software. In this respect, the operating program is also its own development tool.

OPERATION

As in any test procedure, reliable results depend on careful and repeatable experimental technique. Figure 3 outlines the procedure we use.

Contaminant Preparation

A contaminant is selected, the dust cup is packed, and assembled onto the Wright [2] dust feeder. The packing procedure is very specific and unless carefully followed can be a major source of variation. Some contaminants pack better than others (A.C.F.T.D. versus A.C.C.T.D. [3]), while others may pack well but cannot be dispersed in a uniform manner (ASHRAE Test Dust [4]).

Contaminant is added to the empty dust cup 2 grams at a time. It is distributed by inserting the tamping tool and rotating the tool a full turn. 300 PSI is applied to the contaminant via the tool and released.

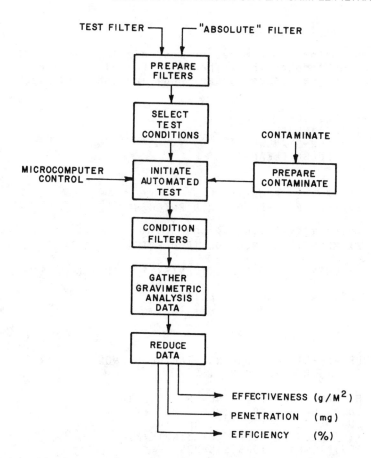

FIGURE 3 -- H/ DUSTRON EXPERIMENTAL PROCEDURE

The tool is turned 60 degrees and the pressure again applied for a total of three times. It is noteworthy that a consistent environment is required for generating predictable dust packs.

Filter Preparation

The test filter and "absolute" filter are conditioned prior to weighting by allowing them to equilibrate to controlled conditions for at least 15 minutes.

After the sample has been properly conditioned, it is clamped in the test fixture with the same side facing upstream that will be used in the actual application. The response of media to exposing its wire side vs. its felt side to contamination is discussed later.

Test Condition Selection

The operator is prompted by menu displays on the CRT
and responds through the keyboard. Typical CRT displays
and operator responses (underlined) are shown in Figure
4a.

```
SELECT TEST TYPE
(1) STANDARD
(2) NON-STANDARD
(3) PULSE CLEAN
? 2

ENTER TEST FLOW
EXAMPLE 8.0 FPM DESIRED ENTER 80 ?  160

ENTER TERMINATING PRESURE°
EXAMPLE 15.0" W.C. DESIRED ENTER 150 ?  300

IF ENTRIES ARE CORRECT ENTER 1 IF NOT 0 ?  1

RESET AND ZERO X-Y PLOTTER
PRESS [RETURN] TO START TEST  _

SENSORS BEING ZEROED
```

FIGURE 4a--DISPLAY DURING STARTUP OF NON-STANDARD
DUST LOADING TEST

```
HI RANGE DP   50    INCHES W.C. X10
LO RANGE DP   49    INCHES W.C. X10
DUST SIZE     1250  MICRON-DIA. SQ.
TEST FLOW     160   FEET/MINUTE X10
TEST TIME     1975  SECONDS


NON-STANDARD DUST LOADING TEST
TEST PROCEEDING

FLOW SET 160
TERMINATING PRESSURE 150

PRESS [ESCAPE] FOR EMERGENCY ABORT
```

FIGURE 4b--DISPLAY DURING NON-STANDARD
DUST LOADING TEST

Test Initiation

From this point on, the filtration test is
automatically controlled by the microcomputer. The
motors are started and flows are maintained at the
selected setpoints. An X-Y plotter records the
differential pressure across the sample with respect to
time and the CRT displays this and other pertinent
parameters (Figure 4b). The CRT display is updated once

each second. At the terminating point, the test is automatically stopped and the final parameters displayed until a new response is keyed into the system.

Condition Filters

Both the test and "absolute" filters are conditioned again in the same way as before the test.

Gravimetric Analysis and Data Reduction

The contaminant weight on the test filter produces an effectiveness (grams per square meter). The contaminant weight on the "absolute" filter produces a penetration (milligrams) and the efficiency (percent) and feed concentration are calculated from the sum of the contaminant weights.

SYSTEM CAPABILITIES

If a test sample has been conditioned and pre-weighed, and the dust pack has not been exhausted (typically 15 tests per pack), it is possible to begin another test within one to two minutes. The maximum turn-around time when removing an exhausted dust pack and installing a pre-packed cup is 10 - 15 minutes. The result is a high percentage of actual test time without compromising quality.

In addition to the normal sample holder, a specially constructed honeycomblike frame may be inserted in the test sample assembly to allow testing of specimens that may easily deform. For example, test samples with the ream weight and strength of facial tissue have been successfully evaluated.

The solid aerosol contaminant concentration is controlled by varying the speed of the dust feeder. A maximum contaminant concentration of 5.36 grams per cubic meter in the airstream (0.150 grams per cubic foot) may be obtained. Any powdered contaminant which will form a stable cake may be packed in the contaminant cup and used. Current use and experience indicate that a particle size of 40 microns and lower is most useful.

Sample restriction is measured by two differential pressure sensors whose ranges are 0 to 70 cm (0 to 27.6 inches) and 0 to 343 cm (0 to 135 inches) water column. The face velocity range is 2 to 15 cm/sec (3 to 17 feet per minute) with the standard 0.0182 sq.M (0.196 square feet) test area. Higher face velocities may be obtained by reducing the sample test area.

The upflow arrangement, mentioned earlier, allows control of the maximum aerodynamic size of particles in the contaminated air stream to be varied independent of the test area's face velocity. The maximum aerodynamic size obtainable is about 80 microns.

The downflow arrangement provides a mechanism for studying reverse and pulse clean performance. There are three options available for pulse clean testing.

    (1)   Terminate when the differential pressure after the pulse reaches a predetermined value (pulsing every preselected time interval).

    (2)   Terminate after selected number of pulse cycles (pulsing every preselected time interval).

    (3)   Terminate when the differential pressure reaches a predetermined value (pulsing when the differential pressure equals a lower pre-selected value or every pre-selected time interval).

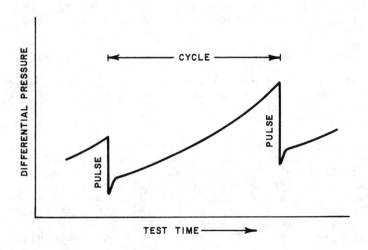

FIGURE 5 -- PULSE CLEAN CYCLE

TEST DATA AND DISCUSSION

Dustron Standard Media

One roll of a typical filtration medium was selected in 1980 and samples from this roll of media are routinely tested on the Dustron. The test parameters are:

Contaminant Type   AC Fine Test Dust
Face Velocity   8.0 FPM (4.06 cm/sec)
Contaminant Concentration   0.017 gms per ft (0.61 gm/M)
Terminal Pressure   15.0 inches W.C. (38.1 cm)

In a previous paper [5], standard data were presented for 138 tests over a three year period. Table 3 tabulates the results since that period of time

TABLE 3--Results of 15 Standard Tests

	Effectiveness (g/M)	Penetration (mg)
Minimum Value	106.	1.9
Maximum Value	132.	5.5
Mean	117.	3.5
Standard Deviation	7.20	0.976
Coefficient of Variation-%	6.2	27.9

The term of effectiveness (gm./sq. meter) refers to that amount of contaminant captured by the filtration medium while penetration (milligrams) refers to the amount of contaminant that passes through the filtration medium and is retained by the more efficient "absolute" material (No. FM-004 Owens Corning Fiberglass). The variation in penetration is quite large, but could be reduced with additional conditioning time. However conditioning time in excess of testing time severely affects test thruput.

Felt Versus Wire

In the filtration field it is widely accepted that paper type products usually possess two distinct surfaces. The smoother more consolidated surface is commonly referred to as the wire side of the sheet. The other surface is the felt side. Under most air flow regimes, the contaminant laden airstream will charge the felt surface. As the contaminant is removed by the fiber matrix, cleaned air exits from the wire side of the material.

When the paper filter medium is oriented incorrectly, the filter assembly (element or cartridge plus housing) can result in a performance loss of 30%. The Dustron performance test demonstrates this characteristic of filter paper. Using the same test parameters and three samples from the same roll of a different paper filter type material, the data in Table 4 were generated.

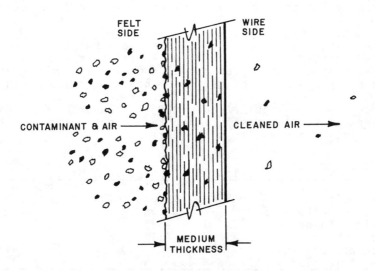

FELT
SIDE

WIRE
SIDE

CONTAMINANT & AIR ⟶

CLEANED AIR ⟶

MEDIUM
THICKNESS

FIGURE 6 -- FELT VERSUS WIRE SURFACE

	TABLE 4--Felt Versus Wire Results		
Sample Number	Sample Orientation	Effectiveness (gm./sq. meter)	Penetration (milligrams)
1	A	93	0.3
2	A	104	0.6
3	A	87	0.3
4	B	45	0.2
5	B	73	0.3
6	B	66	0.2

A - Felt Side Upstream   B - Wire Side Upstream

In like manner, the Dustron Standard was tested and resulted in:

	Effectiveness	Penetration
Felt Side Upstream	108	2.7
Wire Side Upstream	63	4.6

Both sets of data reveal a loss in effectiveness of 35-42% when the medium is oriented so that wire side faced upstream.

Types of Contaminant

A. C. Fine and A. C. Coarse test dusts 3 are two widely used standardized contaminants. A. C. Coarse has a wider range of micron sizes than A. C. Fine and the percentage of each micron fraction differs from that of A. C. Fine. A composite filter material was tested using A. C. Fine and A. C. Coarse test dust under the control parameters of 16 FPM (8.1 cm/sec) and a terminating pressure of 15 inches (38.1 cm) W.C. The results were:

Contaminant Type Dust	Effectiveness (gm/sq meter)	Penetration (milligrams)
ACF	14.3	1.5
ACC	116	4.3

The Dustron revealed quite different results depending upon which type of contaminant was used.

Types of Filter Media

Figure 7 reveals the effectiveness of four entirely different types of filtration medium and figure 8 shows the degree of penetration for the same four media.

All data were generated using the Dustron and the following setup:

Contaminant Type	ACF
Face Velocity	16 FPM (8.1 cm/sec)
Terminal Pressure	3, 5, 10, 15 inches W.C.
	(7.6, 12.7, 25.4 38.1 cm)

At the end of each pressure terminating point, the contaminant laden sample and "absolute" were weighed and returned to the assembly for the next test. Thus the same sample type was analyzed at four different terminating points. The same absolute was used for all 16 performance tests.

The samples are described as:

POLY--------100% polyester fiber
PCG---------Polymer Bonded Cellulose,Polyester,Glass
CELLULOSE---Polymer Bonded Cellulose
GLASS-------93 - 95% Glass fiber

Each sample possesses an air permeability [7] of 20 - 30 CFM/Sq. Ft. No other selection criterion was controlled.

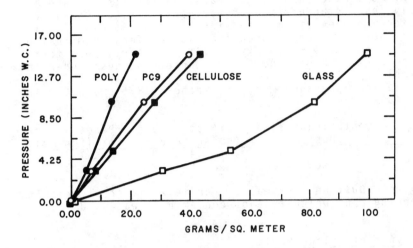

FIGURE 7 -- FILTER MEDIA EFFECTIVENESS

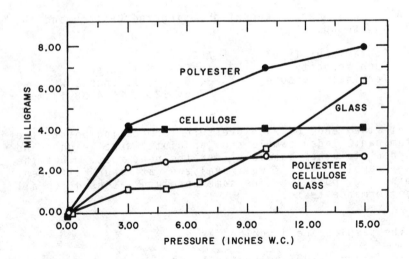

FIGURE 8 -- FILTER MEDIA PENETRATION

The Dustron was able to clearly distinguish between the 100% polyester, the polymer bonded cellulose and the nearly 100% glass medium. While the distinction was not as clear between PCG and cellulose, one could still make a decision especially for those applications that require a 10 - 15 inch terminating pressure.

The other data reduction-penetration (Figure 8)-reveals some interesting trends. At the 3 inch terminating point, the glass sample was the most efficient, but at the 15 inch point, it was the third most efficient. After the initial penetration (between 0 to 3 inches of restriction), both PCG and cellulose allowed no further penetration. The 100% polyester medium continuously allowed contaminant to penetrate the matrix.

CONCLUSION

The Dustron has met the design objectives of versatility, flexibility and accuracy in the performance of testing of flat samples. Through the use of microprocessor techniques, a rather simple sample in/data out procedure has been documented. The Dustron can perform typical air filtration test procedures and because of its unique design, can run a variety of other procedures including pulse clean cycle operations.

Effectiveness measurements on an internal standard over a five year period reveals a coefficient of variation of 7 - 10% with the last three years being between 6 - 7%. Particle retention efficiency has been reduced to a more meaningful gravimetric penetration value.

Data has been presented to illustrate the capability of the Dustron to disperse different solid particle type contaminants, to evaluate one filtration surface from another and one type of filtration medium from another.

REFERENCES

[1a] Davies, C. N., Air Filtration, Academic Press,
     Inc., 111 Fifth Avenue, New York, NY 1973
[1b] Dennis, R., Handbook on Aerosols, 1976, TID-26608,
     National Technician Information Service,
     Springfield, VA. 22161.
[1c] Orr, Clyde,
     Filtration, Principles and Practices, Part I,
     Marcel Dekker, Inc., New York, NY 1977.

[2] Willeke, K, Ed.,
     Generation of Aerosols and Facilities for Exposure
     Experiments , Ann Arbor Science Publishers, Ann
     Arbor, MI, 1980

[3] A. C. Fine Test Dust and A. C. Coarse Test Dust, A.
     C. Sparkplug Division, General Motors Corp., Flint,
     MI.

[4] The American Society of Heating, Refrigerating, and
     Air-Conditioning Engineers, Inc.,
     Method of Testing Air Cleaning Devices Used in Gener
     al
     Ventilation for Removing Particulate Matter, Standar
     d 52-76, ASHRAE Circulation Sales Dept., 345 East
     47th St., New York, NY 10017.

[5] Moulton, J. E., Wilson, J. C., "Dustron--An Air
     Filtration Testing System", presented at the TAPPI
     1984 Nonwovens Symposium, Myrtle Beach, SC,
     available as reprint from Technical Association of
     the Pulp and Paper Industry, Atlanta, GA.

[6] American Association for Testing and Materials,
     "Gas Flow Resistance Testing of Filtration Media",
     1983 Annual Book of ASTM Standards , Vol. 14.02,
     Standard F-778-82, ASTM, 1916 Race St.,
     Philadelphia, PA.

Richard J. Remiarz, Brian R. Johnson, and Jugal K. Agarwal

AUTOMATED SYSTEMS FOR FILTER EFFICIENCY MEASUREMENTS

REFERENCE: Remiarz, R. J., Johnson, B. R., and Agarwal,
J. K., "Automated Systems for Filter Efficiency
Measurements," Fluid Filtration: Gas, Volume I, ASTM STP
975, R. R. Raber , Ed., American Society for Testing and
Materials, Philadelphia, 1986

ABSTRACT: New methods of filter testing have been
developed which eliminate the shortcomings of previous
methods. The systems are fully automated and easy to
use. The systems use a continuous flow condensation
nucleus counter. Filter efficiencies of 99.9999+% can
be measured with these systems. One system uses monodisperse aerosols for the challenge and can measure
filter efficiency as a function of particle size. The
systems incorporate an air-operated filter holder assembly for ease of use. A microcomputer system is used for
automation and data analysis.

KEYWORDS: filter testing, automation, condensation
nucleus counter, monodisperse aerosol generation

INTRODUCTION

In the past, the most commonly used filter testing technique
for high efficiency filters has been the dioctyl phthalate
(DOP)/photometer technique. In this technique the challenge aerosol
is generated by either nebulizing DOP or by quenching DOP vapor.
The particle concentration detector is typically a forward-
scattering photometer employing a white light bulb as the light
source and a photomultiplier tube as the photodetector. The basic
problem with this technique is that photometer sensitivity decreases
sharply as the particle concentration decreases. Hence, while the
photometer is able to accurately measure high concentrations
upstream of the filter, it can not easily measure the low concentrations found downstream of high efficiency filters. To overcome
this low concentration problem very high particle concentrations

Mr. Remiarz is a senior project engineer and Mr. Johnson is an
engineer in the Advanced Technology Group of TSI Inc. Dr. Agarwal
is the manager of the Advanced Technology Group, TSI Inc., P.O. Box
64394, St. Paul, MN 55164.

must be generated upstream of the filter. Also, frequent
adjustments must be made to zero the photomultiplier tube to
compensate for electronic and photometric noise drift. Even then,
this technique is not capable of measuring penetrations less than
0.01%. In addition, photometers are sensitive to the particle size
distribution, and are not sensitive to small particles. Therefore,
the particle size distribution of the challenge aerosol must be
carefully controlled, and it is not possible to measure efficiency
as a function of particle size.

Other methods of filter testing include the sodium
chloride/flame photometer technique and gravimetric techniques. In
the sodium chloride/flame photometer technique, sodium chloride (or
other sodium salt) challenge aerosol is generated and a flame
photometer is used as the particle detector. With gravimetric
techniques, challenge aerosol is generated by dispersing a powder
such as silica dust into the airstream. The aerosol concentration
upstream and downstream of the test filter is determined by
collecting the particles on two separate sample filters and weighing
the filters on a microbalance.

All of these various methods of filter testing have the
following shortcomings. They all need very high upstream challenge
concentrations, are not able to measure filter efficiencies greater
than 99.99%, and are not readily able to measure filter efficiency
versus particle size.

In the past several years, new methods for filter testing have
been developed which do not have these shortcomings [1-3]. Each of
these systems is capable of measuring very high efficiencies and
some are capable of measuring efficiency as a function of particle
size. What all of these systems have in common is the use of a
single particle counting continuous flow condensation nucleus
counter (CNC) as the particle detector.

The CNC first grows submicrometer particles to supermicrometer
alcohol droplets and then measures the alcohol droplet
concentration. This makes the CNC sensitive to particles as small
as 0.01 μm diameter, but insensitive to variations in particle size,
shape, composition, and refractive index. Thus, a filter test
method using the CNC as the particle detector can use virtually any
aerosol as the challenge aerosol. In addition, the CNC can measure
particle concentrations from less than 0.01 particles/cm^3 to greater
than 10^6 particles/cm^3. This feature makes the CNC a nearly ideal
instrument for filter testing with the ability of measuring
efficiencies of 99.9999+%.

This paper describes an automated system for filter efficiency
measurements. The principle of operation of the system is to
challenge the filter with a selected size monodisperse aerosol and
measure the aerosol particle concentration upstream and downstream
of the filter. Then the size of the monodisperse aerosol is changed
and the test is repeated. The CNC is used to measure the
concentrations of particles smaller than 0.5 μm. An aerodynamic
particle sizer (APS) is used to measure the concentration of
particles larger than 0.5 μm. An electrostatic classification

method is used to generate monodisperse aerosols in the size range of 0.01 μm to 0.5 μm and uniform polystyrene latex (PSL) spheres are used to generate monodisperse aerosols in the size range of 0.5 μm to 5.0 μm. This paper also describes variations of the system for different applications.

DESCRIPTION OF THE SYSTEM

A schematic of the system is shown in Figure 1. The system can be broken into 6 major components: 1) aerosol generation system for particles <0.5 μm, 2) aerosol generation system for particles ≥0.5 μm, 3) filter holder assembly, 4) aerosol flow system, 5) particle detectors, and 6) microcomputer system.

Aerosol Generation System for Particles <0.5 μm

Monodisperse aerosols in the size range of 0.01 to 0.5 μm are generated by using the electrostatic classification method first described by Liu and Pui [4] and schematically shown in Figure 2. In this method a polydisperse aerosol generated by an atomizer is brought to Boltzmann equilibrium which, in the submicrometer size range, means that most of the particles are either electrically neutral or carry only ±1 elementary unit of charge. These particles then pass through an electrical field. Particles of a certain electrical mobility exit through a slit in the central rod. By varying the voltage on the collector rod, the size of the exiting particles can be varied. To maximize the particle concentration of the monodisperse aerosol exiting through the slit, the mean particle size of the incoming polydisperse aerosol should be matched to the desired size of the exiting monodisperse aerosol.

In the current system, this is accomplished by using three atomizers. Each atomizer contains a NaCl/water solution, but of differing concentrations. When monodisperse aerosol of a smaller particle size is desired, the atomizer containing a lower concentration solution is turned on. When monodisperse aerosol with a larger size is desired, the atomizer containing a higher concentration solution is turned on. The entire process is automated and con- trolled by the microcomputer system. Also, to facilitate the drying of the polydisperse aerosol generated by the atomizers, a diffusion drier is incorporated into the system. The diffusion dryer consists of two coaxial tubes. The inner tube is a cylinder made of wire mesh. The space between the two tubes is filled with silica gel desicant. As the aerosol passes through the unobstructed inner cylinder, moisture diffuses to the silica gel, drying the particles.

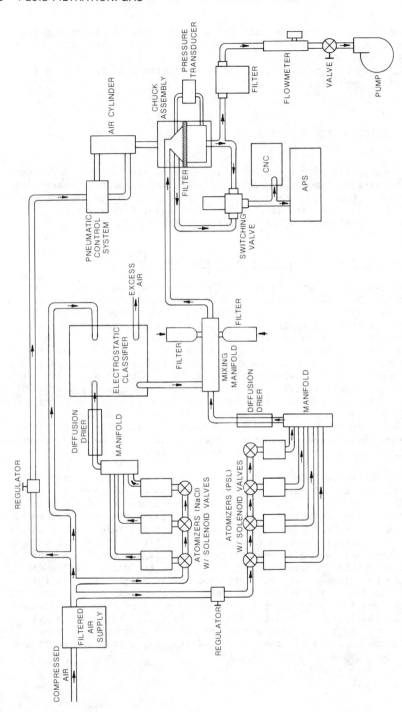

FIGURE 1--Schematic of filter test system.

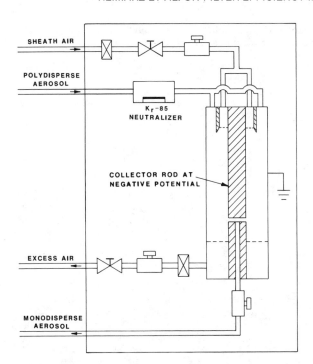

FIGURE 2--Schematic of electrostatic classifier.

Aerosol Generation System for Particles ≥0.5 μm

To measure the filter efficiency for particles ≥0.5 μm, the challenge aerosol is generated using polysytrene latex (PSL) spheres of known sizes. Referring to Figure 1, clean air from the filtered air supply passes through an additional regulator and goes to four atomizers. Each of the four atomizers contains a different size of PSL spheres. The outputs of the atomizers go to a manifold which leads to a diffusion drier. The dried particles then go to the mixing manifold where they are mixed with particle-free air to provide the necessary flow rate for challenging the filter in the filter holder assembly.

Filter Holder Assembly

An air-operated filter holder assembly was designed to hold test filters. The assembly is shown in Figure 3. It consists of two halves. The upper half contains the challenge aerosol inlet, upstream sampling port, and upstream pressure port. The bottom half contains the downstream sampling port and downstream pressure port. A differential pressure transducer is connected between the two pressure ports. An air cylinder is used to raise and lower the top half of the assembly. As shown in Figure 1, air from the filtered air supply passes through a regulator on its way to the pneumatic control system for the air cylinder. To test a filter, the user simply places the filter in the lower half of the filter holder and

presses a button to close the assembly. The assembly then closes,
automatically forming an air-tight seal with the filter.

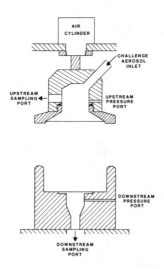

FIGURE 3--Schematic of air-operated filter holder assembly.

Aerosol Flow System

The aerosol from the lower half of the filter holder passes
through an in-line filter, flowmeter, valve and pump. The pump
draws the challenge aerosol through the test filter. The flowmeter
is a thermal mass flowmeter with a range of 0 to 120 1/min. The
in-line filter is used to protect the thermal mass flowmeter from
contamination. The valve is used to control the flow through the
test filter.

The aerosol sample is drawn from upstream and downstream of the
test filter and fed to the particle detectors through a three-way
switching valve. The sample port on the upper half of the filter
holder is connected to an inlet on the switching valve. The flow
from the lower half of the filter holder branches, with one path
going to the switching valve and the other path going to the pump.
The outlet of the switching valve also branches, with one path
leading to the CNC and the other to the APS. With this arrangement,
the switching valve directs the sample flow to the particle
instruments, allowing them to sample both upstream and downstream of
the filter. The sampling lines were designed to minimize particle
loss. Also, care was exercised to make the upstream and downstream
sampling paths similar so that the error due to particle loss is
small. Particle losses in the sampling lines were evaluated by
running efficiency tests without a filter in the filter holder

assembly. Tests were run at particles sizes from 0.01 μm to
2.65 μm. The tests showed that upstream and downstream losses were
equivalent.

Particle Detectors

As shown in Figure 1, the aerosol is simultaneously sampled by
the CNC and APS. The CNC output is used for measuring particle
concentration when the challenge aerosol is <0.5 μm. The APS output
is used when the challenge aerosol is ≥0.5 μm.

Condensation Nucleus Counter: The continuous flow CNC is
described in detail by Agarwal and Sem [5]. A schematic of the CNC
is shown in Figure 4.

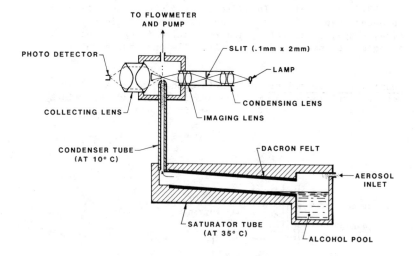

FIGURE 4.   Schematic of condensation nucleus counter.

Aerosol enters the inlet of the CNC and passes through an
inclined heated tube with the inside covered by a thick felt lining.
The lower end of the felt lining dips into an alcohol reservoir
located at the inlet. The felt becomes saturated with alcohol,
providing a larger surface area for alcohol evaporation. The
aerosol, saturated with alcohol vapor, enters a vertical cold
condenser tube. Alcohol vapor condenses on the aerosol particles
causing them to grow to larger droplets. A particle sensor,
consisting of a light source, focusing optics, collecting optics,
and photodetector counts the alcohol droplets.

The CNC has two modes of operation: (1) count mode and (2)
photometric mode. At concentrations below 1000 particles/cm^3, the

CNC counts electrical pulses generated by light scattered from individual droplets. At particle concentrations above 1000 particles/cm^3, the photodetector measures the total light from all the droplets present in the viewing volume at any time. The photodetector output passes through a linearizer circuit which provides a voltage signal linearly proportional to the particle concentration.

The significance of single particle counting is that it eliminates the limit on the minimum concentration that can be measured. The lowest measurable concentration is limited only by the length of the sample time. Because of the dual measurement modes, the CNC can measure the high concentrations found upstream of filters with photometric mode, while measuring the low concentrations encountered downstream of high efficiency filters with single count mode. The CNC can measure concentrations from less than 0.01 particles/cm^3 to greater than 10^6. This allows measuring efficiencies of 99.9999+% accurately. In addition, the CNC is sensitive to particles greater than 0.01 µm and insensitive to variations in particle size, shape, composition, and refractive index. Thus, filter testing can be performed using virtually any challenge aerosol.

Aerodynamic particle sizer: The APS is used to measure filter efficiencies of particles ≥0.5 µm. The APS is described by Remiarz et al [6]. A schematic of the APS is shown in Figure 5.

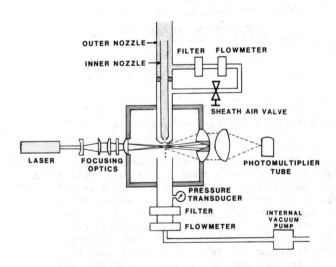

FIGURE 5--Schematic of aerodynamic particle sizer.

In the APS, the sampled aerosol passes through an accelerating nozzle. Because of their inertia, the aerosol particles will lag behind the fluid near the exit of the nozzle. The velocity of the aerosol particles as they exit the accelerating nozzle is measured by a two spot laser velocimeter. At the end of the sample period, the particle size distribution is calculated from the particle velocity distribution.

Microcomputer System

The microcomputer system controls the filter test system and provides data reduction functions. The microcomputer system is based on an IBM PC-XT microcomputer with 256K of memory and two 360K floppy disk drives. The system also includes an Epson FX-85 dot matrix printer and Sony CPD-1201 color monitor. The microcomputer system is interfaced to the rest of the filter test system through a Metrabyte DASCON1 data acquisition and control interface board and Metrabyte STA01 and SRA01 screw terminal and solid state relay accessories.

Figure 6 shows how the microcomputer system is interfaced to the various components of the filter test system. The microcomputer system reads the CNC, pressure transducer, and flowmeter outputs. The microcomputer system also reads and controls the APS. Finally, the microcomputer system controls the electrostatic classifier collector rod voltage, switching valve, and solenoid valves for the atomizers.

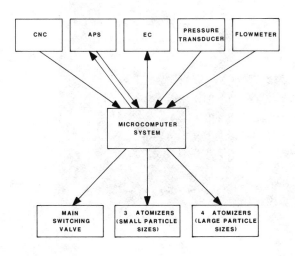

FIGURE 6--Schematic of microcomputer system interfacing.

During a test, for each particle size the microcomputer system turns on an atomizer and sets the electrostatic classifier collector rod voltage. The filter flow rate and pressure drop are then measured. Next, the switching valve is set to measure upstream concentration, and a concentration measurement is made. The switching valve is then set to measure downstream concentration, and a downstream concentration measurement is made. Finally, the microcomputer calculates filter efficiency and penetration. This is repeated for each particle size.

A photograph of the complete filter testing system is shown in Figure 7.

FIGURE 7--TSI filter test system.

OPERATION OF SYSTEM

The filter testing procedure begins with the operator inserting the filter (or filter media) into the filter holder assembly and setting the filter flowrate. The filter flowrate is adjusted by observing the real-time readout of flowrate and pressure drop displayed on the monitor. Once the filter flow is set, the operator presses a key to run the test. The system begins challenging the filter with selected monodisperse aerosols of known size. Upstream and downstream concentrations, percent penetration and percent efficiency are listed for each particle size. A maximum of 14 different particle sizes can be selected to challenge a filter. To improve the precision of filter efficiency measurements, the software enables the user to increase the downstream sample time on the filter. This is especially useful when testing high efficiency filters with very low downstream concentrations.

TEST DATA

Many filter samples have been tested using the system. Table 1 shows the test results of a high-efficiency filter canister. The test results show that at all sizes the filter had an efficiency greater than 99.999%, with the minimum efficiency occuring at 0.15 µm. Efficiencies marked with an asterisk indicate that less than ten particles were counted on the downstream side of that filter, resulting in lower statistical accuracy. This occurs most often at the larger sizes where the upstream concentrations are much lower. To improve the accuracy at these points, the sampling time must be increased. For filters with very high efficiencies, this could require hours. A sample in which no particles were counted downstream is indicated by a greater than sign in front of the efficiency. In this case, the efficiency is calculated from the upstream concentration and downstream sampling time assuming one particle had been counted downstream. Since no particles were counted, the actual efficiency is higher than this amount, and is so indicated by the "greater than" sign.

Table 2 shows test results on a filter media. The media had a much lower overall efficiency, with a minimum efficiency of 99.947% occuring at 0.15 µm. At larger and smaller sizes the efficiency increased as expected, with efficiencies increasing to greater than 99.9999% at sizes larger than 0.5 µm.

TABLE 1--High efficiency canister test results.

FLOWRATE ....: 31.9 LPM
PRESSURE DROP ..: 17.2 MM. H_2O

SIZE (MICRON)	UPSTREAM CONC. (PART/CC)	DOWNSTREAM CONC. (PART/CC)	PENETRATION (%)	EFFICIENCY (%)
0.010	8.57E+03	1.50E-02	0.0001750	99.99983 *
0.050	6.48E+04	3.17E-02	0.0000489	99.999951
0.080	4.08E+04	6.00E-02	0.0001470	99.99985
0.100	1.91E+05	5.88E-01	0.0003074	99.99969
0.150	7.26E+04	4.22E-01	0.0005813	99.99942
0.200	3.73E+04	1.67E-01	0.0004474	99.99955
0.300	9.41E+03	3.50E-02	0.0003718	99.99963
0.540	7.53E+02	5.00E-04	0.0000664	99.999934 *
1.180	7.40E+01	2.50E-04	0.0003377	99.99966 *
2.830	2.44E+01	0.00E+00	< 0.0008192	> 99.99918 *

* - DENOTES LESS THAN 10 PARTICLES COUNTED DOWNSTREAM OF FILTER

MAXIMUM CNC DOWNSTREAM SAMPLE TIME: 2 MIN.

MAXIMUM APS DOWNSTREAM SAMPLE TIME
         PARTICLE SIZE 0.54 :   2 MIN.
         PARTICLE SIZE 1.18 :   4 MIN.
         PARTICLE SIZE 2.83 :   5 MIN.

TABLE 2--Filter media test results.

FLOWRATE ....: 26.3 LPM
PRESSURE DROP ..: 31.1 MM. H_2O

SIZE (MICRON)	UPSTREAM CONC. (PART/CC)	DOWNSTREAM CONC. (PART/CC)	PENETRATION (%)	EFFICIENCY (%)
0.050	5.98E+04	4.31E+00	0.0072160	99.9928
0.080	4.09E+04	1.46E+01	0.0356173	99.964
0.100	2.17E+05	9.44E+01	0.0434415	99.957
0.150	7.88E+04	4.18E+01	0.0530470	99.947
0.200	3.94E+04	1.36E+01	0.0345634	99.965
0.300	9.48E+03	7.23E-01	0.0076253	99.9924
0.540	3.19E+02	9.50E-05	0.0000298	99.999970 *
1.140	1.31E+02	6.67E-05	0.0000509	99.999949 *

* - DENOTES LESS THAN 10 PARTICLES COUNTED DOWNSTREAM OF FILTER

MAXIMUM CNC DOWNSTREAM SAMPLE TIME: 2 MIN.

MAXIMUM APS DOWNSTREAM SAMPLE TIME
         PARTICLE SIZE 0.54 :   10 MIN.
         PARTICLE SIZE 1.14 :   15 MIN.

VARIATIONS OF THE SYSTEM

The system described here was developed for an application where it was necessary to measure filter efficiencies at particle sizes up to 5.0 μm. However, filter efficiency normally increases after reaching a minimum in the 0.1 to 0.3 μm range, and in most cases, it is not necessary to measure efficiencies in the larger size ranges since they are typically orders of magnitude higher than the minimum efficiency. For these applications, the APS and the four atomizers for PSL could be eliminated. The system would consist of the aerosol generation system for particles <0.5 μm, CNC, sampling and flow system, filter holder assembly, and microcomputer system. Using this system, it is possible to measure filter efficiency as a function of particle size from 0.01 to 0.5 μm, which covers the range of primary interest.

In many applications, it is not necessary to measure the filter efficiency versus particle size, but it is important to measure overall efficiency in as short a time as possible. This is true in applications such as quality assurance, where it is important to find defective filters quickly. In these cases, it is important that total test time be minimized so as many filters as possible can be tested. For this purpose, the aerosol generation system would be only an atomizer and diffusion drier. The concentration of the atomizer solution can be selected so the mean size of the aerosol corresponds with the particle size at which the minimum efficiency occurs. By not having to step through the various sizes, the test time is greatly minimized. In addition, the polydisperse aerosol from the atomizer is much higher concentration than the monodisperse aerosol concentration from the electrostatic classifier, and therefore downstream concentration measurements can be made much faster. The system still uses the CNC, so it has the same advantages of being able to measure higher efficiencies with a variety of aerosols.

CONCLUSIONS

By using a CNC for concentration measurements upstream and downstream of the filter, efficiencies of 99.9999+% can be measured. By generating a series of monodisperse aerosols of different sizes, efficiency can easily be measured as a function of particle size. Since the CNC and APS are insensitive to changes in particle composition and refractive index, virtually any challenge aerosol can be used.

Along with measuring filter efficiency, the systems measure flow rate and pressure drop. The systems include a microcomputer system for automation. The microcomputer controls the test procedure and calculates the results. In conclusion, automated systems for filter efficiency measurements have been developed which eliminate shortcomings of previous test methods.

REFERENCES

[1]   Remiarz, R. J., Agarwal, J. K., Nelson, P. A., and Moyer, E.,
      "A New, Automated Method for Testing Particulate Respirators,"
      Journal of the International Society for Respiratory Protec-
      tion, Vol. 2, No. 3, July-September 1984, pp. 275-287.
[2]   Rubow, K. L. and Liu, B. Y. H., "Evaluation of Ultra-High
      Efficiency Membrane Filters," in Proceedings of the 30th
      Annual Technical Meeting of the Institute of Environmental
      Sciences, 1984, pp. 64-68.
[3]   Agarwal, J. K., Sem, G. J., and Remiarz, R. J., "Filter
      Testing with the Continuous-Flow, Single-Particle Counting
      Condensation Nucleus Counter," TSI Quarterly, Vol. 11, No. 1,
      January-March 1985, pp. 3-12.
[4]   Liu, B. Y. H. and Pui, D. Y. H., "A Submicron Aerosol Standard
      and the Primary, Absolute Calibration of the Condensation
      Nucleus Counter," Journal of Colloid and Interface Science,
      Volume 47, No. 1, April 1974, pp. 155-171.
[5]   Agarwal, J. K. and Sem, G. J., "Continuous Flow, Single-
      Particle-Counting Condensation Nucleus Counter," Journal of
      Aerosol Science, Volume 11, No. 4, 1980, pp. 343-357.
[6]   Remiarz, R. J., Agarwal, J. K., Quant, F. R., and Sem, G. J.,
      "Real-Time Aerodynamic Particle Size Analyzer," in Aerosols In
      the Mining and Industrial Work Environments, Volume 3 Instru-
      mentation, Ann Arbor Science, Michigan, 1983, pp. 879-895.

Robert M. Nicholson

A STANDARD TEST METHOD FOR INITIAL EFFICIENCY MEASUREMENTS ON
FLATSHEET FILTER MEDIA

REFERENCE: Nicholson, Robert M., "Std. Test Method for
Initial Efficiency Measurements on Flatsheet Media", Fluid
Filtration: Gas, Volume I, ASTM STP 975, R. R. Raber, Ed.,
American Society for Testing and Materials, Philadelphia,
1986.

ABSTRACT: Reproducible initial efficiency measurements on
flat sheet filter media in air, require well controlled test
aerosols for particle count and particle size resolution.
This Test Standard drafted by ASTM F21.20 Subcommittee on
Gas Filtration utilized the atomization of monodispersed
latex spheres and particle counting by Optical Particle
Counters to measure the initial, particle size efficiencies
of flatsheet filter media. A test system and procedure is
defined that can provide a ±1% reproducibility on initial
penetration in the particle size range of 0.5 to 5 μm at a
95% confidence level.

KEYWORDS: initial efficiency, initial penetration, latex
aerosols, flatsheet filter media, optical particle counter,
aerosol test system.

INTRODUCTION

Over the years the filtration industry has had a continuing
requirement to qualify non-woven filter paper as to product
performance without conducting a full efficiency-life/loading test.
Ideally, one wants to be able to quickly (minutes) compare commercial
grades of filter media against one's own filter media architecture.
The following ASTM recommended test procedure is an adaptation of this
type testing to develop non-woven filter media efficiency
specifications.

As one reviews the materials and techniques used in the standard
it is readily apparent that most of these techniques have been
available for 10 years or more [1]. The successful utilization of
these techniques however, lies in the control and stability of each
instrument into a total system.

Robert M. Nicholson is a Senior Project Engineer in Corporate
Technology of the Donaldson Co., Inc.,1400 W. 94th Street,
Minneapolis, MN. 55440.

Briefly, the system uses the Optical Particle Counter to count the inlet and penetrating aerosol concentrations to determine the efficiency of the filter media. The aerosol particle size is determined by the monodispersed latex (Geometric Standard Deviation less than 1.2) within the particle size range of 0.5 to 5.0 μm. The test system is designed to establish filter media test velocities of 1 to 25 cm/sec. The system can measure initial efficiencies in the range of 1 to 99.9% in less than 2 minutes per datapoint. Reproducibility on a given media sample is expected to be within 1 to 2% of a mean penetration on this or a comparable system. Current Round Robin testing indicates that inter-laboratory testing may not be this reproducible [2].

Some of the controls and stabilizing requirements are more easily stated than practiced, particularly if the system is flexible as to be a compressed air/pushing system or a vacuum pump/pulling system. The critical control is the media velocity or pressure loss across the filter media. When a test system can reproduce this measurement, reproducibility between laboratories has a much higher confidence level, up to (95%)

Incorporated within this testing are two assumptions that cannot be minimized. First, the initial efficiency is defined as the condition where there is no perceptible increase in pressure loss across the filter media with time of aerosol exposure. Second, there must be a stable aerosol count (concentration) at the inlet and outlet sampling probes so as to allow evaluation of counting errors due to sampling line losses and coincidence as a function of particle size.

Therefore, in this standard, initial efficiency is viewed as simplistically as possible such that the efficiency (η) of the filter media at the latex particle size (D_p) is

$$\eta_{D_p} \ (\%) = (1 - P_{D_p}) \ 100$$

and P_{D_p} is the penetration count as defined by:

$$P_{D_p} = \frac{(\text{Penetrating Particle Count})_{D_p}}{(\text{Inlet Particle Count}_{D_p})}$$

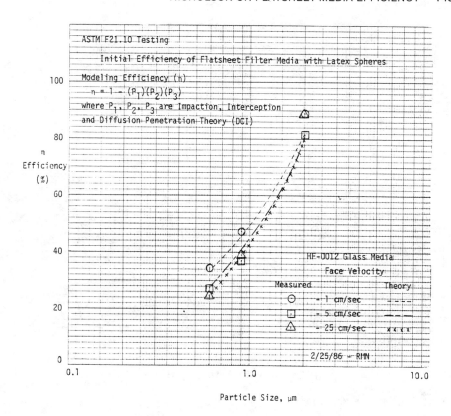

FIG. 1 -- Standard Test Method Versus
Media Modeling

Figure 1 illustrates the use of such testing to verify filter media modeling in theory and formulation. Such modeling includes the media parameters of fibers size, solidity and thickness [3].

A final note of caution, this procedure is not a comprehensive characterization of the performance of a given filter media. Measurements for permeability and loading rate should be included in total filter media specification.

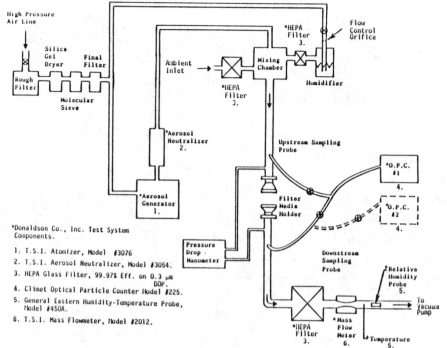

FIG. 2--Typical test system

Test System and Materials

A typical test system is presented in figure 2. The first crucial component is a stable, reproducible aerosol size and count concentration. This standard requires atomization of a solution of suspended latex spheres into a stable aerosol cloud. The aerosol concentrations required are in the range of 10^7 to 10^8 particles/m³ (1 to 10 mg/m³). These concentrations must be held within 1 to 2% of total count for test times greater than 1 hour. The latex suspension can be established with "certified" [4] particle sizes and retained up to 6 months. Guidelines for these suspensions have been established by others [5] and are used here to make dilution of 1000:1 to 10,000:1 of Standard Latex Suspensions [4].

Aerosol conditioning must be maintained throughout the test system to minimize aerosol losses. The aerosol cloud must be electrostatically neutralized and the atomized cloud must be dried and/or diluted to establish 30-50% R.H. through the filter media test holder.

There are mechanical features that require special attention. There must be no flow leakage around the filter media in the sample holder and no expansion or contraction to the velocity profile as it approaches the media cross-section. In some cases of high pressure loss through the filter a support media may be required in the filter holder.

The pressure loss is measured by flush mounted, static pressure taps at 1 duct diameter upstream and downstream of the filter holder. The pressure is monitored at an accuracy of 0.025 cm of water gauge or less.

Geometrically and kinematically identical centerline probes are used to extract representative aerosols from the inlet and outlet sides of the filter medium test section. These probes should have a radius of curvature (R) of 12 cm or R/D (Dia) 20:1 and present a cross-sectional area of less than 10% of the cross-sectional area of the test system ducting. The upstream probe is to be located 8 duct diameters (minimum) downstream of the aerosol injection point and 2 duct diameters ahead of the filter medium specimen. The downstream probe is to be located 3 duct diameters downstream of the filter medium specimen.

Isokinetic aerosol sampling is recommended to minimize probe inlet losses. However, in those cases where isokinetic conditions cannot be met it is recommended that the operation of these probes be ±10% of isokinetic or that the particle Stokes number at the probe inlet be held to less than 1.0 in order to minimize inertial losses at the probe inlet. It is also recommended that the Reynolds number of the sample flow lines be held to less than 2000.

Recommended sampling flowrates for extraction of the counting volume are to be less than 10% of the total test system flowrate.

To minimize aerosol sampling transport line losses due to settling, diffusion and inertia for the aerosol particle size range of this practice the following characteristics of the sampling lines are recommended:

A. Maintain the sampling line flow in the laminar flow regime.

B. Limit horizontal sampling line length to less than 100 cm and the total sample transport line to less than 2 meters.

C. Maintain all radius of curvatures to greater than 12 cm.

The primary airflow is monitored and controlled by an ASME orifice plate or hot wire anemometer mass flowmeter. The aerosol must be filtered out of the flow stream before entering the airflow meters. To resolve differences in flow measurement the ASTM subcommittee has chosen to use a control filter media in its Round Robin testing. The results of this testing indicated a linear regression, correlation coeficient or pressure drop versus flow rate or reproducibility between laboratories of better than 0.9991. Detailed analyses of this testing is being presented by Japuntich in this symposium [2].

Further monitoring for stability of the test aerosol can be provided by using a Multichannel Analyzer (M.C.A.) with the O.P.C. The resolution on the O.P.C. counting output can be measured according to the number of channels available on the M.C.A., but more importantly errors from multiple particle agglomerates and coincidence counting in the O.P.C. can be minimized. It is also advantageous to calibrate the O.P.C. using the M.C.A. to improve peak aerosol counting. A personal computer with appropriate transducers for monitoring and recording pressure drop, relative humidity, temperature, airflow and count data is recommended. The computer can then make the efficiency calculation with corrections and provide hard copy by printer or plotter.

Test Procedure and Results

The Appendix presents the Test Procedure used for ASTM Round Robin Testing. Care should be taken in selecting the filter media sample for testing. Representative media sampling can be treated as per ASTM Std. D-2905-81, Practice for Statements on the Number of Specimens Required to Determine the Average Quality of Textiles. Pre-conditioning is also recommended for each media sample to test duct conditions, i.e., 30-50% ± 5% R.H. and 70°F ± 2°F temperature. It is also recommended that each sample be retained in individual containers to avoid handling contamination and damage.

First, the air supply used for atomizing and drying should be tested for background counting. This can be accomplished by sampling the test airflow with and without latex aerosol. Background counts are to be reduced to less than 1% of total upstream test aerosol count.

The face velocity/pressure losses verification test should be conducted with a control sample.

The aerosol can now be injected into the test system and stabilized until reproducible counts of ± 3% of the mean are made for the upstream and downstream probes over a 15 minute sampling period on each probe. A zero efficiency test is conducted for monitoring sampling line losses and a determination of the purge times through the sampling lines.

A minimum of 100 downstream counts in the particle sizes of interest are required for penetration calculation. The particle sizes of interest must be well defined in the O.P.C. The conditions for valid counting are listed in the Round Robin, Section V., Notes #1 and #2. Typical sample times are on the order of 30 seconds to a minute to provide adequate sampling statistics.

The recommended efficiency sampling strategy is to alternate between upstream and downstream probes if using a single O.P.C. If using two O.P.C.'s upstream aerosol stability is not as critical. Again, a note of caution should be made, that for low face velocities adequate time must be provided to purge the downstream air volume so that true penetration is measured. This standard recommends a minimum of 25 sampling line volume changes occur before counting resumes. For sampling flowrates of 118 cm^3/s in 0.635 cm I.D. lines this is on the order of 10 to 15 seconds.

There are some further points of caution that should be noted. Because of sampling and aerosol transport losses the test procedure has an upper particle size limit of 5.0 μm [5]. However, the latex aerosol size can be extended to smaller aerosol sizes, particularly if a laser O.P.C. is used. Again, it should be emphasized that this test procedure should not be extrapolated to measure loading or life performance of the filter media or media filter elements.

As a final comment on precision and accuracy the following controls must be considered.

Because of the complexity of integrating several aerosol characterizing techniques into a reliable test system, this test method's development of efficiency values for a given filter medium will be influenced by the standardization used with each piece of equipment. For this test method to be reproducible, all adjunct equipment to the test system should be in good repair and calibrated to the manufacturer's specifications.

Although, not covering the full spectrum of particle size and face velocity indicated as applicable in the standard, the following intra-laboratory testing in Table 1 illustrates the reproducibility this procedure can provide.

Table 1 -- Intra-Laboratory Testing of Uniform Glass Media of 30% and 10% DOP Aerosol Penetration Ratings.

MEDIA NAME SAMPLE (#)	TEST #	PARTICLE SIZE micron	FACE VELOCITY cm/sec	DIFFERENTIAL PRESSURE cm of water, gauge	INITIAL EFFICIENCY percent
3M 30% DOP PEN (1)	251200	0.76	10	0.96	88.0
3M 30% DOP PEN (1)	251201	0.76	5	0.51	88.1
3M 30% DOP PEN (2)	251501	0.76	10	0.96	88.3
3M 30% DOP PEN (2)	251500	0.76	5	0.51	88.1
3M 30% DOP PEN (3)	251701	0.76	10	0.99	88.5
3M 30% DOP PEN (3)	251700	0.76	5	0.51	88.6
3M 30% DOP PEN (4)	251901	0.76	10	0.99	88.6
3M 30% DOP PEN (4)	251900	0.76	5	0.51	89.0
3M 30% DOP PEN (5)	252101	0.76	10	0.99	89.1
3M 30% DOP PEN (5)	252100	0.76	5	0.51	89.1
3M 10% DOP PEN (1)	251301	0.76	10	1.65	98.8
3M 10% DOP PEN (1)	251300	0.76	5	0.86	98.8
3M 10% DOP PEN (2)	251400	0.76	10	1.65	98.8
3M 10% DOP PEN (2)	251401	0.76	5	0.86	98.8
3M 10% DOP PEN (3)	251600	0.76	10	1.65	98.8
3M 10% DOP PEN (3)	251601	0.76	5	0.86	98.7
3M 10% DOP PEN (4)	251800	0.76	10	1.68	98.9
3M 10% DOP PEN (4)	251801	0.76	5	0.86	98.8
3M 10% DOP PEN (5)	252000	0.76	10	1.65	98.8
3M 10% DOP PEN (5)	252001	0.76	5	0.86	98.7

For: 30% DOP Pen Media
  Ave = 88.5%
    σ = 1.1%
  @ 95% Confidence,
  Initial Effificency =
      88.5% ± 2.2%

For: 10% DOP Pen Media
  Ave = 98.8%
    σ = 0.1%
  @ 95% Confidence,
  Initial Efficiency =
      98.8% ± 0.20%

In summary if the inlet count can be held to within ± 1 or 2% of mean count on the stability and zero efficiency tests in the Round Robin procedure and the face velocity/pressure losses must be matched on the control sample. Then the initial penetration of a given media can be measured at ± 1% reproducibility at a 95% confidence level for intra-laboratory testing.

ACKNOWLEDGMENTS

The author wishes to express appreciation to the members of ASTM Subcommittee F21.20 for continuing review of this test standard, to Marty Barris and Dick Cardinal of the Donaldson Co., Inc. and Todd Johnson of the 3M Co. for their technical support.

REFERENCES

[1] P. C. Reist and W. A. Burgess, "Atomization of Aquous Suspensions of Polystyrene Latex Particles", J. Colloid and Interface Science, 1967, 24, pp. 271-273.

[2] D. A. Japuntich, "Results of ASTM Round Robin Filter Efficiency Tests Using Latex Spheres", ASTM STP 975, Symposium on Air Filtration, Oct., 1986.

[3] Rubow, K. L. , "Submicron Aerosol Filtration Characteristics of Membrane Filters", Ph.D. Thesis, Mechanical Engineering Department, University of Minesota, December 1981.

[4] Duke Scientific Corporation, 2415 Embarcadero Way, Palo Alto, CA. 94303, Kit #650.

[5] O. G. Raabe, "The Dilution of Monodisperse Suspensions for Aerosolization", Industrial Hygiene Association Journal, 1968, Sept./Oct., pp. 439-443.

[6] P. H. McMurray and J. C. Wilson, "Sampling and Transport", Aerosol Measurement Short Course, Aug., 1978, pp. 1-17.

Other References

Liu, B.Y.H. and Lee, K. W., "An Aerosol Generator of High Stability". AIHA Journal, Dec. 1975, pp. 861-865.

Berglund, R. N. and Liu, B.Y.H., "Generation of Monodispersed Aerosol Standards". Env. Sci. & Tech., Vol. 7, Feb. 1973, pp. 147-153.

Liu, B.Y.H. and Pui, B.Y.H., "Electrical Neutralization of Aerosols". Aerosol Science, Vol. 5, 1974, pp. 465-472.

Fuchs, N. A., "Review Papers on Sampling of Aerosols". Atm. Env., Vol. 9, 1975, pp. 697-707

Bauer, E. J., Reagan, B. T., and Russell, C. A., "Use of Particle Counts for Filter Evaluation". ASHRAE Journal, Oct. 1973, pp. 53-59

Liu, B.Y.H. and Willeke, K., "Single Particle Optical Counters: Principles and Application." Academic Press, 1976, pp. 698-729.

APPENDIX

ASTM F21.20 Round Robin Test Procedure

I. REVIEW EQUIPMENT AND RECORD

a) Flow measurement method; Flow range, accuracy, calibration
b) Pressure drop measurement device; range, accuracy, calibration
c) Particle counter; sampling rate, coincidence limit (maximum count/min.), last calibration, other special controls.
d) Sampling lines; length, I.D. material, sampling probe design, switching valve model.
e) Aerosol generator; air flow rate, solution flow rate.
f) Flow diagram of test system.
g) Schematic of Test Section Geometry.
h) Set-up test system as per figure 4, draft 5 or a comparable test system.

II. SYSTEM START-UP

a) Set main airflow, dilution airflow and aerosol generator airflow to test conditions.
b) Establish airflow controls at required test face velocities as to have air dilution (drying) for aerosol generation of 2:1 or greater.
c) Purge main airflow for 10 to 15 minutes.
d) Warm-up Optical Particle Counter (OPC) for 15 to 30 minutes.
e) After OPC warm-up check built in calibration signal or OPC output signal on an oscilliscope for stable gain on photomultiplier tube or other optical detectors. (Note: This test is conducted with filtered ambient air flowing through the OPC).
f) Switch the OPC into the main airflow and balance OPC air flow and balance OPC airflow against the main airflow.

g)   Record O.P.C. counts for 1 minutes upstream (UPST) and
     downstream (DNST) samples. Conduct counting at lowest
     face velocity for minimum dilution of background.   Note:
     If airflow background count is within O.P.C. electronic
     background extend sampling time until at least 5 counts
     are obtained from airflow background).

III. PRESSURE DROP (ΔP) VERIFICATION OF CONTROL FILTER MEDIA SAMPLE

a)   Insert control sample into media holder of test system for
     ΔP measurement of the required airflow velocity.
b)   Plot the ΔP values for the required test velocities.
c)   Remove the control sample and reestablish system airflow.

IV. AEROSOL DRYING VERIFICATION

a)   Set-up the aerosol generator with a nominal volume of the
     distilled water to be used in the latex dilutions.
b)   Without a filter media sample in the test system establish
     the main system airflow and the O.P.C. sampling airflow
     for the UPST sampling probe.
c)   Sample the UPST and DNST airflow for one minute each.
d)   Verify complete drying of the aerosol generator by
     comparing these counts to counts obtained in II. g.
e)   Record the relative humidity and the temperature of the
     airflow.
f)   Run this drying test for approximately 1 hour, sampling
     every 15 minutes for UPST and DNST counting and record
     aerosol stability and system relative humidity. Measure
     the water consumption of the aerosol generator.
g)   Record any dilution airflow and the required air pressure
     for the aerosol generator.

V. AEROSOL STABILITY AND ZERO EFFICIENCY CHECK

a)   Fill the aerosol generator with the desired dilution of
     latex suspension. Without a filter media sample in the
     test system, close the system and establish the required
     system airflows.
b)   Stabilize the system airflow with the aerosol suspension
     for approximately 5 minutes, then begin successive one
     minute UPST and DNST counts for 15 minutes or until
     reproducible counts are established. Reproducibility is
     based on a ± 0.03) C.O.V. (Coefficient of Variation, $2\sigma/x$)
     limit for the UPST and DNST counts over the 15 minute
     sampling sequence.
c)   Verify that counting is within a 10% coincidence of the
     O.P.C.
     Note #1: The O.P.C. channels to be used for counting
               include channels either side of the maximum
               count channel that produces a count of 50% of
               the maximum channel count or greater.

Note #2:   There must be a clear minimum channel separation in the O.P.C. between any residue particle distribution and the latex particle distribution in the test aerosol. It may be necessary to adjust the gain on the O.P.C. to establish this separation.

VI. EFFICIENCY TEST

a)   Install the test filter sample in the test system and re-establish the required airflows.

b)   Monitor the O.P.C. airflow and adjust for the added filter media ΔP on the sampling flow.

c)   Record the temperature, the relative humidity of the test airflow and the ΔP of the filter media.

d)   Sample and record the UPST and DNST aerosol counts for a minimum of 5 sample counts at each position using a 1 minute sampling time.

Note #3:   If the DNST count is less than 100, extend the sampling time until 100 counts are obtained. However, no count should be conducted longer than 5 minutes to avoid loading the filter media sample.

e)   Monitor the UPST counts, if these counts fall outside the criteria of V.b), stop the test and check the system for aerosol generation instability.

f)   Average the UPST counts and the DNST counts then calculate the decimal efficiency by the following definition:

Efficiency = 1-Penetration

$$= 1 - \frac{\text{Average DNST counts}}{\text{Average UPST counts}}$$

Daniel A. Japuntich

RESULTS OF ASTM ROUND ROBIN TESTS ON FILTER EFFICIENCY
USING LATEX SPHERES AND ON AIRFLOW RESISTANCE

REFERENCE: Japuntich, D. A., "Results of ASTM Round Robin Tests
on Filter Efficiency Using Latex Spheres and on Airflow
Resistance," Fluid Filtration: Gas, Volume I, ASTM STP 975,
R. R. Raber, ED, American Society for Testing and Materials,
Philadelphia, 1986

ABSTRACT: A study was run to determine the appropriateness
of the proposed ASTM filter efficiency standard procedure
[3] at the designated limits of latex sphere diameter and
face velocity. An analysis of variance of filter
performance from six laboratories was performed in the form
of a 2x2 factorial designed experiment with the variable
levels: latex sphere diameter (microns) at .62, .90, 2.02
and face velocity (cm/s) at 1, 5, and 25. Observations
were also made on the precision of the ASTM F778-82 Gas
Flow Resistance Of Filter Media Standard.

KEYWORDS: polymer latex spheres, air filtration, pressure
drop, gas permeability, flow resistance

INTRODUCTION

The last statement on an ASTM Standard Test Method deals with the
test accuracy and precision. In general, for the testing of filter
performance of textiles and nonwovens, it is impossible to make an
accuracy statement since no standard materials exist. Precision,
however, is defined as "the degree of agreement within a set of
observations or test results obtained as directed in a method [1]."
Precision may be designated as single laboratory or as
interlaboratory.

The ASTM F21.20 Subcommittee on Gas Filtration has produced the
test method, "Standard Methods for Gas Flow Resistance of Filtration
Media," ASTM F778-82 and the proposed test method, "Standard Test for
Determining a Singular Particle Size Initial Efficiency of a Flatsheet
Filter Medium in an Airflow." Both of these methods needed studies to
determine their precision.

The F21.20 Subcommittee plan for the Efficiency Test was to run an

---

Dan Japuntich is a Research Specialist at OH&SP Division, 3M
Company 230-BS-06, St. Paul, MN 55144.

interlaboratory analysis of variance replicated twice on the
filtration performance of different flatsheet filter materials at
three face velocities and three particle sizes. At the same time as
this test was run, flow resistance of the materials at the different
face velocities could also be measured.

The Objectives of this Study were:

1. To determine if the filtration efficiency test procedure in the
proposed standard is appropriate at the designated test limits of .5
to 2 microns latex sphere diameter and 1 to 25 cm/s face velocity.
2. To identify test inadequacies and make recommendations for
corrections.
3. To define the testing criteria necessary to establish a
statement on the precision the Efficiency Test.
4. To examine the interlaboratory precision of the F778-82 Gas
Flow Resistance Standard.

EXPERIMENTAL METHODOLOGY

The methodology used to gain the objectives was an analysis of
variance in the form of a 2x2 factorial designed experiment (ANOVA,
see Annex) with the variable levels: latex sphere diameter (microns)
at .62, .9, and 2.02; face velocity (cm/s) at 1, 5, and 25.

The responses chosen for examination were:

    a. % penetration - pooled data, all laboratories
                     - two single laboratories
    b. % coefficient of variation (S.D./$\bar{X}$) - all laboratories
    c. 100% penetration (no filter) % coefficient of variation
                     - all laboratories

Filter Media

Three filter materials were originally tested, but one was
rejected because of high material variability. The two materials upon
which this study is based were made and donated to the subcommittee by
Hollingworth and Voss (H&V): #HE-1021 and HF-0012 fiberglass filter
media. The test specimens were produced by slitting flat sheets of
material from the center of a roll of each medium. The "felt" side of
the media was marked as the challenge side. A "control" sheet of
HE-1021 was included with each laboratory package for face velocity
verification by pressure drop. Table 1 shows the low variability of
these media.

Latex Sphere Challenge Particles

The latex spheres used in the test were purchased from Duke
Scientific Corporation, which subdivided three large lots of .62, .9,
and 2.02 micrometer diameter polymer latex spheres, respectively, into
10 cm^3 samples of 5% solids in an aqueous suspension. Every lab used
the same latex spheres purchased as a set called "Filter Checking
Kit."

TABLE 1--Material Characteristics

	HE-1021	HF-0012
Mean basis wt. (Kg/m^2)	0.078	0.068
% Cov	1.5	1.1
Thickness (mm)	0.61	0.66
Effective fiber diameter (micron) [2]	2.9	5.0
% Solidity	4.9	4.0
Mean Q127 DOP% penetration	38	81
% Cov	1.3	0.69
Mean pressure drop (Pa)	39.0	12.7
% Cov	1.0	3.5

Round Robin Test Procedure

A Round Robin Test Procedure was given with the randomly sampled filter samples to each laboratory with instructions requiring verified set-up procedures, aerosol drying, aerosol stability check, and the test procedure.

Participants

Six laboratories took part, but their identities were not used. Here are the participants in alphabetical order:

Donaldson Company, Inc.
Farr Company
Lawrence Livermore National Laboratory
National Institute for Occupational Safety and Health
Research Triangle Institute
3M Company

A variety of optical particle counters were used (see Table 5), and a variety of atomization aerosol generators were used. Each atomizer used venturi suction as the dilute suspension liquid feed to the atomizing jet. All systems used ionization sources to neutralize particle charge.

Differences between laboratories existed. Filter face area between laboratories varied from 33 to 444 cm^2, although the standard method recommended 20 to 182 cm^2. Two laboratories found that the only way to meet the face velocity criteria was to vary face area with duct cross-section constant. As a matter of fact, only one laboratory out of six had an apparatus which completely met the test criteria as reviewed previously by Nicholson [3]. Five test systems were vertical; one was horizontal. Five test systems used duct mounted samples; one used chamber mounted samples.

Results of Percent Penetration Data Analysis

The analysis of variance yielded:

a. A fitted model equation predicting the response by the levels of the two variables of sphere diameter and face velocity. The degree of fit of the data to the model is a measure of variability and is generally represented by the correlation coefficient squared.

b. Contour plots of the model at different response levels plotted against the two variables call response surfaces.

The variable levels of .62, .9, and 2.02 micron sphere diameter and 1, 5, and 25 cm/s face velocity were transformed into linear design units by taking their logarithms. The response surface contour plots are therefore log-log plots. The logarithm of the penetration fraction was used in the analysis, but the antilog times 100 (% penetration) was reported in the tables and graphs.

Table 2 shows interlaboratory % penetration results on the two filter materials.

TABLE 2--Round Robin Interlaboratory % Penetration Results of the 2x2 Factorial Design

	Sphere Diameter (micron)	HE 1021 Face Velocity(cm/s)			HF 0012 Face Velocity(cm/s)		
		1	5	25	1	5	25
Mean % Penetration	0.62	15.6		18.1	65.2		75.8
	0.90		8.7			60.2	
	2.02	0.49		0.011	21.3		11.4
% Coefficient of	0.62	10.3		5.8	14.2		12.1
Variation:	0.90		7.6			18.7	
ln (% Pen/100)	2.02	10.5		7.3*	26.7		8.3

* Not all data points were used because of zero counts downstream

Graph 1 shows the response surface of the HE-1021 material for percent penetration pooled data from all the laboratories. The isolines denote the decades of percent penetration and, because they are evenly spaced, show % penetration to be an exponential function as theory predicts. Graph 2 shows the same for HF-0012, but with different contour levels because of its lower efficiency. Such response surfaces are signatures of the filter material and, as such, are powerful design tools.

Table 3 shows the degree of fit (R^2) for the models of the pooled data and separate single laboratory data from the lowest variability laboratories. For both filter materials around 14% of the test variability for all the laboratories is unexplained by the model. If the single laboratory data is averaged, 9.7% and 3.9% unexplained variability results, respectively. Therefore, the difference between the interlaboratory unexplained variability and the single laboratory unexplained variability was 4.3% for HE-1021 and 10.1% for HF-0012. Another way of putting this is that 30 to 72 percent of the unexplained variability is because of interlaboratory differences.

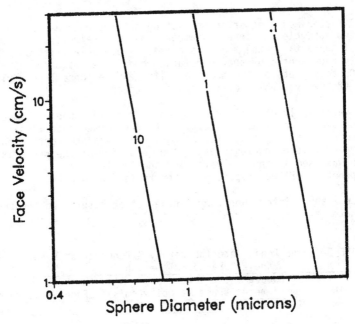

GRAPH 1 - HE-1021 % penetration contours:
sphere diameter vs. face velocity

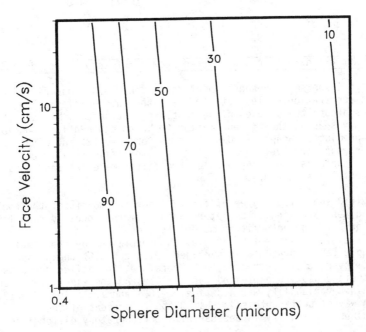

GRAPH 2 - HF-0012 % penetration contours:
sphere diameter vs. face velocity

TABLE 3--Degree of fit to models of percent penetration

H&V #HE 1021	R^2	% Variability Unexplained By Model
All Labs	.860	14.0
Lab 1	.875	12.5
Lab 3	.931	6.9
H&V #HF 0012		
All Labs	.863	13.7
Lab 1	.954	4.6
Lab 3	.969	3.1

A calculation of the percent coefficient of variation of the logarithm of the penetration fraction was taken from the pooled data at each design point. Although this confounds the data, it gives an indication of interlaboratory variability as a function of sphere diameter and face velocity. Graph 3 and Graph 4 show the response surfaces for the two mmaterials. The variability increases as sphere diameter increases and as face velocity decreases. The difference in the level of % C.O.V. between the two models shows that at higher levels of penetration fraction, variability increases, indicating particle concentration variability. In general for this study in the 2.02 micron range, the concentration control was difficult, and the concentration levels were low resulting in penetration calculation error.

It was stipulated in the Round Robin Test Procedure that the desired aerosol counts per sample interval % coefficient of variation be 1.5% giving a 100% penetration variability of .75%. Table 4 shows the average results from each lab of 100% penetration calculations from five consecutive sampling intervals. It would appear that with the atomizer latex sphere aerosol generators used for this testing, the desired count stability is difficult to achieve, especially at the 2.02 micron level. A variance analysis of the 100% penetration % coefficient of variation showed 81% unexplained variability, showing that the concentration variability is random and cannot be modeled. For statistical accuracy the number of upstream/downstream count intervals should be a function of the 100% penetration variability, that is, the more concentration variability the greater the number of intervals for the penetration calculation.

A method of choosing the correct number of sample intervals at a 95% confidence level for a 100% penetration coefficient of variation of 1.5% is to use a statistical sampling plan. Such plans are used every day in industrial quality control. The sampling criteria would be for 100% penetration statistics:
a. The mean should be 100 + 1% penetration.
b. The total range of the 100% penetration data should be within the specified limits of the sampling plan or more data must be taken.

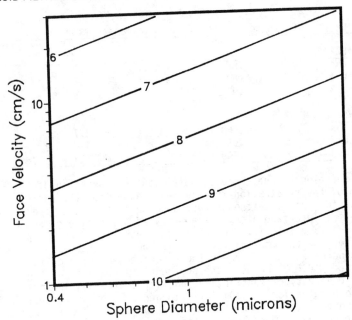

GRAPH 3 - HE-1021 % coefficient of variation contours:
sphere diameter vs. face velocity

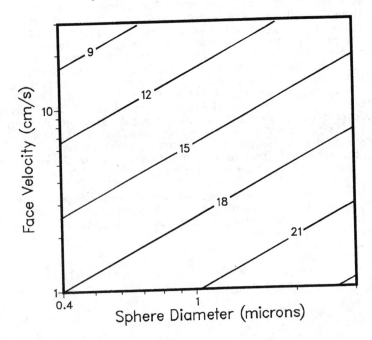

GRAPH 4 - HF-0012 % coefficient of variation contours:
sphere diameter vs. face velocity

TABLE 4—Analysis of 100% Penetration Variability:

5 Consecutive Sample Intervals

Latex Sphere Diameter	.62 μm		.62 μm		.9 μm		2.02 μm		2.02 μm	
Face Velocity	1 cm/s		25 cm/s		5 cm/s		1 cm/s		25 cm/s	
	$\bar{X}$	% Cov	$\bar{X}$	% Cov	$\bar{X}$	% Cov	$\bar{X}$	% Cov	$\bar{X}$	% Cov
Lab 1	98.9	1.6	100.2	0.82	100.2	0.53	97.5	2.7	99.5	4.4
Lab 2	102.4	10.3	102.4	4.7	104.0	5.5	113.3	21.6	98.4	7.8
Lab 3	100.3	0.54	100.4	2.3	104.3	3.8	100.2	2.1	96.8	3.3
Lab 4	99.3	2.7	101.2	1.1	102.3	2.2	106.7	19.5	108.1	11.5
Lab 5	98.8	3.2	96.6	5.6	98.7	1.8	83.5	1.3	93.6	2.0
Lab 6	98.9	2.3	101.4	2.2	97.4	3.3	106.5	6.8	120.0	16.1

TABLE 5—Particle Concentration and Accumulation Per Test for Different Laboratories

Laboratory (Particle Counter)	Particle Counter Sample Flowrate Rate (cm³/s)	Sphere Diameter (microns)	Concentration (Particles/cm³) Face Vel.(cm/s)			Particle Accumulation Per Test (Particles/cm² ×10³) Face Vel.(cm/s)		
			1	5	25	1	5	25
Lab 1 (Climet 225)	118.0	0.62	8.0	28.0	17.9	1.9	33.6	107.0
		0.9	3.3		9.9	0.79		59.4
		2.02						
Lab 2 (PMS–LAS–X)	5.0	0.62	300.0	158.0	260.0	72.0	190.0	156.0
		0.9	150.0		15.0	36.0		90.0
		2.02						
Lab 3 (Climet 208)	118.0	0.62	53.0	8.8	7.2	21.2	17.6	72.0
		0.9	37.0		2.9	14.8		29.0
		2.02						
Lab 4 (Climet 226)	118.0	0.62	17.5	7.8	15.5	28.9	6.4	63.9
		0.9	0.12		0.28	0.19		1.0
		2.02						
Lab 5 (TSI – APS)	472.0	0.62	5.0	1.1	4.1	0.40	0.44	8.2
		0.9	0.26		0.08	0.31		0.60
		2.02						
Lab 6 (PMS–LAS–X)	1.0	0.62	1160.0	483.0	21.7	27.8	580.0	430.0
		0.9	3.3		3.3	0.79		0.79
		2.02						

Particle Concentration and Particle Accumulation

Table 5 shows the concentration in particles per cubic centimeter and particle accumulation per test for each laboratory at each variable level. At lower particle counter sampling rates, greater particle concentration was necessary to obtain a significant number of particle counts per test. This led to greater particle accumulation per test, perhaps forming a particle filter cake lowering percent penetration and adding to the variability. The 2.02 micron spheres were difficult to atomize, giving lower upstream concentrations in most cases. Transport problems also probably affected the large particle counting error.

Results of Pressure Drop Data Analysis

The pressure drop of each media at each face velocity was recorded to give a rough analysis of interlaboratory flow resistance variability. This data alone at 4 runs per face velocity per laboratory was not of large enough sample size to give a precision statement. However, if the interlaboratory data is pooled for each filter material, a plot of a regression curve and its upper and lower 95% confidence limits or pressure drop versus face velocity may be made to give an assessment of the degree of variability at different levels of flow resistance.

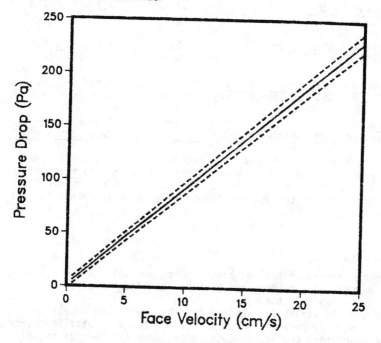

Graph 5 - HE-1021 regression line and 95% confidence
limits: pressure drop vs. face velocity

Graph 5 for the HE-1021 material shows the interlaboratory variability relationships at different pressure drops and face velocities. An increase in interlaboratory percent variability at the low face velocities is caused by a lack of manometer accuracy at these low flow resistances. The ASTM F778-82 Standard predicted this. The variability in the 5 cm/s and 25 cm/s ranges was consistent, but still rather large and an indication that the face velocities in each laboratory were not the same. With this in mind, it is recommended that the F778 Standard include that a plate with evenly distributed flow resistance or a standard reference filter sample be installed in the sample holder periodically to check both the manometer and flow measurement accuracy at the flow resistance level under study.

Table 6 shows some comparisons of variability results on HE-1021 for the pooled interlaboratory data and the mean single laboratory data for HE-1021 and HF-0012. Also included is a separate study of single laboratory flow resistance variability on HE-1021. This separate study was conducted by Frazier Precision Instruments on twenty-five single samples at each face velocity, while LAB 2 ran the same sample number as a check at the 5 cm/s face velocity.

TABLE 6--Comparisons of Pressure Drop
Variability on H&V #HE-1021

	Face Velocity (cm/s)		
	1	5	25
Interlaboratory % Cov	13.0	7.7	7.3
(HF-0012)	(25.2)	(8.6)	(9.0)
Mean Single Laboratory % Cov	5.4	---	6.9
(HF-0012)	(6.9)	---	(4.6)
Frazier Precision Instruments % Cov	2.24	1.03	.94
% variation, 95% confidence level	4.48	2.06	1.88
LAB 2 % Cov	---	1.00	---
% variation, 95% confidence level		2.00	

Since Lab 2 and Frazier Precision Instruments obtained the same precision and since the precision at 5 cm/s and 25 cm/s is essentially the same, a statement can be made that the single laboratory single operation precision at the 95% Confidence level of this method is less than 2.5 percent variation.

CONCLUSIONS:

1. Concentration control of latex sphere aerosols is a large factor in percent penetration variability. Low face velocity measurement is also a factor.

2. Of the 14% variability unexplained by the fitted models to % penetration, 4 to 10 percent is the difference between interlaboratory and single laboratory testing (one-third to three-quarters of the unexplained variability).

3. In single laboratory testing of % penetration, this method gave excellent results over an exceptionally wide range of particle size and face velocity with only 3 to 12 percent of the variability unexplained by the fitted models. Narrower design unit ranges on the designed experiment would have given much lower variabilities. Since this designed experimentation gave only linear fit models, it is probable that curvilinear models would improve the variability.

4. Test variability increased as sphere diameter increased and face velocity decreased.

5. Much of the interlaboratory variability can be directly attributed to differences in test apparatus, face area, concentration control and flow rates.

6. For the ASTM F778 "Standard Method Gas Flow Resistance Testing of Filtration Media," a statement can be made:

"The single laboratory single operator percent variation of this test is less than 2.5 percent at the 95 percent confidence level. No interlaboratory precision statement can be made at this time."

7. The filter efficiency test procedure is appropriate at the designated levels if concentration control and transport particle loss is considered. It is recommended that a new Round Robin in the form of the analysis of variance of a nested designed experiment be run on a statistically significant number of samples to determine the single sample, single laboratory and interlaboratory variabilities at two levels: large sphere diameter, low face velocity and small sphere diameter, high face velocity. This would document statistically the highest and lowest variability regions and generate an appropriate precision statement for the efficiency standard.

Author's Note:

Theoretically, particle efficiency test variability in the large (>.8 micron) particle diameter will always be greater than that in the "most penetration particle" diameter (.1 to .4 micron). The log-log plot of % penetration versus particle diameter in the .1 to .4 micron range gives a maximum, essentially a flat region of change in penetration for a discrete change in particle diameter. In the greater than .8 micron range, however, the log-log relationship has a slope of -2, giving at least twice the variability encountered in the most penetrating diameter range at the same discrete change in particle diameter. In a like manner, the variability measurement of materials tested should also be different, with the larger particle test being more sensitive to filter material structure.

Annex:

An analysis of variance (ANOVA) is a method of examining the variability of a set of statistics. Statistics can be collected and a curve-fit relationship produced by examining how a dependent, measured quantity (response) changes as another independent quantity (variable) is changed. A measure of the degree of fit or confidence level is provided by an analysis of variance of that relationship. The relationship of how a response changes while two variables are changed at the same time may be done statistically by carefully structuring the changes in the variables. One method of doing this is a 2x2 factorial designed experiment, the response results of which may be analyzed by variance analysis to give a curve-fit equation and the degree of fit. In the case of the data analysis for Graph 1 for the HE-1021 filters (d = sphere diameter, FV - face velocity) the equation was in linear form:

$$ -\ln\left[\frac{\% \text{ pen}}{100}\right] = 3.171 + 1.624\left[\frac{\ln 10\, d - 2.197}{.372}\right] + .615\left[\frac{\ln FV - 1.609}{1.609}\right] $$

The degree of fit of such a model can be measured by determining a correlation coefficient (R), which in this case is the degree of dependence of a response to structured changes in the two variables. The square of the correlation coefficient multiplied by 100 is the percent total variability explained by the model, and subtracting this from 100 gives the percent unexplained variability. In the case of Table 3, final conclusions on interlaboratory variability were made by comparing the pooled interlaboratory unexplained variability to the two best cases of single laboratory unexplained variability. A more proper method to separate the single laboratory data variability from the interlaboratory variability is by the use of a nested designed experiment. The coefficient of variation of a normal distribution used in this paper is defined as the standard deviation divided by the mean multiplied by 100.

Acknowledgement

The F21.20 Subcommittee is indebted to Brian D. Johnson of 3M Company for his technical expertise and guidance in statistics and analysis of variance.

References:

[1] ASTM D2906-85 "Standard Practice for Statements on Precision and Bias for Textiles"
[2] Davies, C. N., "The Separation of Airborne Dust and Particles," Institution of Mechanical Engineers, London, Proceedings 1B, 1952 (Assuming Fiberglass Density = 2.6 g/cm^3).
[3] Nicholson, R. M. "Standardized Test Method for Using Latex Aerosols and Optical Particle Counters to Determine Particle Size Initial Efficiency of Flat Sheet Filter Media," Fluid Filtrations: Gas, Volume I, ASTM STP975, R. R. Raber Ed. American Society for Testing and Materials, Philadelphia, 1986.

Applications and Testing: Respirators

Ernest S. Moyer

RESPIRATOR FILTRATION EFFICIENCY TESTING

REFERENCE: Moyer, Ernest S. Moyer, "Respirator Filtration
Efficiency Testing," Fluid Filtration: Gas, Volume I,
ASTM STP 975, R. R. Raber, Ed., American Society for
Testing and Materials, Philadelphia, 1986

ABSTRACT: The existing methods for evaluating
particulate respirator filters as found in 30 CFR
Part 11 are presented. Shortcomings with these
methods are discussed in detail. NIOSH's research
objectives for updating particulate respirator
filter testing are presented along with some
research findings. Conclusions are drawn from
these results and areas of future research in
respirator particulate filter evaluation are
stated.

KEYWORDS: filter efficiency, "worst case" aerosol,
particulate filter testing, particulate filters,
filter efficiency test system

INTRODUCTION

The hazards of inhaling toxic compounds or other toxicants present
in the work environment is of major concern for the well-being of
workers in the modern industrial setting. Recently, much interest
has been focussed on the inhalation of carcinogens and other toxic
hazards associated with industrial processes. To provide adequate
respiratory protection programs to such work environments is an
important and necessary preventive measure against cancer and other
workplace associated diseases.

Clearly, engineering and administrative controls, such as improved
ventilation, process changes, and good work practices, are the
preferred approaches to protecting workers from exposure to airborne
toxic materials. However, the use of respirators is the only
immediate option where engineering controls are unfeasible or not

Dr. Moyer is a research chemist at the National Institute for
Occupational Safety and Health, Division of Safety Research, Injury
Prevention Research Branch, Laboratory Investigations Section, 944
Chestnut Ridge Road, Morgantown, WV 26505-2888

yet available. The same may be true during the installation of engineering controls and during life-threatening emergency situations. Thus it can easily be seen that respirators are frequently needed to supplement inadequate engineering controls or to protect workers under emergency conditions.

Estimates suggest that some 15% of the industrial work force utilize respirators at some time during their work activities. Respirator sales are reported to be on the order of $300 million annually, which translates into tens of thousands of respirators being sold annually. Unfortunately, these respirators may not meet the needs of the workplace, because of improper selection, use, and/or maintenance.

BACKGROUND

NIOSH, in cooperation with the Mine Safety and Health Administration (MSHA), is currently responsible for the testing and certification of respiratory protective devices. NIOSH is mandated by the Occupational Safety and Health Act of 1970 and the Federal Mine Safety and Health Amendments Act of 1977 to conduct research to eliminate on-the-job hazards to American workers. Research into and certification of respiratory protective devices is a significant part of that mandate.

The first bill establishing an agency of the Federal government to oversee and be responsible for the U.S. mineral industries was passed by Congress in 1910. The Bureau of Mines was thus established as part of the Department of the Interior. Over the years attention was given to obtaining equipment which could be used in hazardous mine atmospheres. By World War I, the Bureau had accumulated considerable knowledge in respiratory protection devices. This led to a cooperative effort during the war between the Bureau and the Defense Department to produce air-purifying respirators which were effective against warfare gases.

After the war, the Bureau continued to conduct scientific investigations in respiratory protection to improve health conditions in mines and to impact upon wearer safety. This culminated in 1919 with federal regulations on the respirator testing and approval program. A publication of schedules dealing with permissible gas masks and permissible self-contained mine rescue breathing apparatus resulted. In each case the schedule set forth the procedure by which manufacturers could get approval to have their devices listed among the certified or approved devices.

With the exception of coal mining and a few other relatively small industries, it was not until the promulgation of regulations in 1974 under the Occupational Safety and Health Act of 1970 that the use of approved respirators became mandatory in general industry. The various Bureau of Mines schedules and regulations dealing with respirators were consolidated, amended, and promulgated in 1972 in Title 30, Code of Federal Regulations (CFR) Part 11 which provided for joint certification by NIOSH and the Bureau of Mines.

Although the Federal Government has been involved in testing and certifying respirators for more than 60 years, the complexity, variety, use, and application of respirators have dramatically expanded over the last 15 years. As the uses and applications of these devices have expanded, so has NIOSH's need to assess the adequacy of the application and performance of such devices, and the validity of our testing methodologies. Based upon the many factors involved, we have recognized the need to propose revisions to the certification regulations (30 CFR Part 11) to more properly reflect the requirement for appropriate performance in the expanding use environment and to incorporate recent technological advances in testing methodology.

In the past ten years, NIOSH has become increasingly concerned that the test criteria used for evaluating respirators under the present 30 CFR Part 11 have gradually become irrelevant. Also in question are the relevancy, accuracy, and reproducibility of existing test protocols. The Bureau of Mines, and recently NIOSH, have conducted some research into the development of new criteria for respiratory protection and for the development of test procedures as new respirators were offered for certification. However, little basic research conducted since the middle 1940's has been translated into test criteria or procedures. Existing data need to be incorporated into the respirator testing and certification program, and much more new data need to be developed in anticipation of an accelerated development of respiratory protective devices. A special evaluation of the current NIOSH certification program by a panel of independent experts stated:

"The establishment of performance criteria is a major investment of those charged with developing a satisfactory assurance program. These criteria must be realistic and must be technologically as advanced as the state of the art will permit. Many current NIOSH tests of PPE, in particular, are recognized by manufacturers and researchers in the health and safety field to be lagging in the current state of the art of testing. A satisfactory assurance program must rely on advanced investigatory work by NIOSH. A satisfactory assurance program cannot be one associated with NIOSH performance criteria and tests which are technological anachronisms."

NIOSH has a fundamental responsibility not only to raise its certification programs to a modern level but to simultaneously advance the state of the art and provide leadership in the respiratory protection field. This is why in 1981, NIOSH undertook a project to rewrite and update 30 CFR Part 11.

PARTICULATE FILTER TESTING

The existing methods for evaluating particulate respirator filter efficiency are found in 30 CFR Part 11 Subpart K - Dust, Fume, and Mist Respirators and are as follows:

11.140-4 Silica dust test; single-use or reuseable filters; minimum requirements

11.140-5 Silica-dust test; single-use dust respirators;
          minimum requirements
11.140-6 Lead fume test; minimum requirements
11.140-7 Silica mist test; minimum requirements
11.140-8 Tests for respirators designed for respiratory
          protection against more than one type of dispersoid;
          minimum requirements
11.140-11 DOP filter test; respirators designed as respiratory
          protection against dusts, fumes, and mists having an
          air contamination level less than 0.05 milligram per
          cubic meter and against radionuclides; minimum
          requirements.
11.140-12 Silica dust loading test; respirators designed as
          protection against dusts, fumes, and mists having an
          air contamination level less than 0.05 milligram per
          cubic meter and against radionuclides; minimum
          requirements.

The test procedures and requirements are as follows:

Certification Tests

Dust, fume, and Mist Respirators

(1) Silica Dust Test

     (A) 90 Min Test (4 hrs. for powered)
     (B) Continuous flow 32 liter/minute, 115 liter/minute, or 170
          liter/minute.
     (C) RH 20-80% and 25°C
     (D) 50-60 mg/m^3
     (E) Geometric mean of 0.4-0.6 micrometer
     (F) Standard geometric deviation will not exceed 2
     (G) Unretained material
          (1) < 1.5 mg for air-purifying
          (2) < 14.4 mg tight fitting powered
          (3) < 21.3 mg loose fitting powered
     (H) Single use - breathing machine cyclic flow instead of
          above.

(2) Lead Fume Test (lead-oxide fume)

     (A) 312 Min Test (4 hrs for powered)
     (B) Continuous flow 32 liter/min, 115 liter/min, or 170
          liter/min.
     (C) RH 20-80% and 25°C
     (D) 15-20 mg/m^3 (as Pb)
     (E) Unretained Material as Pb.
          (1) < 1.5 mg for air-purifying
          (2) < 4.2 mg tight-fitting powered
          (3) < 6.2 mg loose-fitting powered

(3) Silica Mist Test

     (A) 312 min test (4 hrs for powered)

    (B)   Continuous flow 32 liter/min., 115 liter/min, or 170
          liter/min.
    (C)   Approximately 25°C
    (D)   20-25 mg silica mist (weighed silica dust) by
          spraying an aqueous suspension
    (E)   Unretained material weighed as silica dust
          (1) < 2.5 mg for air-purifying
          (2) < 6.9 mg for tight-fitting powered
          (3) < 10.2 mg loose-fitting powered

(4) Tests Against More than One Type of Dispersoid

    (A)   Shall comply with requirements for each specific hazard.

(5) DOP Test

    (A)   Challenge time of 5 to 10 seconds
    (B)   Continuous flow rates of 32 and 85 liters per minute (16
          and 42.5 lpm for pairs).
    (C)   DOP concentration 100 micrograms/liter
    (D)   Total leakage for the connector and filter shall not
          exceed 0.03 percent of the ambient DOP.

(6) Silica Dust Loading Test

    (A)   Test as per silica dust test
    (B)   Meet silica dust test requirements
    (C)   Meet airflow resistance tests.

The current certification testing methods were reviewed in
conjunction with the rewrite of 30 CFR Part 11. Significant
shortcomings were identified with the particulate air-purifying test
methodology. For example, the current tests:

    (1)   measure a time averaged, rather than instantaneous
          penetration

    (2)   do not specifically consider the effects of particle size,
          face velocity, or aerosol type on filter penetration

    (3)   appear to lack sensitivity and are non-reproducible

    (4)   do not consider environmental conditions such as
          temperature and relative humidity.

(1) Integrated Versus Instantaneous Monitoring

    The present tests (except for DOP) measure the penetration
    averaged over a time period of 90 minutes or more rather than
    an instantaneous penetration of the filter media at any earlier
    time. As a result, vital information on how much penetration
    occurs at any particular moment is lost. Yet, the
    instantaneous penetration is important because many existing
    respirator filter media have either good initial filter
    efficiencies or good final filter efficiencies due to loading
    effects.

Two different cases which would be detrimental to the respirator wearer are possible. Case I involves a respirator containing an "electrostatic" filter media. Electrostatic media have good initial filter efficiencies, but the filter degrades with increased particulate loading. This loading causes a masking or loss of electrostatic charge (filter degradation) resulting in reduced filter efficiency and increased worker exposure. This is possible since there are no end-of-service-life indicators for such respirators. Note that the longer the wearer continues to use this respirator under these conditions the higher the exposure level.

Case II involves a respirator which contains a "mechanical" filter media, which has poor initial filter efficiency, but where the filter's efficiency increases as a function of filter loading. Should a user be exposed to a relatively low concentration of a contaminant, the filter might take a long time before it loads and the filter's efficiency increases. If this particular contaminant is highly toxic (low permissible exposure level), then the wearer could initially be exposed to hazardous levels.

These two cases represent real world situations which are not simulated in the current performance tests in 30 CFR Part 11. However, methodology which would employ continuous monitoring of filter efficiency with time would result in a means for better predicting these use situations.

(2)   Particle Size, Aerosol Type, Flowrate

Langmuir [1], Ramskill and Anderson [2], and Gillespie [3] have all predicted the existence of a "worst case" aerosol size, against which fibrous filters have minimum efficiency. These predictions are based on theoretical single fiber filtration mechanisms of diffusion, interception, and impaction. The calculations based on single fiber theory predicts an aerosol size of between 0.1-0.4 μm (AMMD) for minimum efficiency. In fact some experimental investigations [4,5,6] have verified these predictions. Thus, we are proposing that filter efficiency tests with two aerosols of the "worst case" type (approximately .3 micrometers) be studied.

To minimize any possible differences due to aerosol types and their degrading behavior on filter media, two aerosols will be used. One will be a solid aerosol and the other will be a liquid aerosol. This is warranted, since it is known that liquid aerosols can affect filter media in a much different manner (coating) than a solid aerosol. However, it should be noted that both aerosols will be neutralized prior to their use as a challenge aerosol.

Filter penetration is a function of face velocity through the filter. Therefore, the filter efficiency tests will be run at two flowrates, one high (85 lpm) and the other low (32 lpm).

(3) Discriminating ability and reproducibility

The existing methods in 30 CFR Part 11 appear to lack
sensitivity. They appear to be limited in their ability to
differentiate between good, medium, or low efficiency filters.
This is substantiated by the ANSI committee data, literature
data, [4,7,8,9] and in-house data which indicate that the
efficiency measured by a "worst case" type aerosol, when
continuously measured, was significantly less than that
obtained by the current certification tests. Also, present
tests have been demonstrated to be non-reproducible between
laboratories. As a result, respirator wearers who rely on
currently certified respirators for protection against certain
particulates, may not be getting adequate protection,
especially if the environment contains a significant quantity
of the "worst case" (those hardest to filter) contaminants.

(4) Environmental considerations

The influence of environmental conditions such as temperature
and relative humidity are not considered in the test regimen in
the present 30 CFR Part 11. This has been addressed in some of
our recent work which was reported at AIHC in Dallas. Filters
were pretreated (ANSI Ad hoc Respirator Committee
recommendation) at 85% relative humidity and 38°C for 24
hours prior to filter efficiency testing. A minimal effect
($\approx$ 2%) was seen, which indicates that longer pretreatment
times (weeks) might be required. Additional work is needed in
this area to determine the best RH testing regimen.

NIOSH Approach

NIOSH's overall objectives were as follows:

(1) support proposed revisions to 30 CFR Part 11

(2) determine "worst case" aerosol challenge conditions

(3) evaluate commercial particulate filters under "worst case"
     conditions

(4) compare filter efficiencies - certification method against
     "worst case" aerosol.

NIOSH started by pursuing work on the determination of the optimum
aerosol particle size for testing particulate respirator filters
(and to see how this varied as a function of test flowrate). This
so-called "worst case" challenge aerosol would give the maximum
filter penetration (minimum filter efficiency). A method utilizing
such an aerosol would be able to differentiate between good, medium,
and low efficiency filters of all types. The important safety issue
involved would be that filters tested against a "worst case" aerosol
would protect wearers' against smaller as well as larger particles.
This is the only way of guaranteeing performance against virtually
all size particles.

The efficiency of a filter medium is dependent on the characteristics of the challenge aerosol as well as the filter characteristics. Challenge aerosol characteristics such as particle size, composition, density, shape, electrical charge and flowrate are influential in determining the filter's efficiency. Naturally, many filter characteristics such as packing density and single fiber diameter, are also critical. These latter characteristics, which are controlled by the manufacturer, are of minimal importance to NIOSH as long as the filters meet certification criteria. NIOSH's interest is with the overall performance of the commercially-marketed filters of all types as a function of testing parameters.

Single fiber filtration theory [1,2,3,10] predicts an aerosol size of maximum penetration, since an increase in particle size will cause increased filtration by the interception and inertial impaction whereas a decrease in particle size will enhance collection by Brownian diffusion. Thus there is an intermediate particle size region where two or more of these mechanisms are simultaneously operating. In this region the particle penetration through the filter is a maximum and the efficiency of the filter is a minimum. This most penetrating aerosol size has been reported to be in the range between 0.1 - 0.4 μm for most fibrous filters [4,5,6]. However, Liu and Lee [11] have shown that at high filtration velocities (100-300 cm/sec.), the most penetrating particle size becomes substantially smaller than 0.3μm.

Although some experimental investigations [12,13] have verified these facts, little work has been done on commercially-available respirator filters to determine the aerosol size at which the maximum penetration occurs. Stafford and Ettinger [14] did evaluate Whatman 41 and IPC 1478 filter paper as a function of particle size and velocity against polystyrene latex spheres and concluded that a reevaluation of filter testing should be considered since a 0.3μm aerosol does not yield minimum efficiencies for all filter media at the different velocities of concern.

NIOSH's study (being prepared for publication) looked at two of the parameters: particle size and flowrate as a function of filter penetration. Particles in the size range from .01 - .30 μm (CMD) were employed in this study to identify the "worst case" aerosol size for testing commercially-available filters of the dust and mist; paint, lacquer and enamel mist; dust, fume and mist; and high efficiency types. Both solid and liquid aerosols have been looked at since reports in the literature [15-19] indicate that differences in filter penetration exist between them due to increased degradation, loading effects, or differences in charging. This report will deal only with results obtained when a liquid DOP aerosol was used.

EXPERIMENTAL DESIGN

Air-purifying filters along with the filter's holder and gasket (where separable) were tested for the initial instantaneous filter efficiency as mounted on the connector in a manner as used on the respirator. When the filter holders were not separable, the exhalation valves were blocked to ensure that valve leakage, if present, was not included in the filter efficiency results. Also, wherever possible, all filters tested were from the same lot to eliminate lot-to-lot variability.

All filters were tested as received from the manufacturer without any kind of preconditioning. Filters were challenged with a liquid DOP aerosol and a solid NaCl aerosol which had been passed through a Kr-85 radioactive source to neutralize the aerosol. The liquid DOP aerosol was produced by an evaporization/condensation technique similar to that described by Liu and Lee [20]. The NaCl aerosol was produced using the following devices in sequence: 1) constant output atomizer, 2) drier, 3) electrostatic classifier, 4) mixer, 5) neutralizer (conditioned with a Kr-85 radioactive source to establish a Boltzmann charge distribution), 6) test chamber (TSI - filter efficiency test system).

A continuous flowrate of 16 - 85 lpm was employed. Room temperature was employed for all the studies. Where possible at least five filters of each type from the same lot were tested and the average efficiency determined.

Filters from different manufacturers were tested against various particle sizes in the range from .01 - .30 μm (CMD). Filters of the dust and mist; dust, fume and mist; paint lacquer and enamel mist; and high efficiency types were employed in this study. In all cases the challenge concentration was maintained at less than 10^7 particles per cm^3 to avoid coagulation. The exact concentration did vary over a limited range but this probably didn't affect the results since both upstream and downstream concentrations were monitored before and after each test. Also, if the upstream concentration at the onset and completion of a run changed by more than 2-3% the test was not considered valid.

Aerosol Efficiency Measurements

The efficiency and penetration was monitored and recorded by means of the TSI Incorporated Filter Efficiency Test System (FETS) which has been described by Remiarz, Moyer, et al [21]. This instrument which was built for NIOSH under contract contains a continuous flow single particle counting Condensation Nucleus Counter (CNC, TSI model 3020). The CNC can measure concentrations as high as 10^7 particles/cm^3 when using the photometric mode, and using its single particle counting ability, can measure concentrations down to 10^{-2} particles/cm^3. When used to measure the particle concentrations both upstream and downstream of a filter, the CNC's large dynamic range allows count filter efficiencies as high as 99.99999+% to be measured. This instrument's sensitivity and dynamic concentration measurement range was necessary in order to detect the differences in filter efficiency as well as to cover the large range of concentrations anticipated in going from upstream to downstream concentrations. This was especially true in the case of high efficiency filters. In addition to testing efficiencies, the system measures respirator flowrates and pressure drops. The three parameters are monitored over time to obtain respirator loading data. Up to three respirators can be tested simultaneously and automatically. The instruments are interfaced to a dedicated microcomputer system that monitors, reads, and controls the instruments. Extensive data analysis and presentation is done by the microcomputer system, including cross-correlations between the various tested parameters.

To date results have been obtained on a complete line of respirator filters from three manufacturers. Some typical results (each point is an average efficiency determined on multiple filters of the same lot) are presented in Figure I (high efficiency filters) and Figure II (Paint, Lacquer & Enamel Mist Filter). The observed ranges of filter efficiency at the point of maximum penetration are presented in the following table:

Table 1--Filter Efficiency at Maximum Penetration for Selected Filter Types.

Filter Type	Range of Percent Minimum Efficiency DOP
Dust, Fume & Mist	91-93 %
Dust & Mist	67-75 %
Paint Lacquer & Enamel Mist	72-82 %
High Efficiency Filters	> 99.97 %

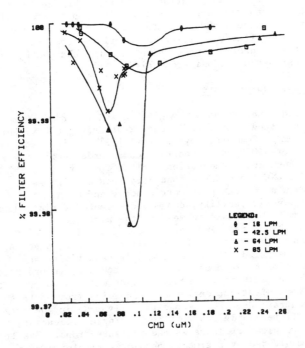

FIG. 1.--High efficiency filter efficiency for DOP.

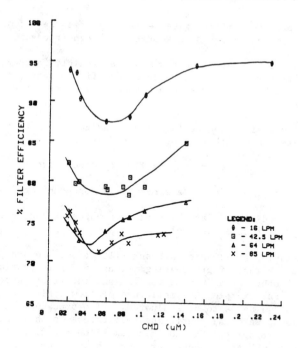

FIG. 2 - Paint, lacquer and enamel mist filter
efficiency for DOP.

It should be noted that these significantly lower efficiencies for
dust, fume and mist; dust and mist; and paint, lacquer, and enamel
mist cannot be explained by detector response differences but indeed
indicate that the "worst case" type test is much more rigorous.
Thus, the present tests (silica dust & lead fume) may not be true
indicators of the filters' performance. Further, it suggests that
the present tests are possibly not discriminating between good and
bad (high, medium, and low efficiency) filters of these respective
types. It must be noted, however, that this study does not consider
loading effects.

The following points can be derived from these studies.

o   A particle size at which a minimum efficiency occurs does exist
     and has been identified for the various filters tested.

o   The particle size of minimum efficiency varies from filter to
     filter even within the same filter type.

o   The particle size at which the minimum efficiency occurs ranged
     from .02 - .29 μm (CMD)

o   High efficiency filters showed minimum efficiencies at the
     largest particle size as compared to the other filter types
     tested.

o   High efficiency filters gave percent efficiencies of > 99.97% when tested at or near the "worst case" conditions.

o   Dust and mist; dust, fume and mist; and paint, lacquer, and enamel mist filters are not nearly as efficient as the present testing methods indicate when tested by a "worst case" type aerosol.

o   Present certification tests are not indicative of performance when determined with the "worst case" aerosol.

o   Filter efficiency decreases as flowrate increases.

o   Particle size of minimum efficiency is dependent on flowrate.

o   The proposed 30 CFR Part 11 revisions appear to be in the range of the "worst case" aerosol.

Work is continuing at NIOSH on aerosol filtration. We are presently looking at the effect of pretreating filters prior to testing. However, many other areas need attention and future projects will be forthcoming. For example, the influence of flow type (constant vs cyclic flow [22]) needs to be evaluated. Also procedures are needed for evaluating powered air-purifying respirators in their normal operating mode rather than at the minimum allowable flowrate, as is presently done. These are necessary because the flow of a powered air-purifying respirator may vary significantly over its four hours of use. For example, a powered air-purifying respirator with a tight fitting facepiece might have an initial flow as high as 400 liters per minute in order to achieve a final flow of at least 115 liters/minute after 4 hours of operation. Since filter penetration is a function of flowrate (face velocity), it is necessary to test these devices under conditions of normal operation in order to obtain realistic filter efficiency data. Additional areas of consideration are aerosol charge, aerosol density, shape effects, and velocity. It is hoped that data collected will ultimately lead to a model for predicting filter efficiency.

REFERENCES

[1] Langmuir, I., W. H. Rodebush and V.K. Lamer, The Filtration of
     Aerosols and the Development of Filter Materials, Report OSRD
     865, 1942.

[2] Ramskill, E.A., and W. L. Anderson, "The Inertial Mechanism in
     the Mechanical Filtration of Aerosols," Journal of Colloid
     Science, Volume 6, p. 416 (1951).

[3] Gillespie, T., "The Role of Electric Forces in Filtration of
     Aerosols by Fiber Filters," Journal of Colloid Science, Volume
     10, p. 299, 1955.

[4] Thomas, J.W., and R. E. Yoder, "Aerosol Size for Maximum
     Penetration Through Fiberglass and Sand Filters," American
     Medical Association Archives of Industrial Health, Volume 13,
     p. 545, 1956.

[5] Rimberg, D., "Penetration of IPC-1478, Whatman 41, and Type 5G
     Filter Papers as a Function of Particle Size and Velocity,"
     American Industrial Hygiene Association Journal, Volume 30, No.
     4, p. 394, 1969.

[6] Lindeker, C.L., R.L. Morgin, and K.F. Petrock, "Collection
     Efficiency of Whatman 41 Filter Paper for Submicron Aerosols,"
     Health Physics, Volume 9, p. 305, 1963.

[7] Lee, K.W. and B.Y.H. Liu, "The Minimum Efficiency and the Most
     Penetrating Particle Size of Fibrous Filters," Journal of the
     Air Pollution Control Association, Volume 30, No. 4, p. 377,
     1980.

[8] Stafford, Ronald G. and Harry J. Ettinger, "Filter Efficiency
     As a Function of Particle Size and Velocity," Atmospheric
     Environment, Volume 6, p. 353, 1972.

[9] Liu, Benjamin Y.H., and K.W. Lee, "Efficiency of Membrane and
     Nuclepore Filters for Submicrometer Aerosols," Environmental
     Science and Technology, Volume 10, No. 4, p. 345, 1976.

[10] Chen, C.Y., "Filtration of Aerosols by Fibrous Media," Chemical
     Review, Volume 55, p. 595, 1955.

[11] Liu, B.Y.H. and K.W. Lee, "Efficiency of Membrane and Nuclepore
     Filters for Submicrometer Aerosols," Environmental Science
     Technology, Volume 10, p. 345, 1976.

[12] Rimberg, D., "Penetration of IPC-1478, Whatman 41, and Type 5G
     Filter Papers as a Function of Particle Size and Velocity,"
     American Industrial Hygiene Association Journal, Volume 30, No.
     4, p. 394, 1969.

[13] Lindeker, C.L., R.L. Morgin, and K.F. Petrock, "Collection
     Efficiency of Whatman 41 Filter Paper for Submicron Aerosols,"
     Health Physics, Volume 9, p. 305, 1963.

[14] Stafford, R.G. and H.S. Ettinger, "Filter Efficiency as a Function of Particle Size and Velocity," Atmospheric Environment, Volume 6, p. 353, 1972.

[15] Stafford, R.G. and H.J. Ettinger, "Comparison of Filter Media Against Liquid and Solid Aerosols," American Industrial Hygiene Association Journal, Volume 22, p. 319, 1971.

[16] Lockhart, L.B., Jr., R.L. Patterson, Jr. and W.I. Anderson, "Characteristics of Air Filter Media Used for Monitoring Airborne Radioactivity", NRL-6054, March 1964.

[17] Posner, S., "Air Sampling Filter Retention Studies Using Solid Particles," Proceedings of the 7th AEC Air Cleaning Conference, TID 7627, p. 43, Brookhaven National Laboratory, Long Island, New York, October 1861.

[18] Lindeker, C.L., R.L. Morgin, and K.F. Petrock, "Collection Efficiency of Whatman 41 Filter Paper for Submicron Aerosols," UCRL 6691, March 1962.

[19] Adeley, F.E., R.H. Scott, and W.E. Gill, "A study of Efficiencies and Pressure Drop Characteristics of Air filtering Media," HW-28065, August 1953.

[20] Liu, B.Y.H. and K.W. Lee, "An Aerosol Generator of High Stability," American Industrial Hygiene Association Journal, Volume 36, p. 861, 1975.

[21] Remiarz, R.J., J.K. Agarwal, P.A. Nelson, and E. Moyer, "A New, Automated Method for Testing Particulate Respirators," Journal of the International Society for Respiratory Protection, Volume 2, p. 275, 1984.

[22] Stafford, Ronald G., Harry J. Ettinger and Thomas J. Rowland, "Respirator Cartridge Filter Efficiency under Cyclic--and Steady--Flow Conditions," American Industrial Hygiene Association Journal, p. 182, May 1973.

Warren R. Myers

QUANTITATIVE FIT TESTING OF RESPIRATORS: PAST, PRESENT, FUTURE

REFERENCE: Myers, W. R., "Quantitative Fit Testing of
Respirators: Past, Present, Future," Fluid Filtration: Gas,
Volume I, ASTM STP 975, R. R. Raber, Ed., American Society for
Testing and Materials, Philadelphia, 1986.

ABSTRACT: The historical development of quantitative
techniques to evaluate the "fit" of a respirator on a
particular wearer is reviewed. The inferences drawn from such
testing with regard to protection or selection of the "best
fitting facepiece" are examined in light of new research data
recently obtained from field and laboratory studies. Areas of
research for improving quantitative fit testing are identified.

KEYWORDS: quantitative fit testing, fit factors, workplace
protection factors, respiratory protection

The amount of protection that a respirator provides can be no
better than its fit to the face. Given that a respirator is
selected and used properly and worn conscientiously, fit is the
crucial determinant of whether adequate respiratory protection will
be possible. If the facepiece does not fit it will not protect no
matter how effectively other components of the respirator work. For
example, any benefits to be achieved by using a high efficiency
air-purifying element that has penetration of only 0.03% would be
quickly lost if the facepiece were not capable of providing a fit of
comparable quality. During the last 30 years a very substantial
amount of respirator research has been focussed on developing
methods to evaluate how well a respirator fits. The purpose of this
paper is to review some of the highlights of that research and
discuss some of the current issues surrounding quantitative fit
testing and what new research needs to be undertaken.

Dr. Myers is an industrial hygienist at the National Institute
for Occupational Safety and Health, 944 Chestnut Ridge Road,
Morgantown, WV 26505.

QUANTITATIVE FIT TESTING: HISTORICAL REVIEW

One of the first reported attempts to quantitatively measure
faceseal leakage was by Guyton and Lense [1]. They developed a fit
test technique involving B. Globigii bacteria. The bacteria was
aerosolized from suspension in saline into a test chamber. The
concentration of bacteria in the chamber was determined by light
scattering techniques. The amount of leakage into the mask was
determined by collecting the bacteria on a cotton filter which was
held in the mouth. Their technique did not utilize extractive
in-facepiece sampling. Instead, subjects were instructed to inhale
through their mouth so that all the breath was filtered as it passed
through the mouth-held filter. Exhalation was done through the
nose. Thus, the collection of bacteria was accomplished only during
the inspiratory cycle. By comparing estimates of the concentrations
of bacteria in the chamber and inside the mask, a quantitative
estimate of penetration (fit) could be made. One significant
problem with this procedure was that the B. Globigii bacteria were
rather large (generally greater than 1 micron) this left some doubt
as to the interpretation of the results for aerosols of smaller size.

Burgess, Silverman and Stein proposed another procedure to
quantitatively test the fit of a respirator which utilized a uranine
dye aerosol [2]. Air samples were obtained by collecting a small
portion of the atmosphere within the facepiece cavity. The water
soluble dye was easily nebulized to produce a test aerosol with a
geometric mean particle size of 0.2 microns and a standard deviation
of approximately 2. Due to its fluorescent properties the uranine
was detectable in quantities as small as 0.1 ng per ml of water by
photometric technique. In addition, the photometric response was
linear over 3 orders of magnitude in uranine concentration (10^{-9}
to 10^{-6} gm/ml). An exposure chamber was designed and built which
was: 1) able to provide a homogeneous distribution of aerosol to
the breathing zone of the test subject; 2) small enough to be
semi-portable yet large enough to allow the test subject to perform
various work activities; 3) designed to permit various connections
to be made to the respirator worn by the test subject; 4) designed
to provide full vision of the test subject; and 5) designed to allow
removal of the test aerosol.

Samples for determining the concentration of uranine inside the
test chamber and inside the respirator cavity were collected on
one-inch, membrane filters in an open-face filter holder. The
uranine aerosol outside the respirator was sampled continuously at a
flowrate determined by a critical orifice; however, the flowrate was
not specified in the experimental methods. The uranine aerosol
which leaked into the mask was sampled only during the inspiratory
phase of each respiratory cycle. This was accomplished with a
respirator sampling detector, which used the output of a pressure
comparator to initiate and terminate sampling in sequence with the
inspiratory cycle. The sampling time was determined by an elapsed
time indicator which was also activated at the start of the
inspiratory cycle. The open-face filter holder used to sample
uranine leakage was placed inside the respirator so as to occupy a
sampling position in the cavity between the facepiece and subject's
face. A hypodermic needle attached to the filter holder was

used to penetrate the wall of the respirator and provide a means to connect the filter holder to a vacuum line. The needle assembly was pushed through the facepiece until the filter holder rested on the wall of the facepiece. The sampling rate used to collect the inside facepiece sample was not specified in the experimental methods.

The authors reasoned that by simultaneously sampling the air in the exposure chamber and in the respirator cavity during inhalation, the performance of a respirator worn by a test subject could be evaluated in a quantitative manner. This was one of the earliest reports of air sampling from inside a respirator using extractive sampling techniques.

Hounan, et al., described a quantitative fit test method in which the wearer of a respirator was tested in an atmosphere containing a sub-micron sodium chloride (NaCl) aerosol [3,4]. The authors note that even though sodium chloride is hygroscopic (a potential limitation with inhalation studies), it is non-toxic and readily detectable with flame photometry. The quantitative assessment of the respirator's performance was based upon estimates of the concentration of NaCl aerosol in the exhaled air. The exhaled air was routed to a 2-liter reservoir where it was mixed with dry air to prevent condensation and to dry the NaCl particles. An air sample was then collected from the reservoir volume and fed to a flame photometer. The test time was about one minute and no exercise was performed during testing. To deal with the problems of lung retention, the lung deposition of NaCl aerosol was measured on each test subject before testing began.

The authors note that protection factors (PF = C_O/C_I, C_O-outside concentration, C_I-inside concentration) as high as 500 could be readily determined with this test system. They also observed that "minor inaccuracies (+20%) occurred due to variations in an individual's breathing cycle" but felt this was not important in a test where the protection factors varied over several orders of magnitude (e.g. 5, 50, 500). The authors also applied this technique to measuring the performance of a respirator worn by bearded individuals. They were able to show that the fit of the respirator deteriorated as beard length increased. In addition to using NaCl particles, these authors also reported on the use of a halogenated hydrocarbon vapor as a challenge agent in quantitative fit testing. They noted similar estimates of faceseal leakage with both agents.

Hans Flyger reported on a quantitative fit test which utilized helium as the challenge agent [5]. Helium was selected as the tracer gas because it has a low solubility in the lungs. Therefore, he reasoned, the helium concentration in the dead space of the respirator would be independent of the breathing cycle so sampling could be done over the complete cycle without corrections for lung retention. As with previous work, the measurement of the ratio between the concentrations of helium in the dead space and the gas chamber represents an estimation of the leakage into the respirator. In-facepiece sampling was done with a hypodermic needle pushed through the body of the respirator. No indication is given as to where the needle was located or what sampling rate was used. In discussing the use of a gas challenge agent, the author also made the observation that "The penetration of particles through a leak is limited by interception, impaction and diffusion of the particles during passage of the leak, while the penetration of helium varies only with the air flow through the leak. The penetration of helium must consequently be considered the upper limit to the penetration of particles."

White and Beal extended the NaCl quantitative fit method described by Hounan, et al., to include various head and facial movements and normal and deep breathing patterns into the test procedure [6]. Air samples were still collected from exhaled air and analyzed by flame photometry.

Burgess and Shapiro reported on a test method developed to evaluate the protection provided by a powered air-purifying respirator against daughter products of radon [7]. The in-facepiece sampling technique was similar to that reported in earlier work [2] with the exception that personal sampling pumps were used to collect the air samples inside and outside the respirator. The sample collected outside the respirator was taken from immediately in front of the respirator.

The test subject was exercised at a 622 kg-m/min work rate. Once his breathing pattern was established, sampling was done for 4 minutes. The radon gas was allowed to attach to cigarette smoke which was reported to have a mean size of 0.25 μm. It should be noted that the in-facepiece sample was collected over the complete respiratory cycle with no corrections for particle loss in the lungs.

The adaptation of using personal sampling pumps was to later become the method upon which all field testing of respirators would be done. The respirator protection factor, as noted before, was determined from the ratio of the radioactivity on the chamber filter to the radioactivity on the respirator filter. This is the first article in which fit test results are used to draw inference about what level of protection the respirator would provide during actual use conditions.

In 1969, Adley and Uhle reported on a quantitative fit method utilizing Freon® [8]. In-facepiece sampling was done by drawing air though a short length of 1/8" O.D. stainless steel tubing inserted in the facepiece. The tubing was sealed in the facepiece with rubber sealant. A 12-foot polyethylene tube was used toconnect the respirator probe to a Halide Meter. Testing was done with the subject standing motionless, talking, breathing deeply to simulate hard work conditions, and lastly moving his head. This series of "exercises" was a forerunner to the "exercises" now commonly used in present-day quantitative fit testing. In-facepiece sampling appeared to be done over the complete respiratory cycle.

Griffin and Longson described yet another quantitative fit method using argon as a gas tracer [9]. Its rate of leakage into the mask was measured by mass spectrometry. Sampling was done only

from the exhaled air. The sampling rate was not specified. During
testing, the subject walked at 4 mph on a treadmill inclined at
2.5° while slowly carrying out a series of head movements.

In 1972 Hyatt, et al. described quantitative fit methods using
an oil mist of dioctylphthalate (DOP) and an improved quantitative
fit method using polydispersed NaCl [10]. The unique aspect of both
test methods was that they used in-facepiece sampling in conjunction
with fast response detectors and recorders to provide a real-time
measurement of leakage into a respirator being worn by a human test
subject. Up to this time, air samples both outside and in the
respirator (with the exception of Guyton) were based upon
time-weighted average (TWA) samples. The methods described by Hyatt
were rapidly accepted and viewed as models for quantitative fit
testing. In the U.S., the oil mist quantitative fit method has
become the method of choice. Oil mists of different substances are
presently being used.

The DOP aerosol used in the testing is polydispersed having a
mass median aerodynamic particle size of 0.7 μm and a
concentration of 25 mg/m^3 of air. The test chamber was 10 feet
long, 5 feet wide, and 10 feet high. It was equipped with a fan to
help disperse and mix the test agent in the chamber. During a test,
an air sample from inside the respirator was drawn through flexible
tubing at a volumetric flow of 8 lpm and passed through a forward
light scattering photometer to measure the penetration of the DOP
aerosol into the interior of the respirator. For in-facepiece
sampling, a probe was attached to the body of the respirator,
generally around its midline and approximately opposite the nose and
mouth of the wearer. Design characteristics of the probe were not
specified. A strip chart recorder attached to the photometer was
used to record the values of aerosol penetration in units of percent
of the concentration of the DOP aerosol in the test atmosphere.
While wearing the respirator in the test chamber, the test subject
carried out the following exercises:

1) normal breathing
2) deep breathing
3) nodding head up and down
4) turning head side to side
5) bending forward and touching toes
6) talking
7) smiling, if the respirator was equipped with a half-mask
   facepiece, or frowning if the respirator was equipped with a
   full facepiece
8) normal breathing

The average of the peak DOP penetrations leaking into the
interior of the respirator was determined for each exercise. The
mean of the average penetration for all exercises was used in
determining a respirator protection factor for each test. A
correction was not applied for lung retention. The authors felt
that to assign a half-facepiece respirator to an individual test
subject, the protection measured by the fit test had to equal or
exceed 10. Similarly, to assign a full facepiece to an individual

test subject, the protection factor measured by the fit test had to
equal or exceed 100. These researchers used the data from
in-facepiece sampling not only to assess the fit of individual
respirators but also to compare the fit provided to a specific
individual by different brands of respirators. As a result, they
felt that the "best fitting" respirator could be provided to an
individual.

In the period of 1973-1974, Dorman describes a significant
modification to the NaCl quantitative fit methods being employed by
the British at that time [11]. The method was modified to allow the
determination of leakage into respirators designed with no
exhalation valves. This test procedure became part of British
Standard 2091 to measure the fit of valveless respirators.

In this test, the respirator was probed with a 12-gauge
hypodermic needle. To prevent condensation of the exhalate on the
needle and transport tube, a flow of 1 lpm of dry air was fed into
the 12-gauge sampling needle by an 18-gauge needle. In-facepiece
sampling was done over the complete respiratory cycle at a sampling
rate of approximately 1 lpm. The in-facepiece sampling probe was
positioned on the side of the respirator. A head harness
arrangement was used to support the probe so it would not alter the
"fit" of the facepiece. A flame photometer was used to detect the
NaCl in the air sample obtained from inside the respirator. It was
connected to a chart recorder to show variations in the
concentration during inhalation and exhalation. Dorman related
the"true" penetration (P_R) that occurred during inspiration
(i.e.,the actual penetration since it would only occur during
theperiod of negative pressure on inhalation) to the penetration
measured by fit test (P_A) by correcting the measured penetration
for lung losses. Using a lung retention estimate of 60%, the
following formula was used to calculate the "true" penetration:

$$P_R = P_A/0.6$$

In this equation, it is assumed that inspiratory and expiratory
times are of equal length. By doing head and facial movements
during in-facepiece sampling, variations in leakage could be
observed when slight displacements of the mask occurred.

In addition to the studies already cited, quantitative fit
testing has been used to evaluate the performance of aircrew and
passenger protective breathing equipment [12], mouth-piece
respirators to protect against chlorine [13], open circuit breathing
apparatus [14] and supplied air respirators [15].

QUANTITATIVE FIT TESTING: CURRENT PROBLEMS AND NEEDS FOR RESEARCH

Recent research studies conducted by Hinton [16] and Dixon and
Nelson [17] have indicated that a correlation doesn't exist between
the level of fit ascribed to a respirator-person combination by
quantitative fit testing and the level of protection achieved in the
workplace when the respirator is properly used and conscientiously
worn.

There are several important questions that can be raised from these studies: (1) Will a respirator which gives the "best fit" in a quantitative fit test be the respirator which will provide the "best protection" in the workplace; (2) Should fit factor data be the basis upon which assigned protection factors are established; and (3) Does the in-facepiece sampling methods normally used in quantitative fit testing provide a representative sample?

The relationship between quantitative fit testing and workplace protection is quite uncertain and has yet to be demonstrated. Based upon these new studies, quantitative fit test data do not appear to be good predictors of workplace protection even when the respirator is used with the most ideal respirator program.

Quantitative fit data also has been the principle data used to establish the assigned protection factors given to different classes of respirators. However, if a relationship can not be shown between these two parameters, assigned protection factors established upon fit factors could be grossly in error. Recent field studies have demonstrated this to be true for powered air-purifying respirators (PAPRs) [18,19,20]. Those studies suggest that assigned protection factors of 1000 to 3000 for PAPRs, which were based upon fit data, may be 20 to 40 times too high. As a result some groups within the respirator community are lowering the assigned protection factor levels for PAPRs. As more field data is obtained the estimates of protection provided by other types of respiratory equipment will no doubt also be revised. A clear research need exists for additional field testing and the development of improved fit test procedures.

The third major question concerning fit testing centers on the accuracy and precision of inboard penetration measurements. It has been commonly assumed that faceseal leakage mixes uniformly and very rapidly within the facepiece cavity. Based upon such an assumption in-facepiece sampling could be considered to provide representative samples. Furthermore, with such mixing sampling strategies would not have to consider issues such as where to locate the sampling probe, how deeply it should be extended into the facepiece cavity, etc.

However, recent research studies, first reported in the United States by Myers, et al. [21] and later in the United Kingdom by Bentley [22], have found that in-facepiece sampling will not provide representative samples. These studies demonstrated that faceseal leakage doesn't mix rapidly or uniformly within the respirator. As a result concentration or penetration measurements made by in-facepiece sampling are subject to large, variable sampling biases. Variations in several parameters of the person-respirator system have been identified to cause significant changes in the sampling bias [21]. They were: 1) location of the probe on the midline of the respirator; 2) depth at which the probe is inserted into the facepiece cavity; 3) breathing being done all through the mouth or all through the nose; 4) area where the faceseal leakage occurs; and 5) an interaction between the area where faceseal leakage occurs and whether breathing is all through the nose or mouth. Based upon these results, it is now hypothesized that faceseal leakage is streamlining within the facepiece cavity [21]. Because of the leak streamlines, in-facepiece sampling produces biased and highly variable concentration measurements.

A key observation from this work was that different brands of half or full facepieces had significantly different amounts of in-facepiece sampling bias associated with them. These differences are thought to be related to the design features of the facepiece. The important point to be drawn is that as a result of different magnitudes of sampling bias, one brand of facepiece may appear to be falsely superior in fit to another, while in actuality having no better fit.

These observations may explain in part why a correlation between quantitative fit results and protection has not been demonstrated. More research is needed to develop an improved in-facepiece sampling strategy for respirators which is not subject to the large and viable sampling biases inharent with current in-facepiece sampling strategies.

REFERENCES

[1]   Guyton, H. G. and Lense, F. T., "Methods for Evaluating Respiratory Protection Masks and Their Components", AMA ARC'H Ind. Health, Vol. 14, 1956, p. 236.

[2]   Burgess, W. A., Silverman, L., and Stein, F., "A New Technique for Evaluating Respirator Performance," Am. Ind. Hyg. Assoc. J., Vol. 22, No. 6, 1961, p. 422.

[3]   Hounam, R. F., "A Method for Evaluating the Protection Afforded When Wearing a Respirator," AERE-R4125, Atomic Energy Research Establishment, Harwell, U.K., 1962.

[4]   Hounam, R. F., Morgan, D. J., O'Connor, D. T., and Sherwood, R. J., "The Evaluation of Protection Provided by Respirators," Ann. Occup. Hyg., Vol. 7, 1964, p. 353.

[5]   Flyger, H., "A Helium Leak-Detection Method for Respirator Control," Health Physics, Vol. 11, 1965, p. 223.

[6]   White, J. M. and Beal, R. J., "The Measurement of Leakage of Respirators," Am. Ind. Hyg. Assoc. J., Vol. 27, No. 3, 1966, p. 239.

[7]   Burgess, W. A. and Shapiro, J., "Protection from the Daughter Products of Radon Through the Use of a Powered Air-Purifying Respirator," Health Physics, Vol. 15, (1968), p. 115.

[8]   Adley, F. and Uhle, R. J., "Protection Factors of Self-Contained Compressed-Air Breathing Apparatus," Am. Ind. Hyg. Assoc. J., Vol. 30, No. 4, 1969, p. 355.

[9] Griffin, O. G. and Longson, D. L., "The Hazard Due to Inward
     Leakage of Gas into a Full Face Mask," Ann. Occup. Hyg., Vol.
     13, 1970, p. 147.

[10] Hyatt, E. C., Pritchard, J. A. and Richards, C. P., "Respirator
     Efficiency Measurement Using Quantitative DOP Man Tests," Am.
     Ind. Hyg. Assoc. J., Vol. 33, No. 10, 1972, p. 635.

[11] Dorman, R. G., Face-seal Leakage, Chemical Defense
     Establishment, Porton Down, U.K.

[12] deSteigure, D., Pinski, M. S., Bannister, J. R., and McFadden,
     E. B., "Aircrew and Passenger Protective Breathing Equipment
     Studies," Report No. FAA-AM-78-4, FAA Civil Aeromedical
     Institute, Oklahoma City, OH, 1978.

[13] Packard, L. H., Brady, H. L., and Scumm, O. F., "Quantitative
     Fit Testing of Personnel Utilizing a Mouthpiece Respirator,"
     Am. Ind. Hyg. Assoc. J., Vol. 39, No. 9, 1978, p. 723.

[14] Hack, A., Trugillo, A., Bradley, O. D., and Carter, K.,
     "Respirator Studies for the Nuclear Regulatory
     Commission-Evaluation and Performance of Open Circuit Breathing
     Apparatus," Report No. LA-8188-PR, Los Alamos National
     Laboratory, Los Alamos, NM, 1980.

[15] Hack, A., Bradley, O. D., and Trujillo, A., "Respirator
     Protection Factors: Part II - Protection Factors of
     Supplied-Air Respirators," Am. Ind. Hyg. Assoc. J., Vol. 41,
     No. 5, 1980, p. 376.

[16] Hinton, J. J., Jr., "Reliability of Quantitative Fit Protection
     Factors in Assessing Face-to-Facepiece Seals," Master Thesis
     presented to the faculty of the University of Texas, 1980.

[17] Dixon, S. W. and Nelson, T. J., "Workplace Protection Factors
     for Negative Pressure Half-Mask Facepiece Respirators," J. of
     the I.S.R.P., Vol. 2, No. 4, 1984, p. 347.

[18] Myers, W. R., Peach, M. J., III, Cutright, K. and Iskander, W.,
     "Workplace Protection Factor Measurements on Powered
     Air-Purifying Respirators at a Secondary Lead Smelter: Results
     and Discussion," Am. Ind. Hyg. Assoc. J., Vol. 45, No. 10,
     1984, p. 681.

[19] Lenhart. S. W. and D. L. Campbell, "Assigned Protection Factors
     for Two Respirator Types Based Upon Workplace Performance
     Testing," Ann. Occup. Hyg. 28(2), 1984, p. 173.

[20] Linaskis, S. H. and F. Kalos, "Study of Efficiency and Current
     Use of Respiratory Protective Devices," INFO-0144, Atomic
     Energy Control Board, Ottawa, Canada, 1984.

[21] Myers, W. R., Allender, J., Plummer, R., Stobbe, T.,
     "Parameters that bias the measurement of airborne concentration
     within a respirator," Am. Ind. Hyg. Assoc. J., Vol. 47, No. 2,
     1986, p. 106.

[22] Bentley, R., Critical Issues Conference on In-facepiece
     Sampling, Gaithersburg, Maryland, U.S.A., January 8-9, 1985.
     Sponsored by the International Society for Respiratory
     Protection and the National Institute for Occupational Safety
     and Health.

Applications and Testing: Filtration for Occupied Spaces

James E. Woods and Brian C. Krafthefer

FILTRATION AS A METHOD FOR AIR QUALITY CONTROL IN OCCUPIED SPACES

REFERENCE: Woods, J.E. and Krafthefer, B.C. "Filtration as a
Method for Air Quality Control in Occupied Spaces," Fluid
Filtration: Gas, Volume I, ASTM STP 975, R.R. Raber, Ed.,
American Society for Testing and Materials, Philadelphia,
1986.

ABSTRACT: Filtration, or removal control, is one of three
methods currently available to provide acceptable indoor air
quality in occupied spaces. While the other two methods,
source and dilution control, are primarily employed for
occupant needs, filtration has conventionally been used for
protection of components within the heating, ventilating, and
air conditioning systems. In this paper, filtration technology is reviewed with respect to current ventilation standards for occupants. The difference between ventilation and
air quality control is discussed in terms of acceptability
criteria and control methods. Parameters that relate these
terms are identified and control strategies are proposed that
can be used to optimize removal and dilution methods for
occupant acceptability and cost-effectiveness.

KEYWORDS: Ventilation, Indoor Air Quality, Control
Strategies, Removal Control, Filtration

Although filtration technology has been applied in heating, ventilating, and air conditioning (HVAC) systems for more than fifty years,
its primary function has not been provision of health or comfort for
the occupants. Rather, filters are usually installed to protect the
components (e.g., coils, fans, ductwork) within the HVAC systems from
dust and particulate loading. Only in special cases, such as critical
care areas in hospitals, have filters been installed in the systems
primarily for the protection of the occupants. Otherwise, ventilation
with outdoor air has been depended upon to dilute particulate and
gaseous contaminants in occupied spaces to concentrations below those

Dr. Woods is Senior Engineering Manager, Honeywell Energy Products
Center, Golden Valley, MN 55422. Mr. Krafthefer is Principal
Research Scientist, Honeywell Physical Sciences Center, Bloomington,
MN 55420.

considered to be deleterious or annoying. However, with the advent of
energy conservation practices of the last decade, ventilation rates
have been generally reduced. As a result, many ventilation systems no
longer provide acceptable control of particulate and gaseous contaminants. The objective of this paper is to describe control strategies
that can be used to optimize filtration and dilution for occupant
acceptability and cost-effectiveness.

VENTILATION CONTROL

A schematic of conventional HVAC systems, Figure 1, identifies the
common placements of filters [1]. In residential systems, filters are
usually placed in the return air duct as these systems normally use
100% recirculated air. If outdoor air is supplied mechanically to the
system, a "pre-filter" may be installed, but usually dilution depends
on infiltration (i.e., uncontrolled air leakage through the building
envelope). Mechanical systems for commercial or institutional facilities may have filters installed either in the return air duct or
the "mixed-air" plenum. Note that the placement of these filters is
upstream of the major mechanical components in the system. This practice is encouraged to achieve protection of the system from contamination. The effectiveness of this practice has recently been documented, and results indicate that significant energy and cost savings
can accrue over the lifetime of the system [2].

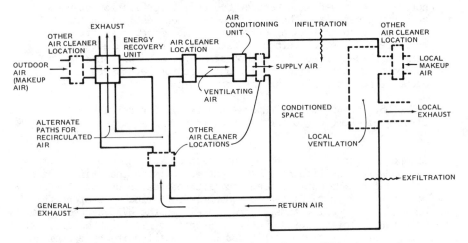

Figure 1. Heating, ventilating, and air conditioning system
schematic indicating variables for indoor air quality control.
(From Ref. 1)

Only in special cases have filters also been installed downstream
of the major components to provide clean air to the occupied space.
For example, in critical areas of hospitals, two filters are specified. The upstream filter (pre-filter) and the downstream (final)
filter must have minimum ASHRAE dust-spot efficiencies of 25% and 85%,
respectively [3, 4].

While common practice indicates that filtration is used for contamination control in only special cases, recognition of its general application for ventilation control has existed for more than a decade. In 1973, the ASHRAE ventilation standard defined ventilation air as: "that portion of supply air which comes from outside (outdoors) plus any recirculated air that has been treated to maintain the desired quality of air within a designated space" [5]. In that standard, minimum and recommended ventilation rates of acceptably clean outdoor air were specified for numerous occupied spaces. That standard also allowed recirculation if adequate temperature and filtration control were employed. For the latter case, the outdoor air requirements could be reduced to 33% of the tabulated values if particulate filters were employed, and to 15% of the values if adsorption or other gas removal equipment were employed. However, two constraints were imposed:

1. the outdoor air quantity could not be less than 2.5 l/s (i.e., 5 cubic feet per minute) per person; and

2. the maximum allowable concentrations of contaminants in the supply air to the occupied space could not exceed the specified values in section 3 of that standard.

The two constraints imposed in those criteria were seldom followed, and as a result, the standard was often misinterpreted during the energy shortages of the 1970s. In 1981, the revised standard clarified the recirculation criteria by providing an equation which relates a selected filter efficiency to a corresponding recirculation rate to provide indoor air quality equivalent to that expected by ventilation with 100% outdoor air [1]:

$$\dot{V}_r = \frac{\dot{V}_o - \dot{V}_m}{\varepsilon}$$

[1]

where:

$\dot{V}_r$ =   recirculation rate (cfm or l/s per person).

$\dot{V}_o$ =   Outdoor air rate specified in Table 3 of ASHRAE Standard 62-1981 (cfm or l/s per person).

$\dot{V}_m$ =   The minimum rate of outdoor air that can be used in the recirculated air ventilating system to provide acceptable indoor air quality, but never less than 2.5 l/s (5 cfm) per person.

ε =   Efficiency of the contaminant removal (air cleaner) device. The efficiency should be derived from the most relevant parameters of the contaminant involved (e.g., ppm, or $\mu g/m^3$).

This revision addressed two short-comings in the previous standard: 1) it defined the air cleaner efficiency, quantitatively; and 2) it specified a recirculation air flow rate as a function of the filter efficiency. In the previous standard, as long as any particulate filter were incorporated into the system, the amount of outdoor air could be reduced and no compensation in recirculation was specified.

Thus, common practice was to reduce the total supply air flow rate to that only required for thermal loads. As energy conservation efforts were increased, lighting and envelope loads were decreased and variable air volume systems became popular. As a result, less supply air was needed for thermal loads and complaints about poor air quality began to increase. Application of Eq 1 requires the recirculation rate to compensate for the reduction in outdoor air, as well as the air cleaner efficiency. Moreover, it requires the total air supply rate, $\dot{V}_S$ (cfm or l/s per person), to be the sum of the recirculation and minimum outdoor air flow rates:

$$\dot{V}_S = \dot{V}_m + \dot{V}_r \qquad [2]$$

For example, ASHRAE 62-1981 specifies the required amount of outdoor air for "meeting and waiting spaces" in offices as 17.5 l/s (35 cfm) per person if smoking is allowed in the space. If it is desired to reduce the outdoor air flow rate to the minimum of 3.5 l/s (7 cfm) per person and to compensate for the smoking by installing air cleaners, the required recirculation rate for a system with an air cleaner efficiency of 100% for "tobacco smoke" (i.e., for particulates, gases, and vapors) would be 17.5 - 3.5 = 14 l/s (35 - 7 = 28 cfm) per person. Thus, the total supply air flow rate would be 3.5 + 14 = 17.5 l/s (7 + 28 = 35 cfm) per person, the same as with 100% outdoor air, but only 20% of the originally required outdoor air would require thermal treatment. For a more realistic air cleaner efficiency of 50%, the required recirculation rate would be 14/0.5 = 28 l/s (28/0.5 = 56 cfm) per person, and the required total supply rate would be 28 + 3.5 = 31.5 l/s (56 + 7 = 63 cfm) per person. In this case, the thermal load for the minimum outdoor air would be the same as for the 100% efficient air cleaner, but the fan power and space requirements for the additional recirculation and total air supply rates would be significantly larger. Thus, Equations 1 and 2 provide criteria for evaluating ventilation options which are expected to provide equivalent indoor air quality.

AIR QUALITY CONTROL

Equations 1 and 2 do not directly address the quality of the indoor air. Rather, they provide an indirect procedure to achieve the same indoor air quality with some recirculation air as would be expected by ventilating with 100% outdoor air, but the value of that quality remains unspecified. In ASHRAE Standard 62-1981, the indirect method is known as the "Ventilation Rate Procedure". Another method is also specified in that standard as the "Indoor Air Quality Procedure." The latter method specifies objective and subjective criteria with which to evaluate the environment, but does not specify the means to achieve the required control. The Ventilation Rate Procedure may be considered a prescriptive standard, and the Air Quality Procedure may be considered a performance standard.

These two procedures may be related by considering a simple one compartment model, Fig. 2. The concentration within this compartment is a function of the rate of contaminant generation within the occupied space, the rate of dilution with outdoor air, the contaminant concentration of the air transported into the occupied space, and the rate of removal of the contaminant by the air cleaning system. Conceptually, this Figure indicates that three methods exist to control the concentration of a contaminant within the occupied space:

o suppression of the generation rate, or source control;

o reduction of indoor air contaminant concentration by outdoor air exchange, or dilution control; and

o removal of indoor air contaminant concentration by air cleaners, or removal control.

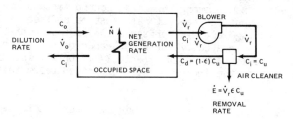

Figure 2. One compartment, uniformly mixed, steady-state model for indoor air quality.

In steady-state within a well-mixed space, a relationship among these three control methods may be expressed as:

$$\Delta C = \frac{\dot{N} - \dot{E}}{\dot{V}_O} \qquad [3]$$

where: $\Delta C = C_i - C_O =$ the difference between the uniformly mixed indoor air concentration, C_i, and the outdoor air concentration, C_O ($\mu g/m^3$).

$\dot{N} = \dot{Q} - \dot{S} =$ the net generation rate of the contaminant in the occupied space, where $\dot{Q}$ is the source strength (i.e., emission rate) and $\dot{S}$ is the sink strength (i.e., settling or sorption rate) ($\mu g/hr$).

$\dot{E} = \dot{V}_r \varepsilon C_u =$ the removal rate of a contaminant in the air cleaner ($\mu g/hr$); where $\dot{V}_r$ is the recirculation rate through the air cleaner (m^3/hr), ε is the efficiency of the air cleaner rated in terms of the contaminant removed (i.e., $\varepsilon = 1 - C_d/C_u$), C_u is the contaminant concentration upstream of the air cleaner (i.e., $C_u = C_i$ for a well-mixed system), and C_d is the contaminant concentration downstream of the air cleaner.

$\dot{V}_O =$ outdoor air flow rate for dilution control (m^3/hr).

In this model, the dilution rate, $\dot{V}_O$, represents infiltration, natural ventilation, or mechanical ventilation with outdoor air. The removal rate, $\dot{E}$, represents the rate at which the contaminant is accumulated in the air cleaner. For the simple case in this model, the removal rate (i.e., $\dot{E} = \dot{V}_r \varepsilon C_u$ could be achieved by fan-filter modules now available as consumer products, or by filtered recirculated air commonly used in residential forced air systems. Note that the removal rate is a function of three factors. Thus, for a given upstream concentration, the same removal rate can be expected from two devices; one which has an air cleaner efficiency of one-half the other if its recirculation rate is twice the other's. Moreover, as the indoor concentration decreases (i.e., C_i and C_u approach zero), the removal rate of a device with a fixed product of $\dot{V}_r \varepsilon$ decreases.

For the case where the air cleaner is located in the mixed air, Fig. 1, the relationship among the three control methods may still be expressed by Eq 3, but the removal rate must be defined as $\dot{E} = (\dot{V}_O C_O + \dot{V}_r C_r) \varepsilon$. The advantage of this configuration is that contaminated outdoor air is also treated by the air cleaner before it enters the occupied space. This feature is especially important when it is necessary to control the indoor air concentration below that of the outdoor air.

Although Eq 3 was derived from a simple model, it serves to identify some basic control strategies and their limitations for indoor air quality control:

o   If removal control is not employed (i.e., $\dot{E} = 0$), the indoor concentration will exceed the outdoor concentration unless the source is removed (i.e., $\dot{N} = 0$) or an infinite dilution rate is provided (i.e., $\dot{V}_O = \infty$).

o   If the outdoor concentration is to be controlled below that of the outdoor air and the dilution rate is finite, the removal rate must exceed the net generation rate (i.e., $\dot{E} > \dot{N}$).

o   Outdoor air required for dilution control may be reduced without affecting ΔC, if the difference between the net generation rate and the removal rate is correspondingly reduced.

o   To achieve an acceptable ΔC economically, a combined strategy of source, removal, and dilution control probably will be required.

ACCEPTABILITY CRITERIA

Only source control can eliminate occupant exposure to a contaminant; dilution and removal control require mixture of the contaminant within the occupied space before their mechanisms become useful. Thus, both of these control methods expose occupants to the contaminants, and should only be used if some level of exposure is "acceptable".

Air quality may be defined, generally, as "the nature of air that affects your health and well-being". In this definition, the World Health Organization's concept of health is implied: "Health is a state of complete physical, mental and social well-being, and not merely the

absence of disease or infirmity" [6]. These definitions offer criteria
for evaluating beneficial, as well as deleterious effects of indoor
environments. However, use of these definitions to ascertain acceptable indoor air quality requires both objective and subjective criteria. Objective criteria may be expressed as quantitative values of
environmental stress which result in measurable physiological or psy-
chological (i.e., behavioral) strains on the occupants. Thus, these
criteria may be used to ascertain compliance with environmental conditions that should not cause measurable disease, disability, or
dysfunction. Subjective criteria may be expressed in terms of af-
fective responses of the occupants (e.g., comfort, annoyance, discomfort). When stress results in a complete state of well-being, a
comfortable (i.e., healthy) strain may exist. The amount of deviation
from these ideal conditions that can be accepted without discomfort or
adverse health effects is dependent upon the occupant's abilities to
adapt to the deviations. As the ability to adapt diminishes, susceptibility to the adverse effects of the stress increases. Thus,
susceptible populations within occupied spaces must be considered when
"acceptable ranges" of environmental stressors are selected.

Traditionally, ASHRAE has defined acceptable environments as those
in which 80% of the occupants find satisfactory:

o Acceptable air quality is defined as "air in which there are no
  known contaminants at harmful concentrations and with which a
  substantial majority (usually 80%) of the people exposed do not
  express dissatisfaction" [1].

o Acceptable thermal environment is defined as "an environment in
  which at least 80% of the occupants would find thermally acceptable" [7].

For purposes of evaluating control strategies, a more technical
definition of indoor air quality has been proposed which incorporates
the above concepts [8]:

The quality of the air in an enclosed space is an indicator of how
well the air satisfies three criteria:

o Thermal conditions of the air must be adequate to provide thermal acceptability for the occupants as defined by ASHRAE Standard 55-1981.

o The concentrations of oxygen and carbon dioxide must be within
  acceptable ranges to allow normal functioning of the respiratory system.

o The concentration of gases, vapors, and particulates should be
  below levels that can have deleterious effects, or that can be
  perceived as objectionable by the occupants.

This definition addresses three important factors needed to
achieve acceptable indoor air quality control. First, it recognizes
that air quality control should be considered to be integral with control of the thermal environment. While some aspects of source control
can be achieved without reliance on thermal control (e.g., product
substitution), suppression of the net generation rate by containment

or isolation requires interface with the thermodynamic state of the air in the occupied space. Moreover, neither dilution nor removal control can be achieved without control of air movement, a thermodynamic process. Second, it suggests that responses to indoor air quality can be quantified in terms of a subjective scale, not unlike those for predicted mean vote (PMV), percent people dissatisfied (PPD) or standard effective temperature (SET*) [9, 10]. Third, and maybe most important, it implies that simultaneous control of the three methods is required, if satisfactory responses are to be achieved.

IMPROVED CONTROL STRATEGIES

A classical conflict has developed between concepts of energy conservation and environmental acceptability. During the energy crises of the last decade, ventilation systems were de-activated, building envelopes were "tightened", and temperatures and relative humidities were allowed to decrease in winter and increase in summer. As a result, environmental quality was degraded, sometimes to the extend that the health of occupants was jeopardized. Terms to describe these conditions are now in literature, such as "Sick Building Syndrome", "Tight Building Syndrome", and "Building Related Illness" [11]. Conversely, data also indicate that this conflict need not exist. Rather, if environmental control is approached intelligently, energy efficient operation will result [12, 13].

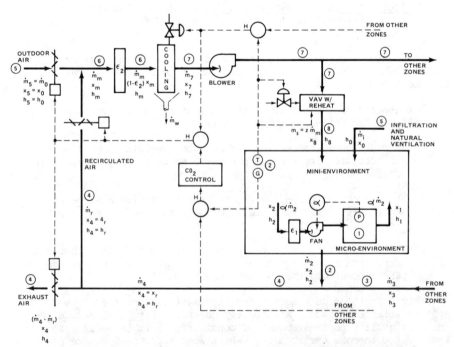

Figure 3. System schematic for micro- and mini-environmental, closed-loop control for thermal and air quality acceptability.

A composite of these control strategies is shown schematically in Fig. 3. In this Figure, the region of primary concern within the room (i.e., micro-environment is shown as Compartment 1, and the remainder of the room (i.e., mini-environment) is shown as Compartment 2. The relationship between the mini- and micro-environments is represented by a filtered room-air supply $(1 - \varepsilon_1)\dot{m}_2$, and a room-coupling coefficient, α, a factor which is similar to the concept of "Ventilation Efficiency" [12, 14].

Room Ventilation Control

The thermal and air quality control interaction between the HVAC system and the mini-environment can be expressed in terms of the following steady-state equations:

o The mass balance of the contaminant in the mini-environment:

$$\dot{m}_s x_s + \dot{m}_i x_o + \dot{N}_{x2} = (\dot{m}_s + \dot{m}_i) x_2 \qquad [4]$$

where:

$\dot{m}$ = mass air flow rate ($g_{dry\ air}/hr$)

$\dot{N}_{x2}$ = the generation rate of the contaminant in the mini-environment.

x = mass concentration of contaminant ($\mu g_{cont.}/g_{dry\ air}$)

o The fraction of mini-environmental supply air flow to system air flow:

$$z = \dot{m}_s / \dot{m}_m \qquad [5]$$

o The mass balance of the contaminant in the HVAC system, downstream from the air cleaner:

$$\dot{m}_m x_m (1 - \varepsilon_2) + \dot{N}_s = \dot{m}_m x_s \qquad [6]$$

where $\dot{N}_s$ is the generation rate of the contaminant in the HVAC system, downstream from the air cleaner which has an efficiency of ε_2. ($\mu g_{cont.}/hr$).

o The mass balance of the air in the mixed air system:

$$\dot{m}_o + \dot{m}_r = \dot{m}_m \qquad [7]$$

o The mass balance of the contaminant in the mixed air system:

$$\dot{m}_o x_o + \dot{m}_r x_r = \dot{m}_m x_m \qquad [8]$$

o And the energy balance of the mixed air system in terms of specific enthalpy, h ($J/g_{dry\ air}$):

$$\dot{m}_o h_o + \dot{m}_r h_r = \dot{m}_m h_m. \qquad [9]$$

In these equations, the subscripts represent:

i = infiltration and natural ventilation rates into the mini-environment (i.e., psychrometric condition 5 in Fig. 3).

m = mixed air in HVAC system (i.e., psychrometric condition 6 in Fig. 3).

o = outdoor air (i.e., psychrometric condition 5 in Fig. 3).

r = recirculated air into HVAC system (i.e., psychrometric condition 4 in Fig. 3).

s = supply air to mini-environment (i.e., psychrometric condition 8 in Fig. 3).

Note that it is necessary to express these balances in terms of mass flow rates of "dry air" rather than volumetric rates, as isothermal conditions can no longer be assumed and changes in air densities must be considered due to the psychrometric processes.

Equations 4 through 9 may be combined to provide an expression for the "Room Acceptability Ratio," K_2, defined as the ratio of room air to outdoor air contaminant concentrations:

$$K_2 = \frac{x_2}{x_o} = \frac{(H + (1 - H)x_r/x_o)(1 - \varepsilon_2) + M + Q_2}{1 + M} \qquad [10]$$

where:

$$H = \dot{m}_o/\dot{m}_m = (h_r - h_m)/(h_r - h_o) \qquad [11]$$

$$M = \dot{m}_i/z\dot{m}_m \qquad [12]$$

$$Q_2 = \dot{N}/z\dot{m}_m x_o \qquad [13]$$

where:

$$\dot{N} = z\dot{N}_s + \dot{N}_{x2}$$

If the contaminant concentration in the mini-environment is identical to that in the recirculated air (i.e., $x_2 = x_r$), Eq 10 simplifies to:

$$K_2 = \frac{x_2}{x_o} = \frac{H(1 - \varepsilon_2) + M + Q_2}{H(1 - \varepsilon_2) + M + \mathcal{E}_2} \qquad [14]$$

Functional relationships from Eq 14 are shown in Figures 4-7 between the "Room Acceptability Ratio," K_2 (i.e., x_2/x_o), and the "Room Contamination Factor," Q_2, with the "Air Cleaner Efficiency," ε_2, the "Passive to Active Air Exchange Ration," M, and the "Outdoor Air Ratio," H, as parameters. In each of these Figures, four sets of graphs are presented for air cleaner efficiencies at 0, 0.2, 0.5, 0.9, at each of four outdoor air ratios of 0.2, 0.5, 0.75, and 1.0. Also shown in each set of graphs is the reference condition of $\varepsilon_2 = 1.0$,

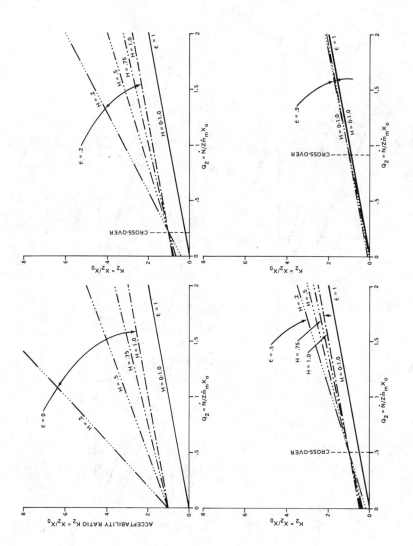

Figure 4. Relationships between the room acceptability factor, K_2 and the room contamination factor, Q_2, as a function of the air cleaner efficiency, $\epsilon 2$, when the passive active air exchange ratio, M, is negligible (i.e., M = 0).

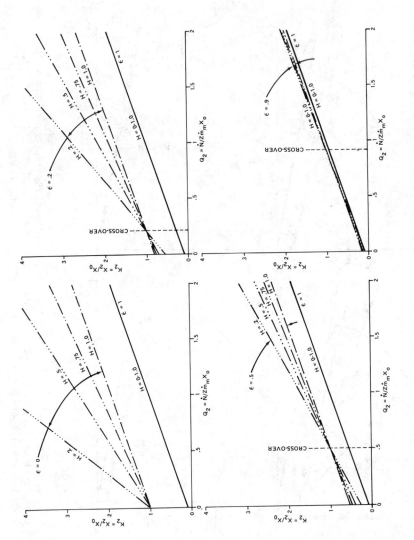

Figure 5. Relationships between the room acceptability factor, K_2, and the room contamination factor, Q_2, as a function of the air cleaner efficiency, ϵ_2, when the passive to active air exchange ratio, M, is low (i.e., M = 0.1).

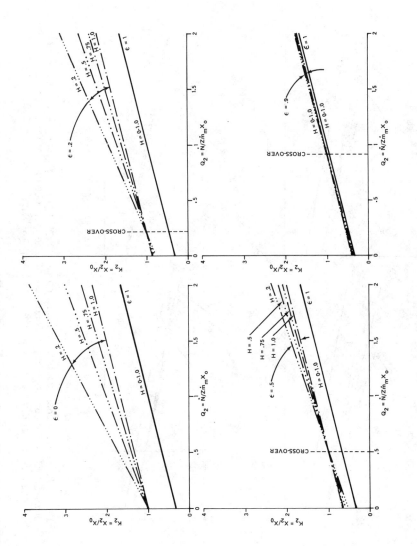

Figure 6. Relationships between the room acceptability factor, K_2, and the room contamination factor, Q_2, as a function of the air cleaner efficiency, ε_2 when the passive to active air exchange ratio, M, is moderate (i.e., M = 0.5).

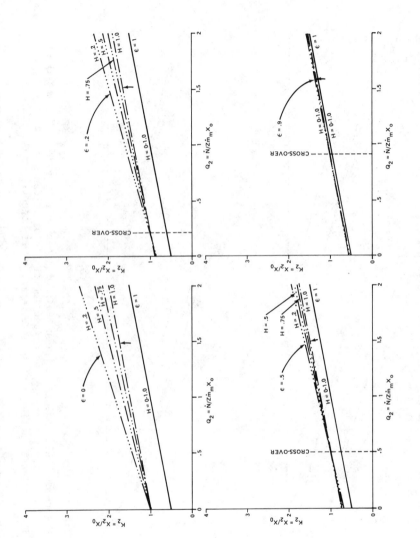

Figure 7. Relationships between the room acceptability factor, K_2, and the room contamination factor, Q_2, as a function of the air cleaner efficiency, ε_2, when the passive to active air exchange ratio, M, is high (i.e., $M = 1.0$).

which holds for all values of H. The graphs in Fig. 4 are presented
for M = 0, which represents interior zones where infiltration and natural ventilation are negligible. Fig. 5 is presented for M = 0.1,
which represents exterior zones with relatively tight envelopes. For
example, if the supply air rate for thermal control is 6 air changes
per hour, the infiltration rate would be 0.6 air changes per hour. In
Fig. 6, M = 0.5, which is representative of exterior rooms that may
have windows partially opened. Fig. 7 represents conditions in which
the rates of passive and active air exchanges are equal, M = 1.0.
This condition may exist in some occupied spaces with large window
areas. Thermal control in this case would be difficult and probably
energy intensive.

The graphs in Figures 4 to 7 identify several performance characteristics that are useful for evaluation of alternative room ventilation control strategies:

o   For all values of M, the Room Acceptability Ratio, K_2, will ex-
    ceed 1.0 (i.e., the indoor concentration will exceed the outdoor
    concentration) for all values of Q_2 greater than zero when re-
    moval control is not employed (i.e., $\varepsilon_2 = 0$).

o   When removal control is employed (i.e., $\varepsilon_2 > 0$), a critical value
    of Q_2 exists for all values of M at which a "cross-over condition"
    exists:

    o   For values of Q_2 smaller than the critical value, lower Room
        Acceptability Ratios, K_2, can be achieved by minimizing the
        outdoor air ratio, H. That is, when the generation rate of a
        contaminant within the mini-environment and HVAC system is less
        than the transport rate of the contaminant from outdoors, the
        ratio of indoor to outdoor concentrations will be reduced by
        reducing the percentage of outdoor air used for ventilation.

    o   For values of Q_2 greater than the critical value, lower Room
        Acceptability Ratios, K_2, can be achieved by maximizing the
        outdoor air ratio, H. That is, when the generation rate of the
        contaminant within the mini-environment and HVAC system is gre-
        ater than the transport rate of contaminant from outdoors, the
        ratio of indoor to outdoor concentrations will be reduced by
        increasing the percentage of outdoor air used for ventilation.

    o   The critical value of Q_2 is increased by increasing the effi-
        ciency of the air cleaner. For example, the cross-over condi-
        tion is increased from $Q_2 = 0.2$ to 1.0 when the air cleaner
        efficiency is increased from $\varepsilon_2 = 0.2$ to 0.9.

o   The relative importance of the outdoor air ratio, H, is greater at
    lower values of M and ε_2. These relationships may be observed by
    comparing Figures 4 and 7.

    o   When M = 0 and $\varepsilon_2 = 0$ (Fig. 4), the only method of ventila-
        tion control is dilution through the HVAC system. Thus, when H
        is changed and Q_2 remains constant, a substantial change in K_2
        will result. Conversely, when M = 1.0 and $\varepsilon_2 = 0$ (Fig. 7), the
        amount of dilution air from natural ventilation and infil-

tration may be equal to (i.e., $H = 1.0$) or exceed (i.e., $H <$ 1.0) that supplied by the HVAC system, thus K_2 is less sensitive to changes in H.

o As the air cleaner efficiency increases for a specific contaminant, the relative importance of H decreases (i.e., see graphs for $\varepsilon_2 = 0.9$) for all values of M.

o The value of Q_2 will vary in a "constant air volume" system only as N and x_o vary. However in a "variable air volume" system, the value of Q_2 will also vary with the thermally controlled values of z and $\dot{m}_m$.

The relationship between thermal and indoor air quality control of the mini-environment is apparent form Equations 10 and 14 and from Figures 4 to 7. To achieve acceptable thermal and air quality, energy consumption, and life-cycle costs, simultaneous control of room temperature, relative humidity, supply and return air rates, and particulate and gaseous concentrations is required. To evaluate the capability of a system to control these factors simultaneously, the following procedure is recommended:

1. Determine the minimum value of supply air flow rate, $z\dot{m}_m$, required to maintain thermal acceptability while the air distribution performance index (ADPI) is maintained at a value of at least 80% [15].

2. For the value of $z\dot{m}_m$, determined in Step 1, the values of x_o expected during the times of minimum supply air flow rates, and the expected generation rates of the contaminants in the room and in the HVAC system during these periods, calculate the values of Q_2 for these contaminants.

3. Specify the acceptable values of K_2 for each of the contaminants of concern.

4. Specify the air cleaner efficiency, ε_2, for each of the contaminants of concern.

5. Determine the outdoor air ratios, H, required for dilution control to compliment removal control specified in Step 4.

6. Compare the minimum value of H, determined in Step 5, to the value of H that will minimize energy consumption during the periods considered in Step 1.

7. Reiterate Steps 4-6 until the environmental acceptability, energy consumption, and life-cycle criteria are optimized.

Personal Exposure Control

Room ventilation control can be effective in maintaining acceptable indoor concentrations of contaminants. However, this control method is usually based on the assumptions that the contaminant is uniformly mixed throughout the controlled space. This assumption implies that the occupants will be exposed to the same contaminant concentrations throughout the controlled space. Uniform mixing seldom occurs and, in some cases, stratification within rooms can be significant. Thus, room ventilation control, alone, is not always sufficient to provide the required air quality for personal exposure control. This phenomenon has long been recognized in hospitals. Examples include:

o Laminar flow clean benches in laboratories and pharmacies.
o Oxygen tents in patients' rooms.
o Biological cabinets in laboratories.
o Portable laminar flow equipment in operating rooms.
o Portable "sterile air" equipment to protect bed-ridden and ambulatory patients that are immunosupressed.

The air quality control interaction between the mini- and micro-environments can be expressed in terms of the following steady-state equations:

o The mass balance of the contaminant in the micro-environment:

$$\dot{m}_2 x_2 (1 - \varepsilon) + \dot{N}_{x1} = \dot{m}_1 x_1 \qquad [15]$$

where $\dot{N}_{x1}$ is the generation rate of the contaminant in the micro-environment.

o The fraction of micro- to mini-environmental mass air flow rate:

$$\alpha = \dot{m}_1 / \dot{m}_2 \qquad [16]$$

where α, the room-coupling coefficient, may be passive (i.e., related to ventilation efficiency) or active (i.e., related to the forced air into the micro-environmental by blowers).

o And the mass balance of air flow in the mini-environment:

$$\dot{m}_2 = \dot{m}_s + \dot{m}_i \qquad [17]$$

Equations 15 through 17 may be combined with Equation 5 to provide an expression for the "Micro-environmental Acceptability Ratio," $K_1 = x_1/x_2$:

$$K_1 = \frac{x_1}{x_2} = (1 - \varepsilon_1) + Q_1 \qquad [18]$$

where:

$$Q_1 = \dot{N}_{x1} / (\alpha (z m_m \dot{} + m_i \dot{}) x_2) \qquad [19]$$

Functional relationships from Eq [18] are shown in Fig. 8 between the micro-environmental acceptability factor, K_1, and the micro-environmental contamination factor, Q_1, with the air cleaner of efficiency, ε_1, as the parameteric function. Because of the expression of Q_1, these relationships are valid for all values of M (i.e., $\dot{m}_i / z \dot{m}_m$). From Figure 8, performance characteristics are identified that are useful for evaluation of alternative personal exposure control strategies:

o A simple linear relationship exists between the micro-environmental contamination factor, Q_1, and the micro-environmental acceptability factor, K_1, with ε_1 as a parameter. A critical value of Q_1 does not exist.

o The micro-environmental contamination factor, Q_1, is affected by:

o The supply air flow rate, $z\dot{m}_m$, and the natural ventilation and infiltration rates, $\dot{m}_i$. For variable air volume systems, or for rooms with operable windows, these changes can be significant (e.g., the room air exchange rate can vary by factors of two or more). Thus, if $(z\dot{m}_m + \dot{m}_i)$ is decreased due to the thermal load in the room controlled by a variable air volume system, the value of Q_1 would increase, and, if α, x_2, and ε_1 remained constant, the value of K_1 would increase, resulting in a degradation of acceptability.

o The room coupling coefficient, α, which may be a passive function of room stratification or ventilation efficiency (i.e., $\alpha \ll 1$). Conversely, α may be an active function of a portable or local forced air device (i.e., fan). In the latter case, the value of α may be much greater than one (e.g., laminar flow clean bench). Note that the air flow rate into the micro-environment, represented by the factor $\alpha(z\dot{m}_m + \dot{m}_i)$, may consist entirely of air recirculated within the mini-environment and need not have been directly supplied by the HVAC system.

o The contaminant concentration in the mini-environment, x_2. Thus, when x_2 is large, the micro-environmental acceptability ratio, K_1, will be more sensitive to the transported contamination than to that generated within the micro-environment. Of particular note, if the value of the factor $(z\dot{m}_m + \dot{m}_i)$ is large and the air cleaner efficiency, ε_1, is high, the value of x_2 may be decreased before steady-state is achieved.

Although the interaction between thermal and indoor air quality control in the micro-environment is not as evident as it is in the mini-environment, a relationship still exists. For example, the heat from a computer terminal is first dissipated into the micro-environment before it is dissipated in the mini-environment and the HVAC system. As a result, both the dry-bulb and mean radiant temperatures in the micro-environment can be significantly higher than in the mini-environment. Thus, simultaneous control of temperature, humidity, particulate and gaseous contaminants is also required within the micro-environment of some functional areas. In others, reductions in energy consumption and operational costs may be achieved by employing micro-environmental control strategies [13, 14]. To evaluate the capability of a system to control these factors simultaneously, the following procedure is recommended:

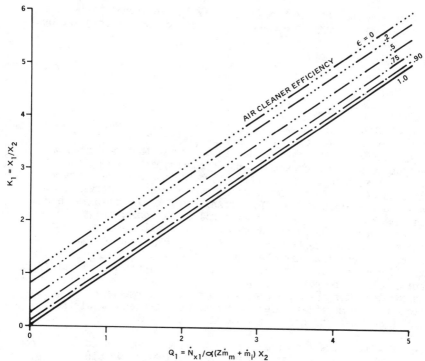

Figure 8.   Relationships between the micro-environmental acceptability factor, K_1 and micro-environmental contamination factor, Q_1 as a function of the air cleaner efficiency, ε_1, for
all values of the passive to active air exchange ratio, M.

1. Determine the minimum value of $z\dot{m}_m + \dot{m}_i$ from Step 1 for evaluation of Room Ventilation Control.

2. Specify the acceptable values of K_1 for each of the contaminants of concern.

3. If passive room coupling is expected (i.e., stratification
   exists), assume α is the ventilation efficiency [11].

4. If active room coupling is to be used, select desired value
   of α.

5. For the mini-environmental air flow rate, determined in Step 1,
   the values of x_2 expected for the contaminants of concern in
   the mini-environment and the expected generation rates of the
   contaminants in the micro-environment during these periods,
   calculate the values of Q_1 for these contaminants.

6. Specify the air cleaner efficiency, ε_1, for each of the contaminants of concern.

7. Compare the resultant K_1 values from Equations 18 and 19 or
   from Figure 8 to those specified in Step 2.

8. Re-iterate Steps 3 through 7 until the environmental acceptability, energy consumption, and life-cycle criteria are optimized.

CONCLUSIONS

We conclude that removal control is vital to the development of
effective control strategies for acceptable indoor air quality.
Removal control is the only method in use today that can provide
assurance that indoor cncentrations of contaminants can be controlled
to values less than those outdoors. Two factors are the primary de-
terminants of the effectiveness of removal control: removal efficiency and air circulation. Yet, standards for evaluating either of
these factors for air quality acceptability are essentially non-
existent. New standards are needed that will allow evaluation of air
cleaners and system performances in terms of the contaminants to be
removed. For example, air cleaner efficiencies could be defined in
terms of characteristic gases, vapors, and particulates; air cleaner
capacities could be evaluated in terms of products of the efficiencies
and air flow rates through them; and system performance could be
evaluated in terms of air cleaner capacities, percent of air recirculation, and effectiveness of room air distribution.

We also conclude that air quality control cannot be isolated from
thermal control of occupied spaces. Stratification within the occupied spaces is highly influenced by convection and system air distribution patterns; psychrometric processes within the system can
result in interactions with the airborne contaminants; and the HVAC
systems, themselves, may act as secondary sources of contamiantion if
they are not properly maintained. These relationships are particularly important to the coupling between zones within an occupied
space, such as "smoking" and "non-smoking" zones in office areas, or
patient and staff zones in chemotherapy treatment areas in hospitals.

Finally, we conclude that control strategies can be developed for
occupied spaces to optimize occupant acceptability with owning and
operating costs. By minimizing contaminant generation rates while
providing the appropriate combination of removal and dilution rates as
functions of thermal and contaminant load characteristics, occupant
acceptabilty can be achieved and costs of operation can be reduced,
thereby increasing productivity within the occupied spaces.

REFERENCES

[1]   ASHRAE Standard 62-1981. Ventilation for Acceptable Indoor Air
       Quality. Atlanta: American Society of Heating, Refrigerating,
       and Air Conditioning Engineers, 1981.

[2]   Krafthefer, B.C., and Bonne, U. Energy Use Implications of Meth-
       ods to Maintain Heat Exchanger Cleanliness. ASHRAE Trans. 92
       (Part 1), 1986 (In Press).

[3]   Guidelines for Construction and Equipment of Hospital and Med-
      ical Facilities. Rockford, MD: U.S. Department of Health and
      Human Services, Public Health Service, Health Resources and
      Services Administration, Publication No. (HRSA) 84-14500, 1984.

[4]   ASHRAE Standard 52-76. Air Cleaning Devices Used in General
      Ventilation for Removing Particulate Matter, Method of Testing.
      Atlanta: American Society of Heating, Refrigerating, and Air
      Conditioning Engineers, 1976.

[5]   ASHRAE Standard 62-73. Standards for Natural and Mechanical
      Ventilation. New York: American Society of Heating, Refrig-
      erating, and Air Conditioning Engineers, 1973.

[6]   "Constitution of the World Health Organizations." Official
      Record of the World Health Organization 2: 100, 1946.

[7]   ASHRAE Standard 55-1981. Thermal Environmental Conditions for
      Human Occupancy. Atlanta: American Society of Heating, Refrig-
      erating, and Air Conditioning Engineers, 1981.

[8]   Woods, J.E., and Maldonado, E.A.B. "Development of a Field
      Method for Assessing Indoor Air Quality in Single Family
      Residences." Final Report: Development of Energy Management
      Program for Buildings in Iowa- Fourth Year. Volume 1.
      Sponsored by the Iowa Energy Policy Council, Des Moines, Iowa.
      Iowa State University, ISU-ERI-Ames 82469, May, 1982.

[9]   Fanger, P.O. Thermal Comfort. New York: McGraw-Hill
      Publishing Co., 1973, p. 15.

[10] Gagge, A.P. "Rational Temperature Indices of Man's Thermal En-
      vironment and Their Use With a 2-Node Model of His Temperature
      Regulation." Fed. Proc., 32[5]:1572-1582, 1973.

[11] Stolwijk, J.A.J. "The 'Sick Building' Syndrome," in Indoor Air
      Volume 1, Recent Advances in the Health Sciences and Technology,
      Swedish Council for Building Research, Stockholm, 1984, pp
      23-29.

[12] Janssen, J.E., Hill, T.J., Woods, J.E., and Maldonado, E.A.B.
      "Ventilation for Control of Indoor Air Quality: A Case Study."
      Environment Internation, 8: 487-496, 1982.

[13] Woods, J.E., Reynolds, G.L., Montag, G.M., Braymen, D.T., and
      Rasmussen, R.W. "Ventilation Requirements in Hospital Operating
      Rooms- Part 2: Energy and Economic Implications." To be
      published in ASHRAE Trans. 92 (Part 2), 1986.

[14] Woods, J.E., Nevins, R.G., and Besch, E.L. "Analysis of Thermal
      and Ventilation Requirements for Laboratory Animal Cage Environ-
      ments." ASHRAE Trans. 81 (Part 2): 559, 1975.

[15] Nevins, R.G. and Miller, P.L. "Analysis, Evaluation and Compar-
      ison of Room Air Distribution Performance - A Summary." ASHRAE
      Trans. 78 (Part 2): 235-242, 1972.

Richard D. Rivers and David J. Murphy

AIR FILTER TESTING: CURRENT STATUS AND FUTURE PROSPECTS

REFERENCE: RIVERS, R. D., and MURPHY, D. J., "Air Filter Testing: Current Status and Future Prospects," Fluid Filtration: Gas, Volume I, ASTM STP 975, R. R. Raber, Ed. American Society for Testing and Materials, Philadelphia, 1986

ABSTRACT: U.S. and foreign test standards for air filters used in general ventilation and as prefilters for high efficiency filters. are described. These existing standards provide means for comparing and ranking filters, but do not provide the complete information needed by air-handling system designers who need to meet specific requirements, or who want to minimize system costs and energy use. Better information about filter service life, quicker tests, and ones which are not dependent on random atmospheric dust are all needed. Committees of the American Society of Heating, Refrigeration and Air Conditioning Engineers (ASHRAE) and the U.S. Department of Energy (DOE) are addressing these problems, and will add particle-size-penetration measurement to the codes.

KEYWORDS: air filter testing, efficiency, dust loading, resistance, penetration, ventilation air filters

The proper air filter for a given application is one that reduces contamination to an acceptable level, has acceptable pressure drop and durability, and provides this performance at an attractive cost. Air filter performance tests attempt to address these requirements. Efficiency tests measure the ability of a filter to remove a specified contaminant. Resistance tests measure filter pressure drop at various flow rates. Dust-holding-capacity or life tests measure the amount of contaminant held by the filter at a specified terminal condition, thus giving at least a rough estimate of the filter's service life. We use the terms "rough estimate" and "some specified contaminant" advisedly. In actual service, contaminant concentrations and chemical and physical characteristics vary with time and location. Filters may operate at constant or variable flow; temperature and humidity of the filtered gas may vary erratically.

Mr. Rivers is vice-president of Environmental Quality Sciences, Inc., 1330 S. Third Street, Louisville KY 40208; Mr. Murphy is president of Air Filter Testing Laboratories, Inc., 4632 Old La Grange Road, Crestwood KY 40014.

Two solutions to the filter evaluation problem are widely used. The first establishes in the laboratory some standard contaminant and test conditions, which presumably mimic typical operating conditions for the class of filters tested. The second method is based on testing what is believed to be the worst performance condition for the filter. ASHRAE 52-76 [1] is an example of the first appproach; the "thermal DOP" test [2], of the second. A third approach has received considerable attention lately: measure performance on contaminants with very accurately known characteristics, then calculate operating performance on an assumed model of the system contaminants and flow conditions [3],[4]. All three appproaches have the same difficulty, that there are too many ill-defined parameters and insufficient theory to allow accurate estimates of real-world performance. In spite of this, existing test methods do allow filters to be compared and ranked for performance. The third approach offers enough information for reasonable estimates of contaminant concentrations produced by filter systems, and also approximate life-cycle costs.

Thus far, we have spoken of "filters" and "contaminants". Air and process gas quality is concerned with both gaseous contaminants and particulates. Most of this discussion will be concerned with testing particulate air filters. We will consider gaseous contaminant filters briefly later in the paper. Our span of interest does not include engine-intake filters, stack-gas cleaning devices, respirators, or High Efficiency Particulate (HEPA) filters, which are dealt with elsewhere at this symposium. The standards discussed evaluate filters useful as prefilters for Clean Rooms of Class 1 to Class 1000, and as final-stage filters for Classes greater than 1000.

U.S. TEST STANDARDS

Considering the wide variety of ventilation-type air filter equipment and the wide range of applications, a surprisingly small number of test standards have been written in the United States. Many methods have been used in reasearch studies, of course, but have not gone through the process of acceptance as standards.

ASHRAE Standard 52-76

This standard [1] was first introduced in 1968, and modified in 1976; both essentially combine the Air Filter Institute Test Code of 1953 and the method developed by the National Bureau of Standards. ASHRAE 52-76 has four main sections:

1. A measurement of filter resistance as a function of air flow.
2. A procedure to load the filter with a standard test dust, a mixture of silica dust, carbon black, and lint, at a fixed air flow.
3. A procedure measuring the percentage of mass of the standard dust captured during the loading test. This percentage is called the Synthetic Dust Weight Arrestance.
4. A procedure to measure the amount of atmospheric dust removed by the filter, in terms of its blackening or staining power. This test, the Atmospheric Dust Spot Efficiency test, is also repeated at intervals during the dust loading process. The existing form of the Dust Spot test samples air upstream and downstream of the filter,

using glassfiber paper targets of equal diameters, and equal sampling flows for both targets. The stain on the targets is measured by change in light transmission. Because this change is not linearly related to the quantity of dust on the targets, the downstream target is sampled continuously, but the upstream target is turned off intermittently to yield equal opacity changes on both targets. The efficiency (in percent) for the filter tested is defined as:

$$E_{ds} = 100(1 - T_u/T_d)$$ (1)

where T_u = total on time for the upstream sampler

and T_d = total on time for the downstream sampler.

In addition to the above, ASHRAE 52-76 also defines a procedure to evaluate roll-type or other self-renewable filters. The loading (or Dust Holding Capacity, DHC) part of this test attempts to simulate steady-state operation of such filters. Arrestance and Dust Spot Efficiency are measured in the same manner as before. Averaging methods for all performance measures are detailed, with special attention given to the possibility that dust-removal performance does not always increase as dust load increases.

Federal Specification F-F-310

This specification [5], dating back to 1957, once called out the "National Bureau of Standards" (NBS) test procedure. The latest version switched to ASHRAE Standard 52-76. Readers interested in interpreting data from the 1950's may refer to the description of the NBS procedure in [6].

MIL-STD-282 ('Thermal DOP' Test for HEPA Filters)

This group of Military Standards [2] was designed to evaluate respirator filters and HEPA filters for use in nuclear safety systems and chemical-biological warfare protective equipment. The 'thermal DOP' test is, however, used for filters of somewhat higher penetration than HEPA filters, such as those used ventilation systems for surgical suites, burn-treatment facilities and lower quality clean rooms. The filtersare challenged with a condensation aerosol of DOP (di-octyl phthalate, or diethylhexyl phthalate). The aerosol has a mean diameter near 0.3 μm and a rather narrow diameter spread (σ_g = 1.2). Concentration upstream and downstream is measured by the amount of white light scattered by the aerosol cloud. Penetrations in the order of 0.001% can be measured in this way. The ventilation-type filters mentioned above typically have penetrations near 5%.

ASTM F-778-82 (Media Resistance)

This standard [7] gives procedures for measuring the pressure drop across flat sheets of filter media. It gives special attention to the means for supporting and clamping the media to minimize leakage and damage to fragile media sheets.

ARI Standard 850-84 (Commercial and Industrial Air Filter Equipment)

This standard [8] by the Air Conditioning and Refrigeration Institute (ARI) is basically an equipment standard, rather than an air filter test standard. It subdivides ventilation air filters into four groups, specifying ASHRAE 52-76 for their evaluation. The groups are panel filters, self-cleaning or self-renewable filters, extended surface filters, and electronic air cleaners. A fifth group is defined as media for use in the first four groups. Detailed limits for test operating parameters are stated. In addition, the standard defines tests of the cleanability of filters, an overpressure test, and a procedure for measuring ozone production by electronic air cleaners.

ARI Standard 680-80 (Residential Air Filter Equipment)

This standard [9] parallels ARI Standard 850-84, but is specific to the types of filters used in residential systems. Self-renewable devices are not included.

UL 900 (Flame Resistance)

This Underwriter's Laboratory (UL) standard [10] defines tests for the flammability of air filters and the dust-adhesive coatings applied to them. The production of flame, sparks and smoke are all evaluated, and two categories of filter (Class I and Class II) are defined by spark, flame and smoke production. The flash point of adhesive coatings may not exceed 163 C when tested by ASTM D92.

TEST STANDARDS USED OUTSIDE THE U.S.

British Standard 2831 (Methylene blue, Alumina Dust Tests)

This standard [11] defines five tests:
1) A resistance-versus-airflow test on the clean filter.
2) A test of the amount of oil or filter adhesive that blows off the filter, where the filter has such a coating.
3) A dust loading procedure, making use of one or both of two fused alumina test dusts. The particle-size distribution for these dusts is given in Table 1.
4) A gravimetric efficiency measurement, obtained by weighing the dust fed to the filter and the test filter itself. This test is repeated four times during the dust loading portion.
5) A dust spot test, similar to the ASHRAE 52-76 method, except that a generated aerosol of methylene blue is used in place of atmospheric dust. The dye stains on the sampling targets are "developed" by exposure to a steam cloud to increase their optical density. This test is repeated four times in the course of the dust-loading. Penetrations in the order of 0.005% can be measured. The standard allows one to substitute the sodium flame test (see below) for the methylene blue test.

British Standard 3928 (Sodium-Flame Test)

This standard [12] specifies a penetration test, using a sodium chloride aerosol with a mass mean diameter (mmd) about 0.6 μm. The concentration upstream and downstream is measured by a flame photo-meter. The upstream concentration is diluted with clean air, and the light in the flame photometer is reduced by neutral density filters. Under these conditions, response is proportional to aerosol mass. Penetrations in the order of 0.0005% can be measured.

TABLE 1 - Characteristics of Test and Atmospheric Aerosols.

Aerosol	Material	MMD μm	Std. Dev.	Density kg/m3	Ref.
AC Fine	Silica	7.7	3.6	2650	1
AC Coarse	Silica	25.0	3.5	2650	*
BS 2831-1	Methylene Blue	0.54	1.5	1260	36
BS 2831-2	Alumina	5.2	1.3	4000	36
BS 3928-69	Sodium Chloride	0.58	1.7	2165	36
AFNOR 44011	Uranin	0.19	1.7	1530	37
DIN 24184 C	Atmospheric/radon	0.37	2.1	1500	16
DIN 24184 B	Paraffin Oil	0.54	1.3	875	16
DIN 24184 A	Quartz	1.84	1.3	2600	16
MIL-STD-282	DOP	0.30	1.2	980	38
Laskin Noz.	DOP	0.77	1.7	980	39
Atmospheric modes:					
Crustal	Soil	18.0	4.8	2500	40,41
Fine	Combustion Prod.	0.4	3.1	1000	40
Nuclei	Salts, Organics	0.02	2.0	1800	40

Standard Deviations are best-fit values to curves which are only approximately log-normal distributions. See References for exact distributions.   *Supplier specification on bottle.

AFNOR NFX 44011 (Uranin Aerosol Test)

This French national (AFNOR) standard [13] measures only filter penetration, expressed as a "purification coefficient" :

$$C_d = U/D \qquad (2)$$

U is the upstream aerosol concentration and D is the downstream. The aerosol used is a dye, uranin, having an mmd of 0.15 um. The aerosol is sampled upstream and downstream simultaneously, captured on a membrane filter downstream and on a medium penetration filter in series with a membrane filter upstream. The purification coefficient of the first-stage sampling filter upstream is measured during each test; this provides a check on the particle size of the aerosol cloud. Concentration measurement is by fluorescence of dissolved uranin dye washed off the sampler targets. Penetrations can be measured down to about 0.0005%.

EUROVENT 4/5 (General Ventilation Filters)

This standard [14] is (with the blessing of ASHRAE) a trilingual version of ASHRAE 52-76 with minor modifications adapting it for use in twelve European countries.

DIN 24184 (Filter Media and Filters for Small-Diameter Aerosols)

This standard [15] was developed at the Staubforschungs-Institute (SFI) in Bonn, and later adopted as a German national (DIN) standard. It covers both filter media and assembled filters, with these tests:
For media only: Mass per unit area, Thickness

For media and assembled filters:
Pressure drop at rated air velocity;
Penetration, using an oil aerosol of about 0.4 µm diameter. The aerosol is generated by aspiration, evaporation, and recondensation of a specified paraffin oil.   Concentration is measured (upstream and down-stream) by a white-light scattering aerosol photometer.
Penetration, using a quartz dust with count-mean diameter about 1.5 µm.   Concentration is measured by the light scattering photometer.
Penetration, using a very fine ambient atmospheric aerosol (filtered through a HEPA filter) which is exposed to radon gas. Some radon progeny, as well as "tagged" non-radioactive particles, are usually present; the count-mean diameter has been measured as 0.07 µm. Radioactivity removed is measured either by sampling or by a rather complicated scheme using two identical test filters in series.
Visual detection of leaks by light-scattering in a darkened chamber. The "thread" of oil aerosol ("oelfaden") downstream of a pinhole leak is surprisingly easy to detect when illuminated by a bright lamp.

DIN 24185 (General Ventilation Filters)

This standard [16] is essentially the German portion of Eurovent 4/5, hence a translation of ASHRAE 52-76.

Australian Standard 1132-1973 (General Ventilation Filters)

This standard [17] mingles the methods of ASHRAE 52-76 and BS 2831. The two alumina dusts and the methylene-blue aerosol of BS 2831 are used in addition to ASHRAE Synthetic Arrestance Test Dust. Procedures for loading static and self-renewing filter devices are included, as are resistance and dust-adhesive blowoff tests. A procedure is defined for locating pinhole leaks in low-penetration filters, using "cold" DOP smoke from a Laskin-nozzle generator and a light-scattering photometer. This is the same test described in the Institute of Environmental Sciences Recommended Practice IES-RP-CC-001-83T [18]. A discussion of Australian test experience, including problems met, is found in [19]. These comments of course apply to the U. S. and British standards from which the Australian was derived.

CURRENT STANDARDS DEVELOPMENT

ASHRAE 52-76: Arrestance, Dust Loading

None of the existing filter test standards provides penetration-vs-particle-size data. Instead, they measure the performance of the filter on specified aerosols, either natural or synthetic. How closely do the aerosols used match those met in actual practice? A look at the data on ambient aerosols is useful here. A great mass of data on this has been gathered; unfortunately, every conceivable particle-size analysis method has been used, making data analysis very difficult. A literature survey indicates that there are four major "modes" in outdoor aerosols:

- Coarse Particle or Crustal Mode. Chiefly soil particles; typically 18 μm mmd, with density about 2500 kg/m3 .
- Fine Particle or Accumulation Mode. Chiefly combustion products, plus some products of photochemical reactions. Typically 0.4 μm mmd, with density near 1000 kg/m3 .
- Nuclei Mode. The result of evaporation and recondensation of organics; also evaporated sea spray and radon progeny. Typical mmd is 0.02 μm, with density from 1000 to 2100 kg/m3 .
- Fibers. Chiefly plant parts. Size varies widely, with length 3 or more times diameters, which are typically 20 μm. Density is approximately 1000 kg/m3 .
- Pollens. These have diameters in the range from 5 to 50 μm, with densities near 1000 kg/m3 .

It is possible to remove the coarsest modes of outdoor dust with very porous coarse-fibered prefilters. It is even possible with dust louvers or multiple small cyclones; these devices are actually used as first-stage filters in air handling systems subject to sandstorms and duststorms. It is wrong, however, to think of successive stages as removing succesively smaller diameter modes. Each filter in a chain removes some fraction of each particle diameter range, though the fraction may be very small. A complete analysis of system performance requires knowledge of the penetrations of all filters in the system over the entire spectrum of particle diameters. This is even more true in the common case of systems which recirculate part of the air flow. In this case, aerosols generated internally must be captured by the filter stages through which the recirculated air passes. The crustal-mode dust recirculated to final-stage filters may be small, but the amount of lint may be relatively high. This is certainly the case for laundries and hospitals, with much handling of fabrics, and for any place where large quantities of paper are handled (printing and data-processing for example.) Recirculatory filters are often subject to high tobacco-smoke loads. Those which are used to control radon progeny must have significant efficiency on particles below 0.02 μm diameter. Thus a fully useful air filter test needs to span the diameter range from 0.01 μm to 30 μm, from radon progeny to lint and pollens.

No one instrument exists to do this, except on a cumulative basis: total mass, total light scattering, or dust-spot opacity. Total light scattering and opacity tend to emphasize finer particle diameters, while mass measurements emphasize larger particles. By using several measures of filter performance, one can get a rough estimate of performance as a function of particle size. ASHRAE Synthetic Arrestance Test Dust is mostly "AC Fine"- Standardized Air Cleaner Test Dust, Fine -hence has a similar mmd about 7.7 μm. A filter's efficiency on a 7.7 μm diameter dust with density 2650 kg/m3 will be close to its ASHRAE Arrestance number. For like reasons, its efficiency on dust of mmd 0.4 μm and density 1000 kg/m3 will be close to its ASHRAE Dust Spot Efficiency. These are rough approximations, however; [20] and [21] discuss how sensitive overall penetrations are to small changes in aerosol distributions, even when mmd remains constant.

The broad acceptance of ASHRAE 52-76 and other tests which do not measure penetration at specific particle diameters is based on the modest cost of these methods, the ease of maintenance of their equipment, and the convenience of having a few performance indices (Arrestance, Dust Spot Efficiency) rather than a set of curves. The backlog of test data and comparative field experience means that reasonable choices of filters can be made to achieve approximate performance levels. This usefulness is reflected in the current revision of ASHRAE 52-76, and the DOE work on MIL-STD-282. Both revisions seek to maintain continuity with the existing standards, so that when particle-size penetration methods are written into codes, users are not suddenly presented with a completely new specification basis.

In the case of the ASHRAE 52-76 revision, a table was prepared listing the virtues of existing sections of the standard, problems that had been noted, and possible alternatives and corrections. Initial discussions showed that writing a particle-size-penetration standard would require much work, but that this could form an independent section of the revision. Effort is therefore concentrated at present on improving the existing parts of the standard; a second section covering particle-size-penetration measurement will be written as soon as the existing parts have been revised.

The Arrestance portion of ASHRAE 52-76 has been criticised on the grounds that the dust used is a poor simulant for aerosols met in actual service. At the time it was first formulated, the dust matched experimental data on the <u>chemical</u> components of atmospheric dust. The percentages of silica, carbon black, and lint are those found in a nationwide survey [22]. These components are not dispersed independently, as they are in the atmosphere; there is a good bit of adhesion between the components. The particle size distribution of the air-suspended mixture is not representative of atmospheric dust as it was in 1954, or as it is today. The dust used was specified because its components could be described, and were readily obtainable with controlled, repeatable properties. (Although [19] correctly points out that the carbon black is specified by a trade term, molocco black, rather than by physical properties). In addition, testing had shown that for a broad range of filter types, dust holding capacities on this test dust matched reasonably well with in-service dust holding capacities. The dust can be formulated and dispersed repeatably. We might like a dust that included tars simulating some combustion products, or that was closer in mmd to typical atmospheric dust. Generating such dusts in relatively heavy concentrations is quite difficult. And heavy concentrations are needed, if the loading is to fit an acceptable time span. Consideration is being given to eliminating the carbon black since it appears mostly as a coating on the silica dust and may not alter DHC greatly. Lint, however, has a strong effect on DHC for many filter media, and a dust for a loading test therefore needs a lint component.

Oddly enough, there is also some complaint that ASHRAE Synthetic Arrestance Dust is too <u>fine</u> to distinguish between the very porous filters used to protect air conditioner coils from lint and coarse-dust plugging. The only ways to deal with this problem would seem to be to define a new coarse-dust/lint combination for this purpose, or to make use of a test giving particle-size-penetration data.

The repeatability of Arrestance tests has been criticized. There seems to be little evidence that the dust itself varies significantly from batch to batch, but there are items in ASHRAE 52-76 which could be more tightly defined to improve test variances. Abrasive wear of the sonic-flow dust aspirator has been observed; the committee is investigating ways to detect this and to limit its impact. Compressed air fed to the aspirator needs to be clean and dry and oil-free; its specification will be tightened. The form of the downstream filter in the present standard allows a rather massive filter; the weight gain of this filter is sometimes in the order of the "readability" of available balances. The filter is bulky and awkward to load onto a balance pan, and "readability" does not consider the effects of air movement, humidity and electrostatics. A considerably lighter downstream filter would reduce all of these effects, and allow use of a balance with better "readability". The revision committee intends to substitute a simple flat sheet of filter media and a suitable holder for the present fabricated filter. Detailed procedures for pre-equilibrating the media to reduce the effects of moisture, and bipolar ionization of the air around the balance to reduce electrostatic effects may be desirable. These improvements should reduce variances while retaining whole-stream sampling, which has many advantages when testing with coarse dusts.

ASHRAE 52-76: Dust-Spot Efficiency Revisions

There are three primary objections to the present dust-spot method: stability, speed, and the guesswork needed to determine test conditions. The method uses ambient atmospheric dust, which varies with location and time. There is evidence that test results at different laboratories are systematically different. This problem can be avoided if a synthetic opaque fine aerosol can be generated at stable concentrations with a stable particle-size distribution. The 52-76 committee hopes for an aerosol which yields dust-spot penetrations about the same as present atmospheric dust-spot penetrations. This means a particle mmd near 0.5 µm. Preliminary studies have used a carbon black dispersion aspirated by a Collison atomizer, and also by a Laskin-nozzle type generator. Building a generator which improves on atmospheric dust is not as easy as it seems. Ref. [23] shows that the output of a Collison atomizer is proportional to the atomizing airflow raised to the power 2.2 . This means that the pressure must be tightly controlled; cleanliness and humidity of the atomizing and evaporation airflow must also be controlled closely if good aerosol stability is wanted. Various studies on aerosol generators [24] [25] [26] [27] give mass fluctuations in the order of 6% . Occasional large droplets must be removed from the generator output by an impactor or louver.

The duration of a 52-76 dust spot test is inversely proportional to penetration. This is the result of the scheme used to compensate for the non-linearity of the opacity-vs-dust load curve for the dust sampling targets. It is a simple task to determine the shape of this curve and to correct for it without the interrupted sampling scheme. (A European Community Standard [28] does just this; in many countries, this curve is used to estimate outdoor air quality. Fig. 1 shows the ECC curve and data obtained in a study made by the 52-76 revision committee). Improvement in the stability and repeatability of the 52-76

photometer will be desirable to implement this, since the alternate
procedure requires measurement of small changes in spot opacities.

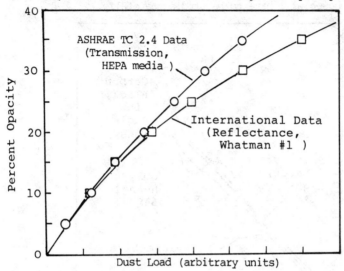

FIG. 1 - Opacity-vs-Dust Load Relation Obtained in U.S. and Europe.

Development of a Particle-size-Penetration Test Under ASHRAE 52-76

As of this date, the 52-76 committee has not chosen an approach
for this portion of the standard. ASHRAE funded a study [29] to evaluate the use of laser aerosol spectrometers (LAS) for this purpose.
The study was, unfortunately, limited to a single form of LAS, but
further studies by these investigators [30] and others [31][32][33]
[34] indicate the constraints to the approach, and its power. The
particle diameter measured by a LAS for a given actual diameter depends on the index of refraction of the particle material. Opaque
particles (such as carbon black) behave quite differently from transparent ones (DOP, petroleum) or translucent ones (polystyrene latex,
silica). In some cases, a substantial range of actual particle diameters are measured as having the same diameter. Fig. 2 (replotted
from [30]) shows these effects for two different LAS and three particle materials. It is apparent that ambient atmospheric aerosols,
with their wide range of refractive indices, create serious measurement uncertainties in a particle-size-penetration method based on
the LAS.

A well-behaved aerosol and aerosol generator are essential parts
of a particle size test method. The useful range of the LAS appears
to be from about 0.2 to 2.0 μm in a single instrument. Aerosol transport problems and the small diameter of the LAS inlet port complicate
matters for particles appreciably larger than 2 μm, even if there were
sufficient numbers of larger particles to provide valid statistics at
larger diameters. The somewhat larger airflows and inlet geometry for
white-light spectrometers do not provide much confidence in their usefulness for larger diameters. Another item concerns diluters, which

are necessary when testing with the LAS. Fig. 3 shows some diluter calibration data, taken from [20] and [35]. The characteristics of the diluter chosen will have to be studied carefully.

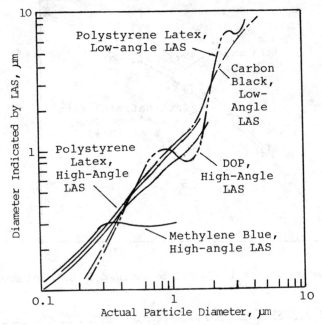

FIG. 2 - LAS Actual and Indicated Particle Diameters

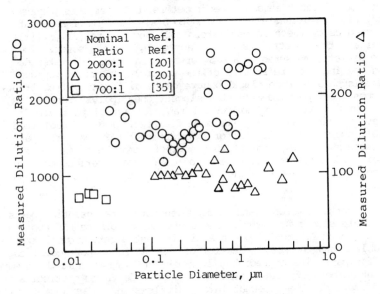

FIG. 3 - Variations in Diluter Ratios

If the test standard is to be concerned with diameters below 0.2 um, entirely different instrumentation will be needed. The combination of an electrostatic particle-size analyzer and a condensation nuclei counter has been used extensively for particles of this size range. This instrumentation, unlike the LAS, does not measure several particle diameters at the same time; it will be slow and rather demanding in operation. It appears that the spectrometer approach is a less obvious solution than it seemed at first, for the wide span of particles of interest to general ventilation filters.

The alternate approach -testing with narrow-band aerosols- has been used effectively for many research studies. Since both the chemical composition and diameter of the particle are known for each size run, there are many more options available for concentration measurement. In some size ranges, total light scattering by the aerosol cloud is an adequate measurement at the penetration levels of interest. For small diameters, the condensation nuclei counter can grow uniform particles to easily detected diameters. For large particles, even counts under the microscope can yield statistically valid data. If a fluorescent aerosol particle is used, this property can be used to measure concentration over very wide ranges. Particles containing sodium or other atoms which produce distinct spectral lines permit the use of the flame photometer for detection.

It is not necessary to evaluate filter penetration at dozens of particle diameters to obtain a useful particle-size-penetration curve. The general form of a filter penetration curve is known, and there is no reason to suppose that sudden breaks occur. Fig. 4 indicates how we might select a few particle diameters to determine the overall form of the curve for a wide array of filter types. Testing at large part-

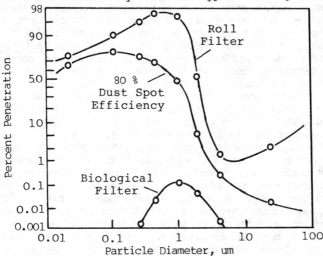

FIG. 4 - Diameters Suitable for Particle-size Penetration Data

icle diameters raises special problems because of the rapid fallout of these particles. Those parts of the national standards listed above

which deal with coarser dusts all have some procedure to cope with
dust fallout in horizontal ducts. A downflow duct is a better solution from the dust-suspension standpoint; the predecessor to 52-76
used such a duct, and the Australian standard allows it. Isokinetic
sampling and good dust distribution must be provided for large particles, and the sampling probe tubing is best kept very short. It may be
necessary to limit particle-size-penetration testing to flat sheets of
filter media.

Problems With Electrostatic Air Cleaners and Room Purifiers.

     Electrostatic air cleaners (two-stage precipitators) create spec-
ial test problems. The adhesion properties of any synthetic test dust
are not likely to match those of atmospheric dusts, nor does the electrical conductivity. Insulators short out far more quickly with a
test dust containing carbon black than is the case with atmospheric
dust. Present habits are to test such filters in the clean state
only; this is fine with conventional plate-type electronic air cleaners, which have almost constant airflow resistance, but is not satisfactory for fiber filters which incorporate electrostatic effects.
Filter units with very low-pressure fans, such as typical "room purifiers", cannot be tested for their real performance in a conventional
test duct. It has been suggested that they be tested in a recirculatory chamber, with air movement provided by their own fans.

GASEOUS CONTAMINANT FILTER TESTING

     There are many gas-cleaning applications where particle removal
is the only concern. However, ventilation of interior spaces is done
largely to make them habitable by people. There has been considerable movement in recent years to reduce outdoor ventilation air flows
in the name of energy conservation. Far too many articles have been
written on the basis that "filtration" can restore recirculated air to
the equivalent of outdoor air for ventilation purposes. Particulates
can be removed to essentially any desired air cleanliness, and frequently with favorable economics. The removal of gaseous contaminants is far more difficult. Regrettably, there are essentiallly no
test procedures which provide the data designers need to determine the
suitability of gaseous contaminant removal materials. Existing standards, such as ASTM D3467 (carbon tetrachloride retentivity of activated carbons) measure the static saturation level of a physical adsorb-
er at a single vapor pressure. Complete isotherm data is more useful,
but hardly transferrable by HVAC engineers into a way to choose air-
handling system components. Designers need kinetic tests run with
gaseous contaminants at concentration levels where toxic effects and
odors actually occur, not thousands of times those levels. Defining
what gaseous contaminants are to be used for testing is difficult;
working out practical ways to determine the real-world performance of
adsorbers, chemisorbers and catalysts will be more so. But the honest
application of filtration to energy reduction in buildings depends on
meaningful performance data, whether we deal with gaseous contaminants
or particulates.

REFERENCES

[1] ASHRAE 52 76: Method of Testing Air Cleaning Devices Used in
     General Ventilation for Removing Particulate Matter. Am. Soc.
     of Heating, Refrigeration and Air Cond. Eng., Atlanta GA (1976)
[2] MIL-STD-282: Filter Units, Protective Clothing, Gas-Mask Compo-
     nents and Related Products- Performance Tests. U.S. Army Chemi-
     cal Corps (1956)
[3] Liu, B.Y.H., Rubow, K.L., and Pui, D.Y.H., "Performance of HEPA
     and ULPA Filters," Proc. 31st Anl. Mtg. Inst. Env. Sci.,
     pp 25-28, (1985)
[4] Rivers, R.D. and Engleman, D.S. "Predicting Air Quality in Recir-
     culatory Ventilation Systems," ASHRAE Transactions 88 Pt.1,
     (1982)
[5] Federal Specification F-F-310: Filter, Air Conditioning, U.S.
     General Services Admin, Washington
[6] Dill, R.S., "A Test Method for Air Filters," ASHRAE Trans. 44,
     pp 379-386, (1938)
[7] ASTM F-778-82: Method for Gas Flow Resistance Testing of Filtra-
     tion Media, Am. Soc. for Testing and Matls., Philadelphia, (1982)
[8] ARI 850-84: Standard for Commercial and Industrial Air Filter
     Equipment, Air Cond. and Refrigeration Inst., Arlington VA,(1984)
[9] ARI 680-80: Standard for Residential Air Filter Equipment,
     Air Cond. and Refrigeration Inst., Arlington VA, (1980)
[10] UL 900: Safety Standard for Air Filter Units, Underwriters
     Laboratories, Northbrook IL, (1984)
[11] BS 2831: Methods of Test for Air Filters Used in Air Conditioning
     and General Ventilation, British Standards Inst., London, 1957
[12] BS 3928: Method for Sodium Flame Test for Air Filters, British
     Standards Inst. London, (1969)
[13] AFNOR NFX 44-011: Method to Measure the Efficiency of Filters by
     Means of Uranin Aerosols, Association Francaise de Normalisation,
     Paris, (1978) (in French)
[14] EUROVENT 4/5: Method of Testing Air Filters Used in General Vent-
     ilation, EUROVENT, Vienna, (1979) (in English, French, German)
[15] DIN 24184: Filter Media and Filters for Small-Diameter Aerosols,
     Deutsches Institut fuer Normung, Berlin, (1974) (in German; sales
     in U.S. by American National Standards Institute, New York)
[16] DIN 24185: Type Testing of Air Filters, Deutsches Institut
     fuer Normung, Berlin, (1980)(see [15])
[17] Australian Standard 1132-1973, Standards Association of Austral-
     ia, North Sidney, N.S.W., (1973)
[18] IES-RPCC-001-83T: HEPA Filters, Inst. of Environmental Sciences,
     Mount Prospect IL, (1983)
[19] Thompson, B.W. and Airah, M., "Air Filter Testing", Australian
     Refrigeration, Air Conditioning and Heating, (Jan. 1986) pp25-33
[20] Bergman, W., et al.,Electric Air Filtration: Theory, Laboratory
     Studies, Hardware Development and Field Evaluations, Lawrence
     Livermore National Lab. (UCID-19952, NTIS) (1984)
[21] Tillery, M.I., Salzman, G.C., and Ettinger, H.J., "The Effect of
     Particle Size Variation on Filtration Efficiency Measured by the
     HEPA Filter Quality Assurance Test," Proc. 17th DOE Nuclear Air
     Cleaning Conf. U.S. DOE, (1982)
[22] Moore, C.E. McCarthy, R., and Logsdon, R.F., "A Partial Chemical
     Analysis of Atmospheric Dust Collected for Study of Soiling
     Properties," Heating, Piping and Air Cond. (Oct. 1954) p145ff

[23] Rappaport, S.M. and Getteny, D.J., "The Generation of Aerosols of Carcinogenic Aromatic Amines," Am. Indust. Hyg. Assn. J. (39)4, pp 287-294 (1978)

[24] Donohue, D.L. and Carter, J.A. "Modified Nebulizer for Inductively Coupled Plasma Spectrometry," Anal. Chem. (50)4 pp686-687 (1978)

[25] Esmen, N.A., Weyel, D.A. and Farrokh, S.R.B., "Design and Characterization of a Low Air Flow Nebulizer," Am. Indust. Hyg. Assn. J. (43)12 pp 934-937 (1982)

[26] De Ford, H.S., Clark, M.L., and Moss, O.R, "A Stabilized Aerosol Generator," Am. Indust. Hyg. Assn. J. (42)8 pp 602-604 (1983)

[27] Collins, G.F., "Measurement of Variation in Output Rate from Aerosol Generators," J. Aerosol Sci. (6)1 pp169-172 (1975)

[28] Ingram, W.T. and Golden, J., "Smoke Curve Calibration," J. Air Poll. Contrl. Assn., (23)/2 pp110-114 (1973)

[29] Liu, B.Y.H. and Pui, D.Y.H., "Response of a Laser Optical Particle Counter to Transparent and Light-Absorbing Particles," ASHRAE Transactions, (1985, Part 1)

[30] Liu, B.Y.H., Szmanski, W.W., and Ahn, K.H., "On Aerosol Size Distribution and Measurement by Laser and White-Light Optical Particle Counters," Proc. Inst. of Envir. Sci. 30th Annl. Mtg., pp1-8 (1984)

[31] Matigot, G., Champsiaux, J., Boivinet, J.M., and Sigli, P., "Application of Laser Spectrometry to the Measurement of Filter Efficiency and Microleak Evaluation," Proc. 7th Intl. Symp. on Contam. Control, paper G3.2 (1984)

[32] Gustavsson, J., "Pleasure and Use of Particle Counters Down to 0.1 um," ibid, paper G2.1

[33] De Worm, J.P. "Laser Detector Methodology for Laboratory and In-Situ Testing of Very Efficient Filters," ibid, paper G2.2

[34] Rivers, R.D. and Engleman, D.S., "Evaluation of a Laser Spectrometer ULPA Filter Test System," Jnl. of Envir. Sci. (29)5 pp 31-36 (1984)

[35] Scripsick, R.C., Soderholm, S.C., and Tillery, M.I., "Evaluation of Methods, Instrumentation and Materials Pertinent to Quality Assurance Filter Penetration Testing," Proc. 18th DOE Nuclear Airborne Waste Management and Air Cleaning Conf. U.S. DOE, (NTIS) (1984)

[36] Hoppitt, H.B., "The Testing and Application Of Filters for Air Conditioning," Filtration and Separation (Dec. 1974) pp 573-580

[37] Dupoux, J., "In-Situ Measurement of the Efficiency of Filtration Installations in the Nuclear Industry by the Uranin Aerosol Method," Proc. 16th DOE Nuclear Air Cleaning Conf., pp17-34 U.S. DOE, (1980) (CONF-801038, NTIS)

[38] Hinds, W., First, M., Gibson,D. and Leith, D., "Size Distribution of 'Hot DOP' Aerosol Produced by ATI Q-127 Aerosol Generator," Proc. 15th DOE Nuclear Air Cleaning Conf. pp1130-1144, U.S. DOE, (1978) (CONF-780819, NTIS)

[39] Prodi, V. and Melandri, C., "Size Spectrometry of DOP Particles Using Low-Temperature Separation and Replica Technique," Staub-Reinhalt. Luft(30)3 pp31-34 (Eng. ed.) (1970)

[40] Sverdrup, G.M. and Whitby, K.T., "Determination of Submicron Atmospheric Aerosol Size Distributions by Use of Continuous Aerosol Sensors," Envir. Sci. and Tech. (11)13 pp 1171-1176 (1977)

[41] Lundgren, D.A., Hausknecht, B.J., and Burton, R.M., "Large Particle Size Distribution in Five U.S. Cities," Aerosol Sci. and Tech. (3)5 pp467-473 (1984)

Robert R. Raber

DEVELOPMENT OF AN ARTIFICIAL DUST SPOT TECHNIQUE FOR USE WITH
ASHRAE 52-76

REFERENCE: Raber, R. R., "Development of an Artificial
Dust Spot Technique For Use With ASHRAE 52-76," Fluid
Filtration: Gas, Volume I, ASTM STP 975, R. R. Raber, Ed.,
American Society for Testing and Materials, Philadelphia,
1986

ABSTRACT: Dust spots comparable to those obtained using the
natural atmospheric aerosol have been obtained using an atomized
and dried aqueous solution of carbon black. Tests which
have been performed during the past two and a half years
have indicated the technique is applicable for the standard
range of dust spots (20% to 95%).

KEYWORDS: dust spot testing, HVAC filter testing

INTRODUCTION

ASHRAE Standard 52-76 [1] is the test procedure in primary
use for rating heating, ventilating, and air conditioning
(HVAC) filters. This paper describes Farr Company experience in
generating a test aerosol to provide an artificial dust spot
that can be implemented with the Standard's existing test
equipment.

First, it should be stated the reason for investigating
an artificial dust spot at all was forced by a changing natural
aerosol in the coastal Los Angeles Basin. The focal point for
the problem was the 25% dust spot extended surface panel filter
which occupies a prominent place in the HVAC market.

These filters typically operate at media velocities on the
order of 0.5 m/s. They have a clean filter particle count
efficiency vs size characteristic on atmospheric dust similar
to that shown in Figure 1. They also occupy an ambiguous
position with regard to the dust spot test, as clean filter dust spot

Robert R. Raber is Research Manager for Farr Co., 2301
Rosecrans Avenue, El Segundo, CA 90245

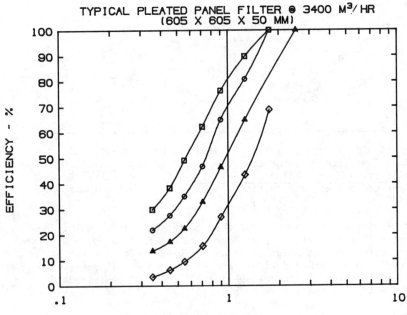

FIG. 1
EFFICIENCY VS SIZE
ON ATMOSPHERIC DUST

TYPICAL PLEATED PANEL FILTER @ 3400 M³/HR
(605 X 605 X 50 MM)

SYMBOL	RUN NO.	ASHRAE DUST FED	FILTER ΔP-"H₂O	DUST SPOT EFFICIENCY
◇	#1	0G	0.31"	16.3%
▲	#2	60G	0.45"	30.4%
○	#3	120G	0.66"	38.5%
□	#4	180G	1.03"	49.4%

is in the range of 11-16%. This is below the range recommended
by ASHRAE 52-76, which establishes 20% as lower limit of test
significance. Ambiguity arises because the average efficiency
is typically 25%, which is within the range of test applicability.

As might be expected from Figure 1, and as was reported by
Bauder in [2], these filters are far more sensitive to differences
in the naturally occurring aerosol than the higher dust spot
efficiency filters. We first began to notice major changes in
the natural dust spots through the clean air winter season of
1983-1984. During this period, routine quality assurance testing
began to show on a frequent basis clean filter efficiencies on
the order of 1-6% and average efficiencies of 16-20% when the values
had been 11-16%, and 23-27%, respectively.

Equipment troubleshooting did not unearth any problems; zero
dust spot efficiency was zero within ± 1 1/2%; the gas meter flows

checked within $\pm$ 1%; the opacity meter readings were not unusual.
It was thought, however, that soiled upstream targets appeared
lighter in color to the naked eye, although without a historical
control sample this was merely speculation.

By the spring of 1984, our test results had deteriorated to
the point where we felt it was necessary to do something to either
supplant or augment the natural aerosol if we were going to
continue to test in our existing location. A program was thus
undertaken to see if something was possible.

INITIAL DEVELOPMENT

Understanding that soiling could primarily be attributed to
carbonaceous particles, and that a one-to-one correspondence had
previously been demonstrated between an artificial sodium chloride
aerosol and the natural atmospheric dust spot by Dyment [3], our
first thoughts centered on doing the following:

   1. cleaning up the air entering the ASHRAE test duct
   2. introducing an artificial carbonaceous aerosol that
could produce dust spot efficiency values similar to those
obtained using the natural atmospheric dust

Because of convenience and a low pressure drop at 2000 cfm,
the first approach taken for cleaning the inlet air was the simple
expedient of adding a 95% dust spot filter ahead of the ASHRAE
duct inlet. For an artificial aerosol, it was decided to atomize
and dry an aqueous carbon black solution using a Collison
atomizer system assembled by Dr. Henry Yu and described in [4].
The dried carbon black aerosol would then be introduced downstream
of the prefilter and upstream of the ASHRAE duct inlet orifice.
The schematic diagram of this original system is shown dashed in
Fig. 2.

Several unsuccessful attempts were made to produce aqueous
solutions of carbon black before we decided to avail ourselves of
the expertise represented in commercial carbon black tinting
products. Our first attempt at this was successful: Pacific
Dispersion's AIT 744 Lampblack. It has proved to be shelf-stable
and produce consistent results batch to batch.

After relatively little experimentation, our first filter
testing produced initial dust spots on the order of 14-18% in
5 to 10 minutes per point. Test variables for this were: 6 ml
of AIT 744 to 300 ml of deionized water; 30 psi (2 bar) supply
pressure to a 4 hole Collison atomizer; a sufficient quantity of
drying air (7.5 m³/hr   drying air for .56 m³/hr   of atomizing
air) to effect a mixed stream relative humidity of 10%. In
addition, 2 drops of Dow Corning Anti-Foam B water dispersible
silicone defoamer were added to each 300 ml solution to eliminate
foaming.

A single successful test is not worth much, particularly if it is not repeatable, so starting in May of 1984 we began to do both natural and artificial dust spots to accumulate some test time and test experience.

Over the next year, the data generated proved to be repeatable and relatively insensitive to concentration and supply pressure variations (see below). In June of 1985, we abandoned natural dust spot testing for development and quality assurance checks, taking the benefit of greatly reduced test time.

In September of 1985 we suddenly and inexplicably saw a large increase in the efficiency numbers being generated. Since this increase was apparently random, it was extremely difficult to trace and the next three months were spent trying to isolate the cause.

As a result of the investigation, 99.95% HEPA filters were added to the duct inlet and a new PO 210 source was installed. This did not cure the problem. Finally, it was noticed the oil trap for the compressed air supply had a good deal of oil in it, indicating a very oily air supply. The solution to this was to re-route the atomizing air line so that it too saw the benefit of full filtering and drying. Since December 1985, the problem has not been encountered again, and the final system now in use is shown in fig. 2.

SUMMARY OF RESULTS

Table I presents a compendium of test data for five different media grades and a development electrostatic air cleaner loaned by Emerson division of White Rogers for evaluation of the artificial dust spot technique.

The data column labeled number of runs indicates the number of two-test runs represented in the data for the same filter. Data in the table were generated using either a four or six hole Collison atomizer, test pressures of either 1 or 2 bar, and AIT 744 concentrations of 4 to 12 ml per 300 ml of deionized water.

SUMMARY OF EFFECTS OF TEST VARIABLES

Without a great deal of elaboration here are our conclusions regarding the system variables:

1. Test Container

    The cut size will be dependent on the shape of the container and whether or not there is a shroud around the atomizer. Limited testing has shown the sloped sides of the 500 ml Ehrlenmeyer flask produce more consistent results and greatly less foam than a straight sided container.

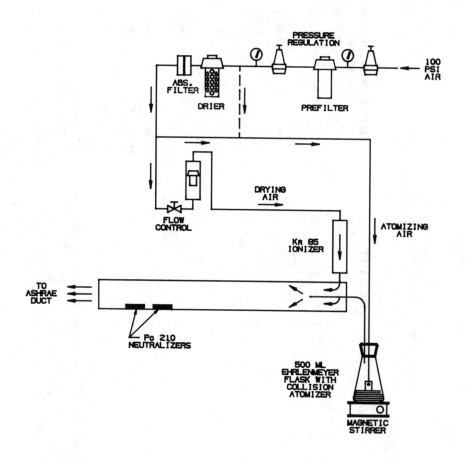

FIG. 2

SCHEMATIC DIAGRAM
OF CARBON BLACK FEEDING SYSTEM

TABLE [I] CLEAN FILTER
COMPENDIUM OF CLEAN FILTER DUST SPOT RESULTS
2.5 m/s FACE VELOCITY (EXCEPT AS NOTED)

NOMINAL MEDIA GRADE (AVG. DUST SPOT EFF)	ARTIFICIAL DUST SPOT			REPRESENTATIVE NATURAL DUST SPOTS	
	No. OF TESTS	AVG.	MAX/MIN OBSERVED	EL SEGUNDO	EASTERN LABS
25%	20	15.8	18.3/12.5	4.3 (4.7/4.0)[1]	13.5 (20/11.3)[1]
40%	3 (2)	23.4	24.2/22.4	8.2	16.9
60%	5 (2)	33.0	35.4/30.4	20.2	34.7
80%	4 (2)	65.8	66.7/64.2	52.9	60.4
90%	12 (2)	83.4	84.7/82.1	75.2	84.4
DEVELOPMENT 70% ELECTROSTATIC AIR CLEANER (2) face velocity					
2.5 m/s	1	73.9	–	67.0	70.0
3.	1	66.4	–	59.5	64.0
3.5	1	59.6	–	50.1	61.0

(1) Natural dust spot data for 25% dust spot filters is for 9 filters; 3 tested in El Segundo and 3 each tested at two Eastern Laboratories; all nine made from same roll of development media. Data in parenthesis are max/min for this data set.

(2) All data are for same air cleaner.

2. Air Flows

Both atomizing and drying streams must be thoroughly clean and dry for consistent results. The effects of different ratios of atomizing air to drying air have not been investigated. The testing reported herein uses (at 2 bar) .56 m^3/hr of atomizing air and 7.5 m^3/hr of drying air on a 4 hole atomizer. Measurements of mixed stream relative humidity indicated a 10% maximum RH. The mixing duct length should be on the order of 20 diameters to insure the drying is complete.

3. Concentration and Supply Pressure

Our limited testing indicated the effects of supply pressure and concentration were negligible, see Table II. We have standardized on 2 bar and usually vary concentrations according to the level of efficiency being measured; 4 ml in 300 for a 15% dust spot efficiency; 8 or 12 ml in 300 for a 90% dust spot.

4. Number of Holes

Collison atomizers can be made with 2 to 6 holes. The bulk of our testing has been with a four hole unit, although recently we have used a six hole satisfactorily.

5. Effect of Different Carbon Blacks

During the period where our artificial dust spots were inexplicably high (see above), we obtained what we believed to be a finer carbon black from another source and the results showed no change. We believe the carbon black dries to certain size agglomerate, and is not materially affected by the original black size. It is more important to have a non-foaming, non-settling, shelf-stable product, and the AIT 744 appears to be most satisfactory in these characteristics.

CONCLUSIONS

1. It is possible using an artificial dust spot technique to produce results consistent with existing values obtainable with the naturally occurring aerosol. In general the artificial dust spot values tend to be slightly higher than natural dust spots for a given media grade.

2. The test does not appear to be extremely sensitive to feeder variables, and is capable of producing results in much shortened times, while still adhering to the minimum 10% opacity change of the downstream target as now specified in ASHRAE 52-76. In general, repeatability at its worst appears to be as good if not better than what

TABLE II

TEST VARIABLE INVESTIGATION
25% AVE. DUST SPOT PANEL FILTER
@ 2.5 m/s FACE VELOCITY

	4:300	4:300	8:300	12:300
AIT 744 CONC. (ml: ml DI WATER)	1	2	2	2
AIR PRESSURE (bar)	17.7	15.3	17.5	18.3
ARTIFICIAL DUST SPOT (%) (Avg. of two runs)	30	10	5	5
TEST TIME (Min)	22	24	20	24
CHANGE IN SAMPLING PAPER OPACITY (%)				

OPTICLE PARTICLE COUNT DATA ON FEED AEROSOL

	4:300	4:300	8:300	12:300
TOTAL COUNT/M^3	1.278×10^6	4.803×10^6	7.725×10^6	13.084×10^6
PER CENT OF TOTAL COUNT IN SIZE RANGE				
.3 - .4 μm	77.3	84.4	83.9	81.9
.4 - .5 μm	17.9	12.9	13.0	14.5
.5 - .6 μm	2.9	1.7	1.9	2.3
.6 - .8 μm	1.6	0.8	1.1	1.1
.8 - 1.0 μm	0.3	0.1	0.1	0.1

is achievable with the natural aerosol.

3. The one test made for an electrostatic air cleaner indicates the technique will also apply to this kind of air cleaner.

ACKNOWLEDGEMENTS

The author wishes to thank Dr. Henry Yu for helping to set up the atomizer system and Mr. Gary dela Cruz for helping to establish the technique.

The assistance of Mr. Joseph J. Fodor, Manager of Technical Services at Pacific Dispersions, Inc., both for his help with aqueous carbon black solutions and for supplying samples of AIT-744 Lampblack, is also appreciated.

REFERENCES

[1] American Society of Heating, Refrigerating, and Air-Conditioning Engineers, Inc; Method of Testing Air-Cleaning Devices Used in General Ventilation For Removing Particulate Matter; Std. 52-76.

[2] Bauder, C.J.; "A Report on The ASHRAE Standard 52-68 Evaluation Test Program"; ASHRAE Symposium Paper AT-78-4 Atlanta, GA, 1978.

[3] Dyment, J.; "Testing Air Filters: The ASHRAE and BS 2831 Tests Compared" Filtration and Separation; Nov/Dec 1978, pp. 546-550.

[4] Liu, Benjamin Y.H., Whitby, Kenneth T., and Yu, Henry H.S.; "A condensation Aerosol Generator for Producing Mono-dispersed Aerosols in The Size Range 0.036 μm to 1.3 μm"; Journal de Recherches Atmospherique, 1966, p. 406.

**Applications and Testing:
Protection of Equipment**

Bruce N. McDonald, Robin E. Schaller, Mark R. Engel, Benjamin Y. H. Liu, David Y. H. Pui, and Todd W. Johnson

TIME RESOLVED MEASUREMENTS OF INDUSTRIAL PULSE-CLEANED CARTRIDGE DUST COLLECTORS

REFERENCE: McDonald, B.N., Schaller, R.E., Engel M.R., Liu, B.Y.H., Pui, D.Y.H., and Johnson, T.W., "Time Resolved Measurements of Industrial Pulse-Cleaned Cartridge Dust Collectors", Fluid Filtration: Gas, Volume I, ASTM STP 975, R. R. Raber, Ed., American Society for Testing and Materials, Philadelphia, 1986

ABSTRACT: An optical particle counting system has been developed for testing industrial pulse-cleaned cartridge dust collectors. The system is applied to measure fractional penetration of a full scale dust collector and to study its short term and long term dynamic behavior. Results show collector efficiencies for dust loadings of 23 g/m³ exceeding 99.999% over the particle size range from 0.5 to 10 μm, with a minimum fractional penetration occurring at approximately 2 μm. The measurement system is shown to be capable of providing particle size and/or concentration measurements on time scales, ranging from 100 ms to many hours. Phenomena from details of individual cleaning pulses to overall collector life and efficiency are demonstrated.

KEYWORDS: pulse-cleaned, cartridge dust collector, optical particle counter, fractional efficiency, time resolved measurements

Pulse-cleaned cartridge dust collectors are gaining increasing acceptance in industry for dust control. The operation of this type of system is shown in Figure 1. While the filtration efficiency of such systems is known to be high, there is relatively little informa-

B.N. McDonald is a Sr. Technical Supervisor, and Dr. R.E. Schaller is Director of Technology Planning & Analysis in the Corporate Development Group and M.R. Engel, P.E., is Product Development Manager in the Industrial Group of Donaldson Company, Inc., Minneapolis, MN 55440; Dr. B.Y.H. Liu is Professor and Director and Dr. D.Y.H. Pui is an Associate Professor in the Particle Technology Laboratory of the University of Minnesota, Minneapolis, MN 55455; and T. W. Johnson is an Engineer in the Medical Products Division of 3M Company, St. Paul, MN.

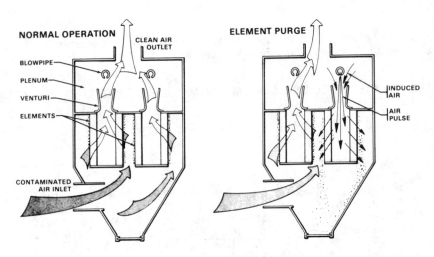

FIG. 1. Operation of Pulse-Cleaned Cartridge Dust Collector

tion available concerning collector efficiency as a function of particle size and as a function of time. Testing of industrial collectors traditionally has been based on source stack sampling methods [1], using a filter to collect dust samples for weighing and overall mass efficiency determination. Although measurement of overall efficiency has yielded valuable information, it falls short of the need of the sophisticated user or equipment designer with an interest in understanding the mechanism of dust collection. The introduction of high efficiency cartridge type collectors has made gravimetric methods impractical because of low downstream concentrations and long sampling times involved. From an environmental perspective, the particle size range between about 0.1 and 10 µm is of particular interest. Moreover, since these systems are dynamic in nature, it is desirable to make such measurements on a temporal basis.

A number of investigators have developed techniques for measuring the effects of particle size on pulse-cleaned baghouse performance. Ensor et al. [2] used cascade impactors on the inlet and outlet of baghouse applied to a utility boiler. Leith and Ellenbecker [3] used sampling filters together with microscopy to analyze particle size flux emissions from a pilot scale baghouse. Ensor et al. [4] used an electrical aerosol analyzer and cascade impactor to measure baghouse particle size penetration in the size range from 0.02 to 10 µm. Klingel and Loffler [5] used an optical particle counter to measure the time dependency of fractional efficiency for a new filter bag in a pilot scale pulse jet lab fixture.

The object of this test program was the development of a new test method for determining the performance of full scale industrial cartridge dust collectors. This paper briefly reviews the equipment and method developed, together with its application to a full scale system. The time resolved measurement capabilities are demonstrated for time scales ranging from 100 ms to hundreds of hours. For a more detailed description of the test system, see Liu et al. [6].

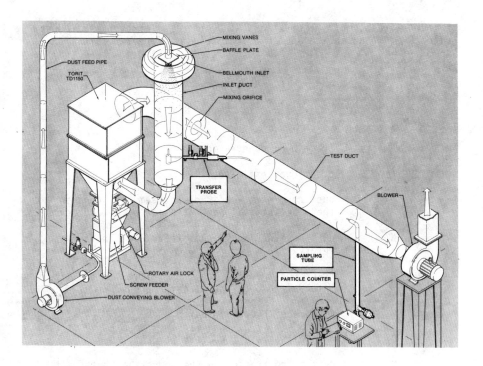

FIG. 2.   Schematic Diagram of Experimental System

DESCRIPTION OF THE EXPERIMENTAL SYSTEM

System Overview

A schematic diagram of the experimental system developed is shown in Figure 2. A dust feeder introduces test dust at a steady rate to the cartridge dust collector. A 60 cm diameter, 6.7 m long test duct is located downstream from the collector. An optical particle counter is used to measure the dust concentration downstream from the collector. By the use of the "transfer probe", a sample of the highly concentrated upstream dust can be extracted and injected downstream, thus sufficiently diluting the upstream dust concentration so that it can be measured by the same optical particle counting system as the downstream concentration at the same fixed point in the test duct.

The fractional penetration of particles through the dust collector, P, and the collector fractional efficiency, η, are:

$$P = 1 - \eta = \frac{N_{dc}}{N_u} = P_p \frac{Q_p}{Q_t} \frac{N_{dc}}{(N_{do} - N_{dc})} \qquad (1)$$

where:

> P   =   fractional penetration,
>
> η =   fractional efficiency,
>
> N_{dc} =   dust concentration in downstream test duct in the size interval with transfer probe closed,
>
> N_u =   dust concentration in upstream duct in the size interval,
>
> P_p =   fractional penetration of particles in the size interval through the transfer probe,
>
> Q_p =   flow rate through transfer probe,
>
> Q_t =   total flow rate through test duct,
>
> N_{do} =   dust concentration in the downstream test duct in the size interval with the transfer probe open.

In this test program, Q_t = 5100 m³/h and Q_p = 1.7 m³/h, providing a dilution ratio Q_t/Q_p of 3000 to 1. It should be noted that the above approach of diluting upstream dust with downstream air is practical only because the efficiency of the dust collector under test is sufficiently high and $Q_p \ll Q_t$.

Calcium carbonate sold under the trade name of Microfil by the Calcium Carbonate Co. was used as the experimental test dust. The volume median diameter of the dust is 5.0 μm.

Instrumentation for Time Resolved Measurements

The combination of the optical particle counter and a multi-channel analyzer (MCA) provides considerable power and flexibility in measuring the performance of the pulse-cleaned dust collectors. A Climet Model 225 Optical Particle Counter was used. The MCA consisted of a Canberra Model 85 with a Model 8529 Signal Processing Module. A small laboratory computer, a Digital Equipment Corp. MINC, was used to acquire and analyze the results from the MCA. The optical particle counter outputs a string of electrical pulses with amplitudes dependent on the size of the individual particles passing through the counter. The MCA provides several methods of analyzing the string of pulses. Two common modes of operation are the pulse height analysis (PHA) mode and the multi-channel scaling (MCS) mode. The heart of the MCA is a block of memory locations or channels in which counts are accumulated.

In the PHA mode, each successive memory location corresponds to a pulse height. An analog to digital converter converts the pulse height to a digital number proportional to the pulse height. That number is used as a memory location and the contents of that memory location are incremented one count. After a period of operation, the memory contains a histogram of number of pulses as a function of pulse height. Using the optical particle counter calibration curve of pulse height -vs- particle size, this data represents the number of particles counted in the period of operation as function of particle size for a number of size increments. While the data is collected in a large number of channels, typically 2048 for these measurements, it is more convenient to group the channels into a smaller number of regions of interest (ROI). Eleven ROI were established, defining 11 geometrically spaced particle sizes from 0.5 to 10 μm. The time periods used for collecting data in the PHA mode ranged from 1 to 15 minutes while both shorter and longer times are possible.

In the MCS mode, each successive memory location or channel corresponds to an increment of time. All the pulses occurring in one time increment are counted in a channel. The pulses occurring in the next time increment are counted in the next channel and so on. After a period of operation in this mode, the memory contains a history of counts per time increment as a function of time for all particle sizes measured. Hence, a time history of particle concentration being sampled by the optical particle counter is obtained.

The start of the counting in the first time increment can be under program control or can be triggered by an external event such as the beginning of a cleaning cycle on the dust collector. The advance to the next channel can be done using an internal time base or by using an external event such as the cleaning pulse to each element in the dust collector. When the MCA has stepped through all of the channels in the assigned memory block once, it has completed a sweep. The process can stop with a single sweep, or can continue by starting over again at the first channel adding to the counts already accumulated for multiple sweeps.

The pulses from the optical particle counter are processed by a single channel analyzer before entering the multi-channel scaler. The single channel analyzer outputs one pulse to the MCS for each input pulse that has an amplitude between specified lower and upper limits. For some tests, the lower and upper limits were set such that all particles from 0.5 to 10 μm were counted. In other tests, the limits were adjusted so that the only particles counted were within a specific size range corresponding to one of the regions of interest described above. For the measurements reported here, dwell times per channel from 10 ms to 3 s were used. Single and multiple sweeps were utilized frequently using the beginning of a cleaning cycle to initiate the sweeps.

Uncertainty of Measurements

At a 95% confidence level, it is estimated that flows Q_t and Q_p are known to ± 5%. The transfer probe penetration P_p is estimated to be known to ± 5%. The uncertainty in the particle counts can be estimated using Poisson statistics to be $2 * \sqrt{N}$ to a 95% confidence level. When these error estimates are combined using the root sum of squares method and Equation 1, the error estimates for the particle penetration of collector range from 5% to 15% (95% confidence) for small to large particles, respectively, for penetrations of 10^{-5} to 10^{-6}. The error bars shown in Figure 4 are calculated in this manner. Monitoring the count rate over an eight hour period indicated that the uncertainty in the downstream counts on the order of ± 40% which is greater than that used above. (See Recovery from Change below). With this error estimate, the uncertainty in penetration through the collector is on the order of ± 50% for 95% confidence. Note that if the penetration is for example 1×10^{-5} ± 50%, then the efficiency is between 99.9985% and 99.9995%.

MEASUREMENTS OF CARTRIDGE DUST COLLECTOR PERFORMANCE

Description of Collector Tested

The experimental system described above was used to measure the performance characteristics of a TD-1150 Pulse-Cleaned Cartridge Dust Collector manufactured by the Torit Division of the Donaldson Company. The collector contained six filter cartridges, each containing 21 m² of filter area, for a total filter area of 126 m². The rated capacity of the collector is 5100 m³/h at a pressure drop of 4.5 cm H₂O. The filter cartridges are cleaned sequentially by a reverse flow pulse of air, each pulse lasting approximately 100 ms, and the time interval between pulses being 10 seconds. Thus, all six cartridges are cleaned once every minute.

The measurements presented below were obtained over a period of time during which several improvements to pulse-cleaned cartridge dust collectors were introduced. Advances have been made in the collector design, filter element design, and the filter media used in the elements. So, while these test results represent pulse-cleaned cartridge dust collector performance, they do not necessarily represent the performance of the latest equipment.

Time Scales of Pulse-Cleaned Dust Collectors

In a pulse-cleaned cartridge dust collector such as the ones described above, there are a number of natural time scales (Table 1).

TABLE 1 -- Time Scales

Order of Magnitude	Event Description
Years	The time for elements to wear out or become plugged
Hours (hr)	The time for new elements to reach "steady state" operation
Minute (min)	The time to complete a cycle of cleaning all elements in unit
10 seconds (s)	The time between cleaning successive elements
100 milliseconds (ms)	The duration of an individual cleaning pulse

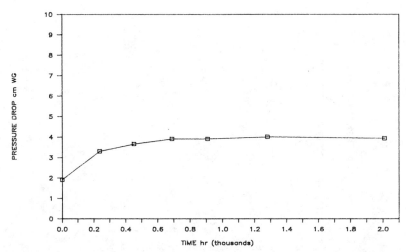

FIG. 3. Life Cycle of Pulse-Cleaned Cartridge Dust Collector

The life (Figure 3) of a pulse-cleaned cartridge dust collector begins with an initial pressure drop. Early in the life cycle, the pressure drop rises for a short time. Thereafter the pressure drop remains nearly constant for a long period while the pulse-cleaning removes dust at essentially the same rate as it is deposited. The period during the initial rise in pressure drop is called "seasoning". Eventually the pressure drop may become excessive or the filter medium may wear out. While this cycle and the associated scales have been described in terms of time, the amount and type of dust are critical factors in determining filter life. The efficiency of the dust collector has a similar life cycle. Throughout the life of the filter, the pulse-cleaning cycles represent actions on shorter time scale that have a large impact on the overall system performance. In the section below, the discussion progresses from the long time scales to the short time scales.

Long Time Scale Measurements

Steady State Efficiency: The measurement system was first used to measure the average performance characteristics during steady state operation when the pressure drop is essentially constant. The MCA was used in the PHA mode with data collection times of 1 to 15 min. It was found that 15 min samples provided stable average efficiency data. Hence, in slightly more than 30 mins, this measurement system is able to measure the upstream and downstream particle size and concentration, and present fractional penetration results (Figure 4).

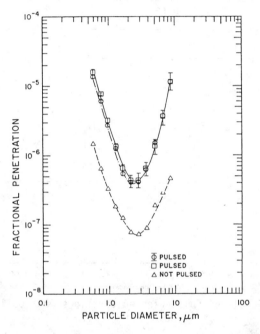

FIG. 4. Fractional Penetration Through Dust Collector at Upstream Concentration of 23 g/m³

The penetration is seen to first decrease with increasing particle size, reaching a minimum of 4 X 10⁻⁷ at a particle diameter of 2.0 μm, and then increase to the upper limit of 10 μm of the optical particle counter. The initial decrease in particle penetration with particle size is consistent with the particle collection mechanisms of interception and inertial impaction. However, the exact cause for the increase in particle penetration beyond the minimum at 2.0 μm is uncertain. A similar trend is reported by Ensor et al. [2] in full scale tests with baghouses on utility boilers. This may be due to the coarse particle bounce through the dust layer and fibrous structure of the filter cartridges. Large particle penetration was attributed by Carr and Smith [7] to "redispersion of particulate agglomerates that bleed through the fabric with the flexing action that occurs during cleaning". This phenomenon is explored further in the short term measurements below.

Similar data for a light inlet dust loading of 0.34 g/m³ show a minimum penetration value of 4 X 10^{-6} at 2 μm. That is approximately a factor of 10 higher than that for the case of the higher dust loading [6].

Data for a special experiment during which the cleaning pulses were stopped are also included in Figure 4 for comparison with the data with cleaning pulses. These data indicate a decrease in penetration through the filter by a factor of greater than approximately 4. This result is consistent with the observation below that the cleaning pulse can cause a momentary increase in downstream particle concentration.

Start Up or Seasoning: Another relatively long time scale measurement is the investigation of the seasoning of the filter elements. The pressure drop and efficiency as a function of time or dust fed are presented (Figures 5 & 6).* The ability of optical particle concentration measurement to make measurements quickly, makes it possible to demonstrate that the efficiency rises very rapidly. The efficiency exceeds 99.99% after only 145 g/m² of contaminant were fed to the filter media. The pressure drop data would imply seasoning taking about 10 times longer.

Recovery From Change: During the testing it soon became obvious that "steady state" operation was not as easy to define as might be expected from a quick glance at a pressure drop -vs- time curve. When a change in operation was introduced, such as a change of dust concentration or an overnight shut-down, the pressure drop curve would make an adjustment in a few minutes to an hour, but the downstream concentration seemed to take much longer to stabilize. For example, the following test was undertaken to determine stability of the downstream concentration.

The MCA was used in the pulse height analysis mode. Sequential samples of 5 minutes duration were taken consecutively for 8 hours. Five minute samples were chosen to average over several of the 1 minute cleaning cycles. The number counts in each region of interest

---

*This set of results was obtained on a smaller pulse-cleaned cartridge dust collector, Torit Model TD-162, using a GCA Environmental Instruments Real-Time Aerosol Monitor 1, Model 1190, to measure the total dust passing the collector. The challenge dust was "5 μm Min-u-sil" from Pennsylvania Glass Sand Corp. The volume median diameter is 1.7 μm. All other results presented are obtained from the test system described in this paper and in Liu, et al. [6].

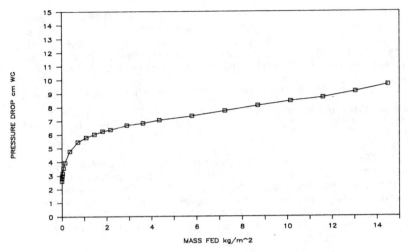

FIG. 5. Seasoning of Collector: Pressure Drop

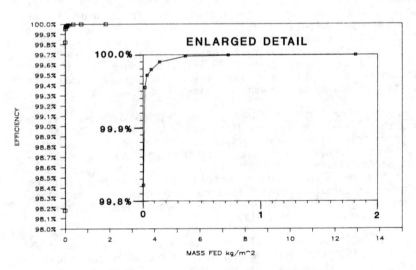

FIG. 6. Seasoning of Collector: Efficiency

for each sample period were stored in the computer. Only downstream readings were taken since the upstream concentration had been
found to be stable. The collector had been operated for 70 hours
since the dust feed rate had been reduced to 0.34 g/m³ from 23 g/m³.
The collector had been operating for several hundred hours total.
Immediately prior to the test, the collector was not operating overnight (16 hours). The collector was operated for 30 minutes prior
to taking data to "stabilize". The results, shown here (Figure 7)
as number of counts in the region of interest representing particles

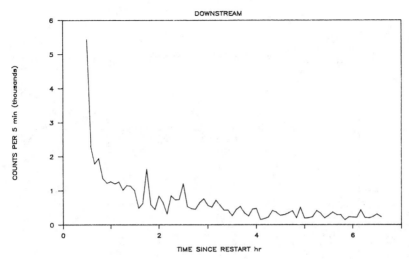

FIG. 7. Downstream Count Rate after Restart

from 4.2 to 5.6 μm, were very similar for all particle sizes. The predominate feature is the trend towards lower counts. Obviously, the collector was not stabilized in over 7 hours of operation after a restart. The particle rate counts in Figure 7 represent penetrations on the order of 10^{-4} to 10^{-5}. This data was used to estimate the uncertainty in the downstream count rate. To minimize the impact of the overall drift toward lower downstream counts, a line was fit to the last 4 hrs. of the data. The variance about that line yielded the 40% estimate for the uncertainty in the data used in the Uncertainty of Measurements section.

Short-Time Scale Measurements

To make measurements with time scales short enough to resolve details of element cleaning cycle, the multi-channel scaling mode was used. The first results (Figures 8 & 9) were for all particles 0.5 to 10 μm and a single sweep. The dwell time per channel was 100 ms.

The time history while the transfer probe is in the open position (Figure 8) indicates that there is a small (±10%), short (3.5 seconds) cyclic fluctuation in the upstream particle concentration, apparently caused by the dust feeding system. A similar measurement of the downstream dust concentration was made with the transfer probe closed (Figure 9). The result shows that there is an increase in the downstream particle count following the initiation of a cleaning pulse and thereafter the count gradually decreases until the next cleaning pulse is initiated. The count rate is low so there is substantial scatter in the data. To see more detail of pulse cleaning cycles, the MCA was set up to start its sweep when the number one element was pulsed. With multiple sweeps (Figure 10), the total counts are increased, thereby reducing scatter due to

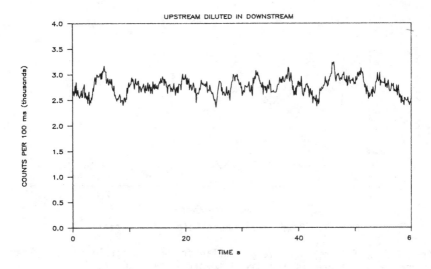

FIG. 8.   Short Term Fluctuations in Upstream Particle Count

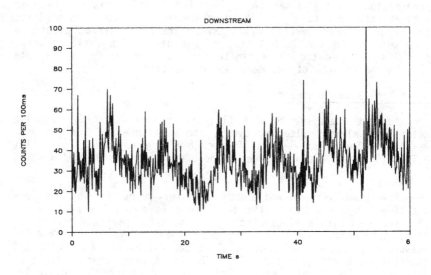

FIG. 9.   Short Term Fluctuations in Downstream
Particle Count: Single Sweep

random event counting statistics. The triggered sweep mode of operation also has the effect of averaging out fluctuations not correlated to the trigger event, the cleaning cycle in this case. In this mode, the increase in downstream particle count following the cleaning pulse is very evident. The pattern of maximum counts being different for each element was quite repeatable. It was not clear if this is due to differences in the elements or if it was due to differences in the element location such as differences between cleaning pulse strength.

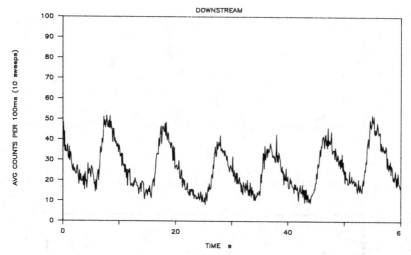

FIG. 10. Short Term Fluctuations in Downstream Particle Count: Multiple Triggered Sweeps

While the precise mechanism causing increased outlet concentration following cleaning is uncertain, it is likely that contributing factors include physical shaking causing collected dust seepage as described by Leith and Ellenbecker, [8], as well as reduced dust layer thickness on the cartridge momentarily reducing efficiency. Also, the concentration immediately upstream of the filter cartridge is increased by the pulse. Loffler, [9], using pilot and full scale pulse jet filters has also shown that outlet concentration increases significantly for a short period following pulse cleaning.

Additional tests were undertaken to investigate the increase in particle concentration downstream due to pulse cleaning and to investigate the increase in particle penetration for particles larger than 2 μm. Triggered multiple sweeps of multi-channel scaling were used with the single channel analyzer adjusted to detect only pulses due to particles in a narrow size range. This test was repeated for six of the eleven previously defined particle size ranges. To obtain better counting statistics, the dwell time was increased to 200 ms and more sweeps were taken. Up to 450 sweeps

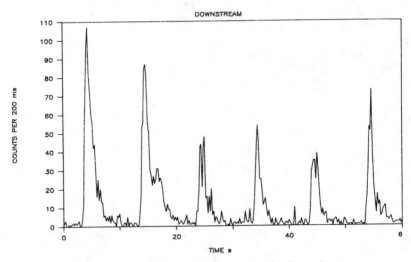

FIG. 11.   Short Term Fluctuations in Downstream
Particle Count: Narrow Range of Particle

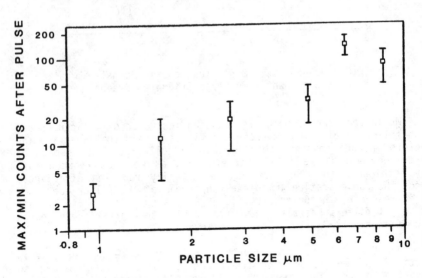

FIG. 12.   Effect of Particle Size on Increased
Penetration Due to Pulse-Cleaning

were taken, representing 15 hours of data gathering. The results shown (Figure 11) are typical. Again the characteristic pattern of peak heights for the 6 different elements is obvious. To determine if large particle penetration was affected more by pulse cleaning than small particle penetration, the ratio of maximum particle penetration per cleaning pulse to minimum particle penetration was plotted -vs- particle size (Figure 12). The maximums and minimums were determined by taking the 10 contiguous channels, i.e., 2 s, with the maximum or minimum total count for each cleaning pulse. The results are the averages for all six elements. The penetration of larger particles is affected more by the pulse cleaning than the small particles. It is not possible to conclude that the increase in penetration for particles larger than 2 µm is due solely to the pulse cleaning because even the non-pulse cleaned measurement (Figure 4) shows increased penetration for the larger sizes. More detailed study is needed to elucidate the mechanisms involved in this phenomenon.

CONCLUSIONS

The result of this test program shows that modern aerosol instrumentation, such as the single particle optical counter and multichannel analyzer, can be used effectively to study the characteristics of industrial dust collectors under realistic conditions. A system has been developed to apply the instrument to measure the fractional penetration of a full scale dust collector, and to study both its short term and long term behavior. Results obtained in applying this system to a high efficiency industrial dust collector show that penetration for an inlet dust loading of 23 g/m³ is less than 1×10^{-5} and the collector efficiency is in excess of 99.999% over the entire size range of the optical counter of 0.5 to 10 µm. Further, there is a minimum in the fractional penetration curve at 2.0 µm. This minimum is present whether or not pulse cleaning is used. The results suggests that the principal particle collection mechanisms in these filters are interception and inertial impaction, and that the increase in penetration above 2.0 µm may be caused by particle bounce; that is, the dry dust particles impacting on the collecting element but not to completely adhering. The particle bounce is enhanced by pulse cleaning. The results also show that the particle concentration downstream of the filter increases immediately following the cleaning pulse followed by a rapid decrease until the next pulse is initiated.

ACKNOWLEDGMENTS

The authors would like to thank the following individuals for their contributions to this program: Richard Cardinal, Mark Gogins, and Bob Nicholson, Donaldson Company, Inc., Minneapolis, MN 55440.

REFERENCES

[1] Bloomfield, B.D., "Source Sampling, in Air Sampling Instruments for Evaluation of Atmospheric Contamiants", ACGIH, Cincinnati 1972, pp. B1-B10.

[2] Ensor, D.S. and Markowski, G., "Size Dependent Penetrations from 0.01 to 10 Micrometers through Several Modern Particulate Control Devices on Utility Boilers", in B.Y.H. Liu, D.Y.H. Pui, H.J. Fissan (eds.), Aerosols. Elsevier, New York, 1984, pp. 639-642.

[3] Ellenbecker, M. and Leith, D., "Dust Deposit Profiles in a High Velocity Pulse-Jet Fabric Filter". J. Air Pollution Control Association Vol.29, 1979, pp. 1236-1240.

[4] Ensor, D.S., Comen, S., Shendrikar, S., Markowski, G., Waffinden, G., Pearson, R., and Scheck, R., "Kraemer Station Fabric Filter Evaluation". RP1130-1 Final Report CS-1669. Electric Power Research Institute, Palo Alto 1981.

[5] Klingel, R. and Loffler, F., "Dust Collection and Cleaning Efficiency of a Pulse-Jet Fabric Filter". Filtration and Separation, May/June 1983, pp. 205-208.

[6] Liu, B.Y.H., Pui, D.Y.H., and Schaller, R.E., McDonald, B.N., and Johnson, T.W., "An Opticle Particle Counting System for Testing Industrial Pulse Cleaned Cartridge Dust Collectors", to be published in Particle Characterization.

[7] Carr, R.C. and Smith, W.B., "Fabric Filter Technology for Utility Coal-Fired Power Plants: Part 1; Utility Baghouse Design and Operation". J. Air Pollution Control Association Vol. 34, 1984, pp. 80-88.

[8] Leith, D. and Ellenbecker, M., "Dust Emissions from a Pulse-Jet Fabric Filter", Proceedings of the Filtration Society July/August, 1983, pp. 311-314.

[9] Loffler, F., "Separation Efficiency and Pressure Loss of Filter Materials of Different Structure at Differing Conditions", Staub-Reinhalt, Luft 30, 1970, pp. 27-31.

R. Bruce Tatge

FILTER TESTS TO SUPPORT GAS TURBINE APPLICATIONS

REFERENCE: Tatge, R.B., "Filter Tests To Support Gas
Turbine Applications", Fluid Filtration: Gas, Volume I,
ASTM STP 975, R.R. Raber, Ed., American Society for Testing
and Materials, Philadelphia, 1986

ABSTRACT: Gas Turbines are in worldwide use as prime
movers, facing tremendous ranges of environment and inlet
air quality. Inlet air filtration is widely used to extend
the life of gas turbines and to maintain their output and
heat rate. In designing an air filtration system, pressure
drop, serviceability, and dust-holding capacity are almost
as important as efficiency. Filter tests performed in the
laboratory generally are idealized to a degree, and do not
always reflect the service that filters actually experience
in the field. Recommendations are made for tests which
would address these concerns, particularly for
self-cleaning filters, which are of increasing importance
in this industry.

KEYWORDS: gas turbines, inlet air filters, self-cleaning
filters, filter testing.

INTRODUCTION

It would be difficult to find a class of machines which are
subjected to a greater variety of environments than gas
turbines. Arctic cold, the heat and humidity of the tropics,
dusty deserts - all must be accommodated. Depending upon the
rating of the machine, the airflow may be very large, such as
several hundred kilograms per second.   It is perhaps not
surprising that the inlet air filter may be one of the largest
components of the power plant, much larger than the gas turbine
itself.

Mr. Tatge, is Principal Engineer, Plant Arrangement and
Inlet Systems, at the General Electric Company, Turbine
Technology Department, Bldg. 53-402, Schenectady, N.Y. 12345.

FIG. 1 - Typical Gas Turbine Power Plant

One of the keys to reliable operation in difficult conditions is clean air, and it is for this reason that most modern gas turbines use inlet air filters. There was a time, some years ago, when that was not the case. Then, fuel was cheap and many gas turbines were in peaking service, only infrequently run. Today, with fuel expensive and with many applications placing a premium upon the ability to run for weeks or months without shutdown, the priorities have changed. At the same time, technological improvements in design and materials have increased output and improved efficiency – changes which also place a premium on clean air if the user is to enjoy high reliability and performance.

EFFECT OF AIR POLLUTANTS

If air quality is inadequate, damage to the gas turbine may result. Problems typically fall into one or more of the following categories: erosion, corrosion, cooling-passage plugging, and fouling. Erosion is the physical wearing-away of blading due to impact with dust particles. When you consider that the compressor tip runs near Mach one, you realize that impact forces can be substantial. At first, erosion may result only in roughening of the surface. This lowers the aerodynamic efficiency, thereby decreasing air flow and degrading performance. In extreme cases, blading may wear more severely until there are major losses in output and efficiency, and there is hazard of blade fracture with disastrous consequences.

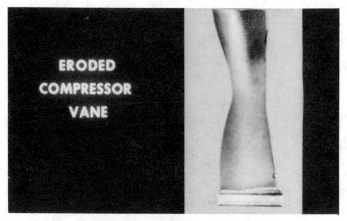

FIG. 2 - Severely Eroded Compressor Vane

Erosion is mainly caused by ingestion of relatively large particles, 10 microns or more in diameter. Smaller particles lack sufficient energy to cause damage.

Corrosion is generally a problem in the turbine section of the gas turbine. There, various metallic elements, in combination with sulfur, can break down protective coatings in the turbine blades.

FIG. 3 - Hot Corrosion of Turbine Bucket

AMBIENT AIR QUALITY

Sulfur generally comes from the fuel, but the corrosive alkali metals sodium and potassium may be present in the air. Desert soils, fertilizers, ocean spray, and road-salt are all potent sources of these metals, but almost any soil may have significant quantities.

TABLE 1 - Alkali Metals in Soil

Location	Total sodium and potassium, %
Kuwait	1.3%
Oman	2.1%
Australia	2.0%
Saudi Arabia	3.3%
U.S.	4.7%

Tests typically show that, in various soil fractions, sodium and potassium percentages increase as particle size decreases.

TABLE 2 - Alkali Metal Variation with Particle Size

Sample	% Sodium and potassium in total sample	Sieve size, microns	% sodium and potassium in sieved fraction
1.	1.2%	37	2.5%
2.	0.6%	74	1.0%
3.	1.8%	74	2.3%

Many modern gas turbines pump air through air passages in the turbine blading for cooling purposes. Dust in the air can accumulate and plug these passages, weakening materials and encouraging corrosion. Cement dust is particularly prone to this sintering action.

All of the above processes are irreversible, in that damage can only be corrected by replacing the damaged parts. Fouling, the last problem to be mentioned, is reversible. Fine particles, vapors, smokes, and fumes can adhere to compressor blades, reducing their aerodynamic efficiency. Unlike erosion, however, cleaning can remove the fouling material and restore performance.

From the above, it can be seen that particle-size distribution and dust chemistry are very important to the gas turbine. Also important is the dust concentration, because this relates to the effect of the dust on the gas turbine, and also to the life of filters which may be used for protection. In much of the United States and other developed countries the dust concentration runs only about 0.05 ppm, but in desert dust storms it can be four orders of magnitude higher.

AIR FILTRATION SYSTEMS

Depending upon the quantity and quality of the dust, gas turbine filtration systems typically use one or more of the elements found in the following table.

TABLE 3 - Typical Gas Turbine Air Filters

---

o Inertial separator

o Prefilter

o High-efficiency filter

   o Conventional

   o Self-cleaning

In designing an air filtration system, a number of criteria must be balanced. Parameters of importance are given in Table 4.

TABLE 4 - Filter System Parameters

---

o Efficiency

o First cost

o Maintenance Cost

o Service interval

o Ease of service

o Pressure drop

Depending upon the function of the gas turbine, different users put different values upon these factors. In continuous service as part of an industrial process, a long service interval might be the primary concern. In peaking service for an electrical utility, first cost would be more important than operating cost, simply because the gas turbine would probably operate for relatively few hours per year. As part of a cogeneration plant, pressure drop is very important because it affects both fuel cost and available output. It is fair to say that the performance-related factors of pressure drop and filter life (service interval) are more important today than they were in the past, due both to rising fuel costs and to increased usage of gas turbines in mid-range and base-loaded applications.

FILTER TESTS

Given this complex interplay of filter characteristics and customer requirements, what can ASTM Standards do to improve the match between the equipment which is delivered by the

manufacturer   and   the   customer's   expectations?   The
opportunities lie mainly in four categories:

o Filter configuration

o Filter aging

o Effect of contaminant characteristics

o Special concerns associated with self-cleaning filters

FILTER CONFIGURATION

Laboratory test of filters are usually made under somewhat
idealized conditions. First, because of practical limitations,
they generally involve a relatively small number of filter
elements. For example, fan capacity may limit the test to only
a few filters, perhaps in the range of 1 to 10, whereas the
actual gas turbine may require use of several hundred filters.
The simplest assumption is that there is no interaction between
elements, so that results from the laboratory tests of isolated
elements may be extrapolated to larger installations. In fact,
there may be significant interaction. We have found, for
instance, that the spacing of inertial separators affects
pressure drop by as much as 50%. Also, in laboratory tests
incoming airflow is almost always smooth and uniform, and
normal to the filter face. In fact, second-stage filters may
encounter turbulence due to upstream elements, and filters may
be arranged in VEE-banks to increase the number of filter
elements per unit of frontal area. As a result, flow is no
longer normally incident. This is generally done because of
practical constraints such as shipping limits, limited space at
the user's site, and disparities between the velocities in
different parts of the inlet system, which can be difficult to
match.

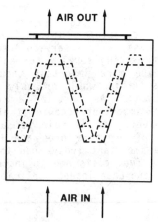

FIG. 4 - Plan View of Inlet Filter Compartment

In a typical filter compartment utilizing inertial separators and high-efficiency filters, the inertial separators run at a face velocity of about 6 m/s, whereas the high-efficiency filters use a face velocity of only about 2.5 m/s. The VEE-bank arrangement for the high-efficiency filters solves the mechanical problem, but current tests do not provide a clue as to the impact on filter performance.

FILTER AGING

It is widely recognized that, as a filter ages, its filtration efficiency tends to improve due to development of a filter cake. However, a point may be reached when a filter reaches its ultimate dust-holding capacity, beyond which the filter may "unload" dust, so that its efficiency suddenly drops. Standard tests to provide information on efficiency vs. dust load would help the user to avoid the hazards associated with either very young or very old filters which might have an efficiency or arrestance below the minimum tolerable level.

Some filters sold for gas turbine service employ a gel to coat the filter media, thereby increasing efficiency. However, it is our experience that the gel may migrate from the filter to the gas turbine's compressor. There, due to its adhesive properties, the gel causes dust build-up on the blading, which is exactly opposite to the goal of the filtration system. Again, a standard could be written to provide a measure of carry-over as a function of velocity and ambient temperature.

Finally, some filters are comprised only of materials which are inert and stable. Others, however, employ media which can be deteriorated by adverse conditions including chemical contaminants, heat, and sunlight. The consequences can vary from deterioration in arrestance to outright physical collapse. We have experienced the latter in plastic parts which were not ultraviolet inhibited. Reoccurrences could be eliminated by suitable qualification procedures.

CONTAMINANT CHARACTERISTICS

Filter efficiency is commonly defined by the use of standardized test dust, usually Arizona Fine and sometimes Arizona Coarse. Actual dusts may differ significantly in particle size distribution from either of these. This is particularly true in multi-stage systems, where initial filter stages classify the dust, so that the material which penetrates to later stages is much different from the ambient dust. What would be more generally useful would be data on efficiency as a function of particle size. However, different test methods for measuring the dust quantity and size distribution upstream and downstream of filters do not always agree because of variations in dust particle size, shape, and specific gravity. A uniform test procedure would be helpful in allowing the user to make meaningful comparisons between different designs.

Many gas turbines are exposed to dusts or aerosols containing water-soluble salts such as sodium chloride. Under humid conditions, they acquire moisture from the atmosphere and go into solution. The airflow through the filter tends to cause the solution to migrate to the downstream side. Then, when humidity drops, the salts re-crystallize on what should be the clean face of the filter, and they may re-enter the airstream. Since salts are generally potentially damaging, there is a need to quantify this procedure. While there are various user specifications which address this problem, it would be better if it were the subject of a recognized Standard.

SPECIAL CONCERNS ASSOCIATED WITH SELF-CLEANING FILTERS

So-called self-cleaning filters, which use a short reverse-blast of air to knock off dust accumulated on the upstream side of a barrier filter, have had a revolutionary impact upon gas turbine air filtration, to the point where this is now the design preferred by most customers.

CLEANING ACTION

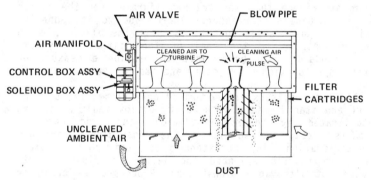

FIG. 5 - Self-Cleaning Air Filter

However, there are no industry-wide recognized tests which can be used to compare the offerings of different manufacturers. The key to such filtration is, of course, cleanability, or the ability of a filter to maintain pressure drop by rejecting accumulated dust. Laboratory data, accumulated over a relatively brief period of time, usually indicates that immense quantities of dust can be handled without significant increase in pressure drop. Field experience, however, accumulated over months or even years, shows a gradual upward trend of pressure drop with time, apparently due to gradual embedding of particles in the filter matrix. This is very important to the user since it affects filter life, but is currently completely unpredictable. Dust chemical and mechanical characteristics and ambient relative humidity are thought to be significant influences.

Also of importance are possible changes in filtration efficiency with time, as a dust cake accumulates. Sodium and potassium which may penetrate a filter when its efficiency is low can coat hot sections of the turbine and have an effect on hot corrosion long after the penetration actually occurs. It is important to ensure that filter efficiency is always sufficiently high to be protective. Standardized tests for measuring efficiency as a function of particle size and dust accumulation would allow users to compare competitive designs.

CONCLUSION

Gas Turbine users recognize that inlet air filtration is a good investment to protect an expensive piece of equipment. They also recognize that filtration has a cost in lost efficiency-due to pressure drop, and in maintenance-due to down time and need for replacement elements. New standards which would help in the evaluation of costs and benefits would be welcomed by both manufacturers of gas turbines and their world-wide customers. These are not high-tech questions, but tend to be simple, straightforward and practically oriented. Given the many ways in which gas turbine filters are used, that does not mean that the answers are easy.

Robert M. Nicholson and Lloyd E. Weisert

A REVIEW OF THE USE OF SAE STANDARD J726 IN HEAVY DUTY ENGINE
AIR CLEANER TESTING.

REFERENCE: Nicholson, R. M. and Weisert, L.E., "A Review of the
Use of SAE Standard J726 in Heavy Duty Engine Air Cleaner
Testing", Fluid Filtration: Gas, Volume I, ASTM STP 975, R. R.
Raber. American Society for Testing and Materials,
Philadelphia, 1986.

ABSTRACT: A brief review of the SAE J726 Standard for Heavy
Duty Engine Air Cleaner Testing is conducted. Some limitations
of the standard test dust, AC Coarse and AC Fine are presented.
Recommendations for revising the Standard to include fractional
or grade efficiency are made.

KEYWORDS: air cleaner testing, test dust, fractional
efficiency, simulated environments.

INTRODUCTION

The acceptance and degree of use of a standard test procedure is
usually a reflection on how accurately and thoroughly it reflects
real world performance. Typically, the test procedures actually used
are user driven; that is, they are tailored to a specific users
application or concern. If the user feels a standard test procedure
adequately challenges the product, it will often be used without
variation. The SAE J726 Air Cleaner Test Code [1] is the most
commonly used standard for heavy duty engine air cleaner testing.

The perfect air cleaner system would not allow any particulate
contamination to enter the intake system. It would accomplish this
without adding any restriction burden to the engine intake system,
and would perform in this manner for the useful life of the engine.
Of course real world air cleaner systems do not attain this level of

Robert M. Nicholson and Lloyd E. Weisert are Senior Project
Engineers in Corporate Technology of the Donaldson Co., Inc.,
1400 W. 94th St., Minneapolis, MN 55440.

performance. The performance they do attain, however, is essentially a comparison to this "standard".

Testing to this "standard" does not always provide final product definition. For example, in over-the-highway heavy duty air cleaning the cleaner will often be exposed to a singular, unique contamination episode, such as acid fume ingestion. The air cleaner user may not log the incident or even be aware of it. Depending on the length of cleaner exposure and the relative dust loading at the time the cleaner will start to pass more dust. The customer then will demand more and better Quality Control (Q.C.) using J726. This will not always correct the problem. The solution is to use the J726 standard as a control for baseline testing, then analyze and improve the system components to make the air cleaner more versatile.

The SAE J726 test standard has much of its origin from military vehicle testing at Yuma, Arizona, in the post World War II era. The AC Coarse (ACC) and Fine (ACF) test dusts specified in the standard are processed fractions of Yuma dust that are used to simulate the most abrasive and penetrating dust clouds found in the vicinity of fast moving, close order, heavy duty military vehicles. The two dust fractions produced by AC Spark Plug, Division of General Motors, have been formulated to challenge different aspects of engine protection. The coarse fraction represents the more abrasive and mechanically clogging particle size spectrum. The fine fraction ("Float Dust") is the more penetrating particle size for air filters and the greater visibility restrictor [2].

The test J726 standard incorporates the SAE practitioner's approach to utilizing fluid flow control measurements to test a non-homogenous fluid of air and particles. Specifically the test standard "measures airflow restriction or pressure drop, dust collection efficiency, dust loading capacity, and air cleaner-air flow integrity" [1].

The test standard assumes a well controlled, stable airflow induction system, it measures pressure drop or restriction in terms of static pressure across the system or components. Dust collection efficiency is measured by dust recovery of that collected by the air cleaner and that penetrating, such that:

$$\text{Efficiency, \%} = \frac{\text{Increase in wt. of unit}}{\text{Increase in wt. of unit + increase in wt. of absolute filter (ABS.)}} \times 100$$

or Efficiency, % = (1-Penetration) 100

$$\text{where Penetration} = \frac{\text{Increase in wt. of ABS. Filter}}{\text{Weight of Total Dust of Inlet}} \quad [1]$$

Dust capacity is usually limited by a maximum air cleaner restriction or pressure drop; or where there is a breakdown in fundamental collection mechanism such as the dust reentrainment out of the pre-cleaner or dust migration through the paper element. Air cleaner integrity is based on element media collapse, gross leaks in system seals or the paper element seals and is stable through a temperature range of -40°F to 250°F.

Typical air cleaner flow rates, efficiencies and pressure drops measured in this standard are:

       o Air Flow   - 0.2 to 300 M³/Hr.
       o Efficiencies - 10 to 99.99%
       o Pressure Drop - 100 to 7500 Pa.

Repeatable system efficiency is measured in two phases. First, the system is challenged for a limited time (usually 30 minutes) at a minimal pressure drop increase to measure initial efficiency. The second phase of the efficiency measurement is to load the air cleaner to its dust loading capacity or terminal pressure drop for the life efficiency. Double stage cleaners may require separate measurements on efficiency particularly if the first stage alters the particle size distribution seen by the second stage. For example, Figure 1, illustrates the concentrating of the ACC test dust into a narrower particle size distribution which the second stage must operate against. This is acceptable if the second stage is designed for this particle size spectrum.

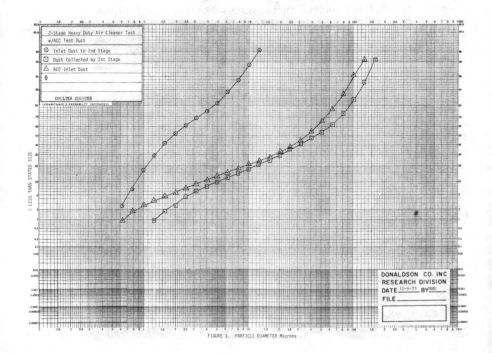

FIGURE 1. PARTICLE DIAMETER Microns

REPRODUCIBILITY

Sampling for a representative air cleaner element is usually the greater challenge in establishing a total test system. If adequate Q.C. is maintained on the respective components, the cleaner performance depends on the sealing integrity and air flow transition from inlet to outlet. Leaking and blow-by must be avoided particularly, if fractional efficiencies are to be considered, as in Figure 2.

Often taken as a "given" is the reproducibility on the dispersion and stable inlet concentration of the standard test dust. It has been the Donaldson Company's experience that this testing aspect is the greatest variant in the standard. Over the years our concern has been with the specific properties of the standard test dust, ACF and ACC. Although these dusts are well quantified in the test standards, there are some subtle aspects of the dust that affect system analyses, particularly for fractional efficiency testing.

The standard test dust has been found to be not log normal in particle size distribution when measured at particle size resolutions greater than that provided by the supplier, AC Spark Plug [1]. Figure 2 illustrates the difference in efficiency developed by an air cleaner with a distinct separation capability in the 1 to 2 μm when the proportion of particles below 5 μm can not be well quantified. The reconstructed ACF is a mixture of several batches of air classified dust, mixed in proportion to the weight distribution specified by the J726 Standard.

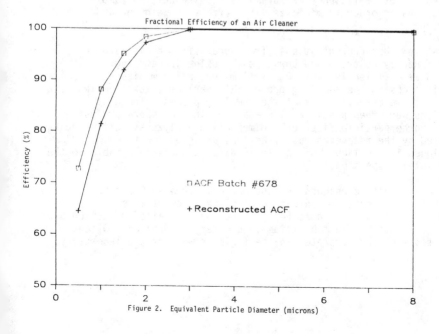

Figure 2. Equivalent Particle Diameter (microns)

A second aspect is the variance in particle shape over the particle size spectrum [3]. The angular shape of the dust greater than 5 μm relative to the more uniform shape of particle size less than 5 μm can have a significant impact on overall penetration and bulking property of the dust cake. We have observed that the more angular the dust, the higher the penetration of the air cleaner. Conversely the more uniform particle shape of the less than 5 μm particle size exhibits better efficiency than an angular dust such as ground silica powder. Because ACC and ACF test dusts often needs enough impaction energy to disperse the fine portion of the particle size distribution, attrition can occur in the larger particle sizes.

However, even though these dust size distributions are less than ideal, they still represent the real world for a wide spectrum of "dusty" environments. Therefore, the real challenge is to reproducibly disperse the dust into a uniform concentration of particle size and mass.

ACCURACY AND INITIAL EFFICIENCY

The J726 standard is oriented toward steady state efficiency and appears adequate for Q.C. product specification but is inadequate for air cleaner development.

In filter media development and modeling, initial efficiency under conditions of low dust concentration, is the primary interest because it provides critical information about filter media structure and avoids the unpredictability of particle interaction under heavy loading. These efficiency measurements require on-line sample extractions in order to minimize large errors in weighing and handling.

At low dust concentrations, fractional efficiencies are established by upstream and downstream isokinetic sampling. Fractional efficiencies assume a well mixed, uniform size distributions at the sampling probes. In earlier times, samples were drawn onto membrane filters at the sampling probes. The dust fractions were then particle size analyzed in the laboratory using a liquid borne particle sizer or sedimentation analyzers. Efficiencies measured by the recovered dust fractions are good for these techniques if the laboratory analysis dispersion is comparable to the test system dispersion.

More recent techniques use the Optical Particle Counters (O.P.C.'s) with dust concentration diluters to count in-situ performance of dust cleaners [4]. These counters allow detection of small variances and instabilities in cleaner performance. Often these tests can be completed in the field as well as the laboratory.

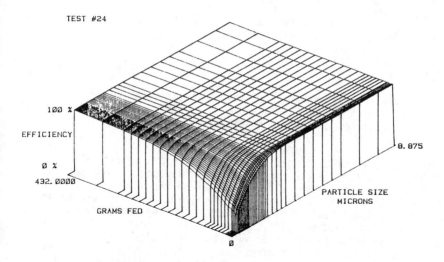

TEST #24

100 %

EFFICIENCY

0 %
432. 0000

GRAMS FED

8. 875

PARTICLE SIZE
MICRONS

0

Figure 3.   O.P.C. Fractional Efficiency for ACF Loading
of a Heavy Duty Engine Air Cleaner

Figure 3 illustrates the Donaldson Co., Inc. (DCI) use of OPC's
in evaluating air cleaner system performance. We have employed this
type of instrumentation in conjunction with computerized data
acquisition and analysis in the laboratory and the field with
universal customer acceptance. In the field it has enabled us to
characterize environments (particle size, distribution and
concentration), locate optimum inlet locations, correlate particle
size with specific engine component wear, as well as evaluate system
performance. This type of testing allows "standard" heavy duty air
cleaners to be modified to specific applications while retaining
their reliability established by the J726C testing. Incorporating
current instrument technology and data acquisition into development
of a new or modified test standard appears due.

A need to characterize this degree of penetration in the
transition stage of initial filtration to cake filtration requires a
greater resolution than provided by the SAE J726C standard. A
fractional efficiency is required that tracks this process. The
DCI system does just that, characterizing the change in particle size
penetration and the change in particle concentration with change in
ΔP of the system.

A system such as the DCI system, allows the study of problems in
filter media pore distribution (leaks) and in dust cake stability and
dust migration.

In the development of filter papers for heavy duty air cleaners,
it is known that the pore rating of given media will have specific
effects throughout the collected dust particle size spectrum. An
O.P.C. system can track the fundamental filtration mechanisms of

inertial interception and impaction during initial loading and into the transition to dust cake filtration.

Special Adaptations

There are several areas of critical interest to heavy duty air cleaner users which are not addressed by the J726C standard. Most of these involve system performance in simulated application environments.

One such area of testing often requested before specification is vibration. Most air cleaner applications are exposed to some sort of vibration excitation in use. The amplitude and frequency range depends on such criteria as on or off-highway applications, engine or chassis mounting and isolation, stationary or moving application, rubber-tired or tracked vehicles, engine RPM range, etc. The parameters most often tested under vibration conditions are structural (fatigue) and seal integrity. The seal integrity test may involve feeding dust to the system and monitoring inefficiency with an optical particle counter while the system is maintained at an extreme of high or low temperature, and subjected to a specified vibration condition.

Another area of interest to manufacturers and users is the performance under wet or humid conditions. Barrier filter media most commonly used in heavy duty engine applications has a cellulose base, which swells and loses structural integrity when exposed to water. This swelling may cause the typical pleated configuration to buckle or deform, rendering a portion of the filter ineffective. We have conducted tests where the life, or dust loading capabilities of a filter are tested at conditions of 115°F and 50% relative humidity. This is a severe test, but has been very useful in evaluating filter structural integrity under humid conditions.

Other special tests conducted, involve dust loading using specific contaminants such as salt or carbon, or imposing conditions such as pressure pulsations or flow oscillation.

ACF and ACC test dusts are representative of a wide spectrum of dusty environments and they will generate extensive wear on metal surfaces. However, these dusts do not represent all environmental dusts that shorten filter life. There needs to be an added component of "soot" or submicrometer carbon or carbonyl aerosol. Such dust has been used for spot testing in air conditioning filtration but not for penetration testing. A problem with such a test dust is maintaining discrete dispersion of the carbon without extensive agglomeration.

RECOMMENDATIONS

The SAE J726C standard is due for some revision. Here are some specific areas. First, a more universal test dust should be incorporated into the standard.

The dust should have the following properties:

1. Be homogenous in chemistry throughout the particle size distribution.

2. Be uniform in particle shape throughout the particle size distribution.

3. Be readily dispersible and stable in dust concentration for the duration of the test requirements.

4. Be normally distributed in particle size about a mean size, such that for any particle parameter of measurement, i.e., number, volume, dynamic shape the size is characterized by a mean size and a standard deviation.

Second, the standard should develop a test technique for establishing fractional efficiency in the range of 1 to 10 μm particle size. This would provide information on the relative penetration of critical particle sizes that damage precision engine parts and gas turbine surfaces.

Third, other tests simulating typical environmental stress to the air cleaner should be considered.

ACKNOWLEDGMENTS

The authors wish to express appreciation to Charlie O. Reinhart of the Donaldson Co. for his review suggestions for this paper.

REFERENCES

[1] SAE Recommended Practices 24.58, Air Cleaner Test Code-SAE J726, May 1981.

[2] Drob, F. T., "Survey of Dust Problems in Tank Automotive Equipment", Detroit Arsenal Report #4242, Center Line Michigan, Nov. 1958.

[3] Smyrloglou, E. I., "A Study of the Shape of AC Fine Test Dust", Fluid Power Testing Symposium, Stillwater, Oklahoma, August 1976.

[4] Reinhart, C. O. and Weisert, L. E., "Measurement of Engine Air Cleaner Efficiency using Airborne Particle Size Analysis", SAE Technical Paper Series #831262, Milwaukee, Wisconsin, Sept 1983.

Applications and Testing:
Protection of the Environment

John D. McKenna and Leighton H. Haley, Jr.

INDUSTRIAL FABRIC FILTER BAG TEST METHODS: USEFULNESS
AND LIMITATIONS

REFERENCE: McKenna, J.D., and Haley, Jr.,
L.H., "Industrial Fabric Filter Bag Test
Methods: Usefulness and Limitations," Fluid
Filtration: Gas, Volume I, ASTM STP 975, R.R.
Raber, Ed., American Society for Testing and
Materials, Philadelphia, 1986.

ABSTRACT: Troubleshooting of a malfunctioning
baghouse can often be cost-effectively guided
by laboratory testing of the filter bags.
Quality assurance programs of fabric can be
successfully and fruitfully conducted in the
lab. Bags made from different fabric candidates
can be given long term, side by side, in plant
comparisons, by periodic laboratory tests.

The recent recognition of the above and the increased use of fabric and bag lab testing has
led to some controversy over the usefulness and
limitations of the lab as a tool for better
baghouse operation.

In the paper which follows a description of the
tests employed, a profile of a few case histories and a discussion of the usefulness and
limitations of lab testing is presented.

KEYWORDS: baghouse, troubleshooting, fabric,
test methods, fabric filter

FABRIC TEST METHODS COMMONLY EMPLOYED

Laboratory tests are often employed by the fabric
filter industry as a diagnostic aid for baghouse

Messrs. McKenna and Haley were at the time of this
writing President and Lab Manager, respectively, at ETS,
Inc., 3140 Chaparral Dr, Suite C-103, Roanoke, VA 24018.
Mr. Haley has since changed affiliation and is currently
with Norfolk Southern Corporation, Alexandria, VA.

troubleshooting, for monitoring or predicting bag life, for ensuring bag quality through quality assurance programs and for new fabric development. The following describes the most commonly used fabric test methods.

Permeability

The effectiveness of fabric filters is greatly affected by the porosity of the fabric employed. If the porosity is too high the fabric may not collect the smaller size fractions of the particulate and if the porosity is too low, the operational power requirements can become prohibitive. The porosity is measured by the air permeability test (ASTM D757-75) which measures the

rate of gas flow, at a specified pressure differential, through a piece of fabric. The pressure differential across the fabric is generally specified to be 12.7mm (0.5 in.) of water.

The test equipment consists of a fan that draws air through a known area, a vertical manometer for measuring the flow rate through the fabric, a clamp for holding the specimen, a method for adjusting the fan speed and nozzles of various diameters. After verifying the operation of the machine using a calibrated test plate, the specimen is clamped in the machine and the fan speed adjusted to provide a pressure differential of 12.7 mm (0.5 in.) across the fabric. The vertical manometer is read and the flow rate through the fabric is determined from tables supplied by the manufacturer.

This procedure can be performed on dirty fabrics to determine if the gas flow channels have been restricted or plugged by the dust particles or if the fabric structure has opened up thereby allowing dust to bleed through. Also, the effectiveness of various cleaning methods can be checked by this method and compared to new fabric test data.

Tensile Strength

The tensile test determines the breaking load and or elongation properties of fabric filters. A specimen of known size is clamped between the jaws of a machine so that the test direction is as nearly parallel to the direction of the load application as possible. The force required to break the sample is recorded by a suitable maximum recording device.

ASTM D1682-64 allows for the use of three different testing machines for this procedure: (1) the constant rate of extension machine (CRE) allows the rate of increase of the specimen length to remain constant over time, (2) the constant rate of traverse machine (CRT) allows the rate of increase of the load to be dependent on the extension characteristics of the fabric and (3) the constant rate of load machine (CRL) applies the increase of load uniformly over time after the first three seconds. This procedure recommends that all of the machines be operated at speeds that ensure breaking the sample within 20 ± 3 seconds. However, the results obtained may vary depending upon the machine used and if there is a controversy, ASTM states that results from the CRE shall prevail. Without knowledge of which machine and test parameters have been used, results from different laboratories can loose meaning.

Mullen Burst

The Mullen Burst test (ASTM D3786) measures the pressure required to rupture a fabric specimen by a force applied perpendicular to the plane of the fabric. A fabric sample is placed between the jaws of the burst machine and pressure applied by a diaphragm at a constant rate until the diaphragm bursts through the fabric. The pressure required for bursting is measured by a maximum recording gage. The proper overall operation of the machine should be periodically verified by the use of aluminum sheets having known bursting values.

The Mullen Burst results coupled with the tensile strength can provide an indication of the strength of the fabric and its ability to withstand the rigors of various cleaning methods.

Flex Test

The MIT Flex test (ASTM D2176) is a method originally designed for testing paper products but has been adopted by the fabric filter industry. This test provides insight (especially for glass fabrics) as to the ability of various fabrics and weaves to withstand self abrasion. A strip of fabric 12.7 mm (0.5 in.) wide is placed between the jaws of the machine so that the test direction is as nearly perpendicular to the jaws as possible. A 1.8 kg (4 lb.) load is applied through the upper jaw. A spring is placed between the jaw and load

to dampen the vibrations and to ensure a uniform load.
The sample is flexed through an angle of 3/2 π radians
(270 degrees) until failure.

The results vary from sample to sample and can be
affected by differences in test machines. The relative
humidity of the testing room reportedly affects the
results and a compensating equation has been proposed
for glass fabrics [1]. One drawback to the test is the
fact that it cannot be conducted in the fabrics use
atmosphere or with a continuous dust loading. However,
this test is sometimes performed on glass fabrics after
exposure to heat and or acid conditions.

Efficiency Testing

The filtration efficiency of a particular fabric
can be determined for a specific application. ASTM
D2986 describes a method for evaluating the efficiency
of particle collecting filters used in air assay work;
however, this procedure is not generally used by the
industrial fabric filter manufactures or testing
facilities. Generally the fabric filter is tested using
a specific dust and the testing procedures vary according to the testing agency. A typical procedure for
testing filter efficiency follows.

A sample of the test fabric is placed in a filter
holder and inserted into a gas stream. A given quantity
of dust of known particle size or distribution is slowly
added to the gas stream. Any dust passing through the
test fabric is collected by a back up filter. From the
initial weight of the dust and the weight of the dust on
the backup filter, the efficiency for a particular
fabric/dust system can be determined by Equation 1.

$$E = \frac{I-F}{I} \times 100 \tag{1}$$

Where:
    E = Efficiency
    I = Initial Dust Weight
    F = Dust Weight from the Backup Filter

Microscopic Examination

Microscopic examination of filter media, both new
and used, has proven to be beneficial. New fabrics,
particularly glass, can be examined to determine the
degree of coverage of a coating, the quality of any

texturization process and the effects of heat treatments in setting the crimp. The examination of used fabric filter media can reveal the size and shape of the collected particulate. One can also gain insight as to whether the particulate is being collected on the fabric surface or is penetrating into or through the fabric.

Chemical Analysis

Fabric filter media have been known to lose strength more quickly than anticipated thereby increasing operating costs due to bag replacement. These failures are frequently caused by excessive baghouse temperature excursions or process changes which generate dust types or gases which are detrimental to the fabric filter employed. Chemical analysis of the fabric can often distinguish the type of degradation, either chemical or thermal, that has affected the fabric. Also, solubility parameters and infrared analysis have proven useful in ensuring that the end user is receiving the type of fabric filter desired.

Visual Bag Inspection

Fabric filter bags can be visually examined in the laboratory to help determine the mode of failure of a bag or to identify potential problems. Bags are generally inspected for holes, rips or tears, areas that have been abraded and the condition of the collection and non-collection side of the bag. The spacing of the cage impressions on pulse jet bags can reveal if the bag and cage fit together properly. In reverse air applications, the flex lines can indicate improper tensioning. Dust found on the non-collection side may indicate dust bleed through, leakage around the tubesheet or holes in other bags. The condition of the dust cake could indicate moisture problems which may lead to increased pressure drops and increased operating costs. The visual inspection may prove to be the most useful test in an overall troubleshooting program.

Laboratory procedures other than those described above are commonly employed in quality assurance programs. Included here would be such tests as weight, construction, fabric thickness, twist and count. Chemical analysis of the particulate and gas in contact with the fabric filter along with particle size data help provide a total picture of the filters environment when trying to troubleshoot baghouse problems.

LAB TESTING OBJECTIVES

The purpose of testing either filter media or filter bags in the lab can be any of the following:

1. New Media Development
2. Fabric Quality Assurance
3. Bag Quality Assurance
4. Baghouse Troubleshooting
5. Filter Media Screening and Selection
6. Bag Life Forecasting

The use of lab analysis is almost invariably an initial part of a new fabric product development. Fabric testing is used here to determine and optimize the physical characteristics of new materials. Lab tests also provide a basis for comparison between new and state-of-the-art commercial fabrics.

Quality assurance programs are normally associated with the acquisition of a new bag set. While most bag manufacturers will replace faulty bags, within the warranty period, they normally only absorb the material cost of the bag replacement, thus the user is liable for any removal and installation costs, plus any loss of production. A new bag set quality assurance program is thus often economically very justifiable.

Since the bag is the heart of the baghouse, it is most often involved in the baghouse malfunctioning. Bag testing is often a low cost way to initially assess the nature of the baghouse problem. A standard battery of tests can usually be conducted on the bag for a few hundred dollars, whereas a plant visitation by either an "in-house" company baghouse expert or an outside consultant can often cost thousands. Outputs of this preliminary analysis include bag strength and flow loss assessment, and also bag coating loss. Subsequent microscopic and chemical analysis often provide further information aiding in the problem definition.

Along these same lines, a program of periodically removing and testing bags from an on-stream baghouse is of merit. Fabric strength monitoring is generally used to aid in the determination of rate of deterioration of bags, and to identify any major change that has occurred to the bag set as a result of a past system upset. This information can act as an alert to solve a system problem before the bag set becomes permanently destroyed.

Anticipating the end of bag life has proven useful in some large systems where wholesale bag changeout would create a long downtime and should be scheduled in advance. Fabric monitoring is also utilized when bag life guarantees are at stake.

Long term testing programs of a number of fabric candidates by periodically removing and lab testing bags made of alternative materials has demonstrated merit for screening and selection purposes. While the use of the lab tests for this purpose appears to be fairly straight forward, the use of the lab for forecasting bag life has proven to be much trickier.

QUALITY ASSURANCE PROGRAMS

In order to insure that the proper bags and related components are obtained, it is necessary to first develop a complete and detailed specification of all important parameters and then incorporate this specification into a purchase order. Once this has been done, the buyer must then implement a testing and inspection program which determines that the specification is met. Without such a procedure, the buyer has no assurance that he is not receiving an inferior bag or even a product rejected by another buyer.

Once a bag specification has been chosen, the normal procedure is to have the bag supplier purchase a sufficient amount of fabric to make the bags. This material, when received at the bag fabricating plant, should be tagged to designate its specific order number. Samples are then taken from this designated lot according to the pre-determined sampling plan as defined by the quality assurance program. These samples are shipped to the fabric testing laboratory where they are subjected to a battery of tests, which may include tensile, flex, burst, permeability and finish. Based on the results of the fabric tests vs. acceptable limits, the fabric lot is either rejected, subjected to additional tests, or accepted for bag fabrication.

The objective of a quality assurance (QA) program is to verify that the bag construction meets the specified quality. Since all major bag manufacturers have their own quality control procedures, these can be used or modified to fit the bag user's specific requirements. A QA program would most likely include reviewing the bag fabricators quality control records and monitoring the production line on site while the bags are being fabricated.

Specific items that should be checked in a bag QA program are the following:

General quality and workmanship of the bag - particularly the sewing and seams.

Fabricated length and diameter of the bags - they should be large enough as to fit into place without stretching, but not too long to cause them to fold over at the bottom.

Length under tension - large bags covered with dust will weigh as much as fifty pounds or more. Bags that stretch too much under tension will fold and cause creases that will eventually cause holes.

Cuff to thimble and cap mate - bags should be checked to see if the cuff fits securely over the thimble and also to see if cuff-to-cap connections are secure and neat.

COMPONENT QUALITY ASSURANCE PROGRAM

A component QA program would have similar design procedures as a fabric and bag QA program. In this case, the components such as caps, rings, bands, or cages should be randomly checked to see if any flaws exist. A QA program should include checking the bag cages for roundness, welds on rings and cages for sharp edges that could tear the fabric, snap bands to see if they fit snugly in place, and caps to see if they are round and do not have sharp edges. Other components such as venturis, valves, diffusers, etc., should also be checked if they are to be employed in the baghouse.

The tests which can be incorporated in the fabric quality assurance testing are shown in Table 1. The frequency of sampling and testing depends upon the fabric production process and the quantity of cloth being produced and purchased. If the fabric is produced or coated in batches then the sampling and testing frequency will have to address this. Three levels of testing are shown with a few key parameters being tested on all samples and all tests being performed only on the fewest number of samples. One employed level frequency is to do Level 1 on 10% of the samples. Level 2 on 25% of the samples and Level 3 on all the remaining samples, another frequency level is Level 1 once on each production batch, Level 2 on 10% of the samples and Level 3 on 25% of the samples.

For planning and scheduling purposes, it should be noted that it is possible that at least a portion of the fabric does not meet specification, and therefore production and installation schedules should have some time built in to accommodate the delay such an occurrence causes.

TROUBLESHOOTING

Lab testing has proven to be a useful tool for troubleshooting.

In the case of pressure drop problems it is usually possible to distinguish between ineffective cleaning and bag blinding. In the case of poor bag life, lab analysis can distinguish between mechanical, chemical and thermal attack. The lab data, while it sometimes may be sufficient to allow one to identify and solve a problem, more often is used in conjunction with field data or helps to identify what field information is needed to troubleshoot the problem.

It should be noted that one major limitation of lab results for flow and strength is that direct correlation between field and lab numbers has not yet been established. For example, it is not possible to take lab permeability data as currently measured with a Frazier machine and indicate what the absolute values of the bag pressure drop will be in the field. It is possible; however, to get a good indication of whether or not there is excessive pressure drop and whether it is due to bag blinding or ineffective cleaning. In order to help clarify the usefulness and limitations of lab testing in troubleshooting a few case histories follow.

Case History #1 - Nomex Felt - Collecting Clay Emissions

As indicated in Table 2, a pulse jet baghouse was applied to clay emissions coming off of a Nichols Hershoff furnace. The gas stream from the furnace was cooled to 400° F by bleeding in ambient air prior to the baghouse. The bags were 5 inches in diameter, eight feet long and made of 14 oz. Nomex felt. A set of bags suddenly failed and had to be replaced. Three bags were sent to the lab for testing. The objective of this investigation was to determine the cause of failure so that the problem could be avoided in the future if possible. The laboratory conducted Mullen Burst and Tensile strength tests and these indicated greater than 90% loss in strength. No signs of heavy abrasion or wear were present and the felt appeared to be completely

intact.   It was, however, embrittled and much less
flexible than the new felt had been.   It was then
decided to conduct an "Amine Ends" test which would
help clarify whether the felt had suffered acid or
thermal attack [Note 1].   The plant personnel had ruled
out thermal attack.   The furnace was gas fired and
preliminary interrogation of the plant personnel had not
turned up any significant acid gas source.   The "Amine
End" test indicated that the fibers had seen excess
temperature and no acid attack.   This was reported back
to the plant which first rejected the data but when the
tests were repeated on additional bags and confirmed the
original results, the plant then went back into the
maintenance records and discovered that the bleed in
damper had stuck in a partially closed position for a
few hours during one night shift.   Temperature records
also indicated that the bags had seen temperatures
greater than 600°F.   The Nomex maximum temperature limit
for continuous operation being 425°F.

Case History #2 - Baghouse System Upgrade

   The fabric finishing division of a Fortune 500
Company has a baghouse operating on a coal-fired
boiler.   They have been plagued with high system
pressure drop.   The causes for excessive pressure drop
are numerous including insufficient bag cleaning, poor
coal quality, improper boiler operation and at full load
conditions, excessive gas-to-cloth ratio.   In an effort
to alleviate the high pressure drop, improvements were
made to the cleaning system.   Bag tests (permeability
profiles) were utilized to determine the effectiveness
of these modifications.

   Alternate fabrics were being screened to determine
their capability to operate effectively at extremely
high gas-to-cloth ratios (>8/1).   Their performance was
being tracked by periodic permeability profiles and
scanning electron microscopic analysis.   Permeabilities
provide indication of the clean-down potential of the
fabric while the SEM enables one to explore the interstices of the fabric, determine the extent of dust
penetration and provide amplification on its clean-down
potential.

   As a result of the combined field and lab effort,
the system is now able to operate at full boiler load
conditions.

Note 1:   This test was conducted by the fabric manufacturer.   The test employs solubility and IR parameters to
indicate the polymer chain breakdown resulting from
thermal rather than acid attack.

Case History #3 - Troubleshooting - Pressure Drop

As profiled in Table 3, this case history involved a shaker baghouse applied to particulate emission control of a furnace. The baghouse is preceded by a chemical injection system to provide gaseous emission control. The gas stream leaves the furnace at about 1200°F and after a number of cooling steps the gas is delivered at 140°F to the baghouse. Excessive pressure drop across the bags led the plant to undertake bag analysis. Initial work indicated that the cake was not being cleaned off the bags. A review of the overall system led to the suspicion that condensation and crystal formation might occur. In order to confirm this, microscopic analysis of the filter medium cross - section was conducted. Photomicrographs clearly showed the presence of crystals in the interstices of the fabric (see Figure 1). These particles are different from much of the dust collected on the surface of the bags. This discovery now provides the insight needed to understand the fundamental cause of the pressure drop problem. Other testing which helped in this analysis in addition to the microscopic analysis included permeability, chemical analysis and visual inspection of the bags.

Other relevant case histories can be found in an earlier publication [2].

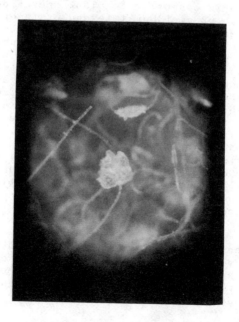

FIGURE 1
CRYSTAL FORMATION WITHIN THE FABRIC

CONCLUSIONS

While the fabric filter baghouse market has not been in a dynamic growth mode the last few years, the promise of future growth has led to the ongoing development of numerous new filter media. This fact combined with the use of baghouses in new and often difficult applications has expanded the need for laboratory techniques which evaluate the fabrics condition and characterize its flow and strength. The laboratory when properly employed can serve as a relatively low cost tool to be employed when troubleshooting a baghouse. Along these same lines lab testing of fabrics for preliminary screening of new fabrics has grown as more and more fabrics are developed. This serves the purpose of providing data for preliminary rejection or fabric improvement before the more costly field tests are performed.

In the case of large baghouses with thousands of bags and in the case of continuous plant operation where proper baghouse operation is critical for maintaining plant production, the use of bag quality assurance programs, wherein the lab testing is a major ingredient, are becoming more and more essential.

Finally, there is one additional need for lab testing which has increased in some industries where the baghouse operation is essential to plant operation. This is the need to predict bag life and/or functionality. In the areas of new fabric screening, quality assurance and troubleshooting, there are numerous successful case histories which demonstrate the usefulness of lab bag and fabric testing. This is not true however when it comes to the use of lab testing as a tool for making predictions and forecasts. In fact, while there have been some successes reported, there have also been notable failures. Perhaps the data base is often not yet in place to allow for forecasting. More long term monitoring programs such as that reported by Ohio Edison personnel [3] are needed to provide the trend data. Certainly proper forecasting cannot be based on a single data point or a snap shot in time. Long term trends need to be established if we are to use the test methods currently available.

The usefulness of fabric testing labs to 1.) provide an independent verification that fabric and bag specifications have been met, 2.) aid in baghouse troubleshooting and, 3.) as an R & D tool has been well established; however, the use of the lab as a vehicle for forecasting bag life has not yet been firmly

established.   The data base is limited and in some
instances has been misapplied.   The forecasting process
is still too much of an art rather than a science.

Limitations of fabric testing include the areas of
lab methods, reporting and data spread and interpretation.

Regarding lab methods, the ASTM Standards in some
cases are not explicit enough.   The fact that they have
been originally developed for another purpose compounds
this.

With respect to lab reporting, it is extremely
difficult to compare the data from one lab to another
because of differences in the test procedural details,
test machinery and in some cases lab conditions.   In
order to make this more concrete, specific tests are
mentioned here.   Firstly, in the case of Mullen Burst
testing, it is noted that it is often important to note
which fabric face is placed up during the test i.e. in
the case of clean woven fabric and even in the case of
used felts, it should be noted if the fabric is tested
"collection side" up or down.   Table 4 shows the
possible impact.   In the case of permeabilities of dirty
fabrics, there is a need to unify and quantify any
cleaning procedures employed.   For example, if the
fabric is vacuumed, the method and pressure employed
should be measured and described.   With respect to
tensile and flex testing more detail needs to be
specified with regards to the sample preparation.

Regarding data interpretation, when it comes to bag
life projection this cannot be done using data for a
single point in time.   One must have data over an
extended period so that trends can be established and
even then it is very difficult.   Unfortunately, a few
incorrect forecasts have been made based on insufficient
data; thus, all labs have lost some credibility.

In order that the fabric testing labs better serve
the roles discussed, it appears that the following needs
should be met:

1.   More explicit test methods and reporting
     formats need to be established.

2.   Testing of used and soiled fabrics needs to be
     addressed.

3.   Data spreads for generic fabric types need to
     be established.

4. Development of additional methods aimed at forecasting bag life should be undertaken.

5. Further study should be devoted to how many bags need to be tested in order to provide representative data for the entire baghouse unit, especially in cases of very large modular units.

The above needs are much easier to identify than to meet but the effort required to meet them is certainly economically justifiable and worthwhile.

TABLE 1--Testing Level

	1	2	3
# Samples Tested	1	6	15
Weight	X	X	X
Thickness	X	X	
Tensile Strength	X		
Machine	X	X	
Across	X	X	
Mullen Burst	X	X	X
Permeability	X	X	
MIT Flex	X	X	
Machine	X	X	
Across	X	X	
Microscopic Exam	X	X	
Scrim			
Weight	X		
Weave	X		
Construction	X		
Fiber	X		

TABLE 2--Case History 1
Nomex Felt - Collecting Clay Dust Emissions

Baghouse Type:	Pulse Jet
Source:	Nichols Hershoff Furnace
Dust:	Kaolin Clay
Volume:	35,000 ACFM
Temperature:	400° F
Bag Dimensions:	5" dia. 8' long
Fabric:	14 oz. Nomex Felt
Problem:	Bag Failure
Gas/Cloth:	3.7/1

TABLE 3--Case History 3
Shaker Felt - Furnace Particulate, and Gaseous Emissions

Baghouse Type	-	Shaker
Source	-	Furnace
Dust	-	Silica and Limestone
Volume	-	100,000 ACFM
Temperature	-	140° F
Fabric	-	Polyester Felt
Problem	-	Pressure Drop
Gas/Cloth	-	2/1

TABLE 4—Comparison of Burst
Unused Woven Glass Fabric Warp vs. Fill Up

Burst lbs/in² Warp Side Up		Burst lb/in² Fill Side Up	
463	497	353	492
482	492	456	455
490	531	353	390
470	498	526	459
533	517	427	435
494	532	446	481
489	464	507	462
525	498	470	434
550	499	469	456
483	468	498	468
523	516	445	438
536	510	403	435
443	494	454	438
566	475	395	442
507		444	
High	566	High	526
Low	463	Low	353
Mean	502	Mean	446

Values are for the same 29 fabric specimens tested
on Mullen Burst in one case sample positioned warp side
up and in one case Fill side up.

REFERENCES

[1] Boudreau, R. J., Hoffman, W., McCluskey, S. C.,
"The Manufacture and Quality Assurance Testing at
Fiberglass Filter Fabrics," Proceedings: First
Conference of Fabric Filter Technology for
Coal-fired Power Plants, EPRI Symposium, Denver,
Co, 1981.

[2] Mycock, John C., Ross, John, Greiner, Gary, "Lab
Analysis as a Tool for Improving Baghouse Operation
and Maintenance," ETS, Incorporated, Roanoke,
Virginia.

[3] Boley, Vaughn V., Dulovich, Joseph R., "Experience
with Solving Pressure Drop Problems on Four
Identical 754,000 A.C.F.M. Fabric Filters," Annual
APCA Meeting, June, 1985.

Larry G. Felix, Dan V. Giovanni, Robert P. Gehri, and Wallace B. Smith

THE NEED FOR STANDARD TEST METHODS FOR FIBERGLASS FABRICS USED ON
UTILITY COAL-FIRED BOILERS

REFERENCE: Felix, L. G., Giovanni, D. V., Gehri, R. P., and
Smith, W. B., "The Need for Standard Test Methods for Fiber-
glass Fabrics Used on Utility Coal-Fired Boilers," Fluid
Filtration: Gas, Volume I, ASTM STP 975, R. R. Raber, Ed.,
American Society for Testing and Materials, Philadelphia,
1986

ABSTRACT: Baghouses are no longer new to the utility industry. As bags need to be replaced and the causes of bag failures determined, utility baghouse users have needed to learn
more about fiberglass filter bag technology so they can write
intelligent bid specifications, determine quality control
procedures, and learn how to test new and used fiberglass
filter fabrics. It has been found that no standard test
methods exist to characterize new or used finished fiberglass
fabrics used for filtration. The development of these standards for new finished fiberglass fabrics is urgently needed.
The development of proper standard test methods for used
fiberglass filtration fabrics is contingent on an understanding of the mechanisms which cause fiberglass filter media to
fail while in service.

KEYWORDS: standard test methods, woven fiberglass
fabrics, fabric filtration, electric power generation, air
pollution control

INTRODUCTION

There are three air pollution control technologies available to the
electric utility industry: electrostatic precipitators, wet scrubbers,

Mr. Felix is a Senior Physicist, and Dr. Smith is Associate Director of
Environmental Sciences Research, Southern Research Institute, P. O. Box
55305, Birmingham, AL 35255-5305. Mr. Gehri is a Senior Research
Specialist, Southern Company Services, Inc., P. O. Box 2625,
Birmingham, AL 35202. Mr. Giovanni is a consultant with Electric Power
Technologies, Inc., P. O. Box 5560, Berkeley, CA 94705.

and baghouses. Of the three, baghouses are generally recognized as having the highest overall collection efficiency, as well as an insensitivity to the resistivity of coal fly ash, which limits the effectiveness of electrostatic precipitation [1]. Wet scrubbers are not suitable for the control of particulate emissions, although they have been used after baghouses and electrostatic precipitators for flue gas desulfurization.

As the name implies, a baghouse is a collection of compartments of bags which filter particles from flue gas. Pressure drop across a baghouse is controlled by periodically cleaning the collected dust or ash from each compartment of filter bags. During cleaning, compartments are isolated from the inlet flow of dirty gas, and the bags are cleaned by a reversed flow of clean gas, which collapses the bags, or by shaking.

From a utility perspective, baghouses also offer the attractive capability of flue gas desulfurization by dry scrubbing. In this process, powdered calcium or sodium-based sorbents are injected into the flue gas upstream from the baghouse. SO_2 may be removed by sorbents in the gas stream or in the dustcake. Pilot and full-scale demonstrations have shown that some dry-scrubbing processes can remove almost all of the sulfur oxide emissions from a typical utility boiler [2].

Baghouses were first used in the domestic utility industry in 1973, at the Pennsylvania Power & Light Company's Sunbury Station. Since then, as of October 1985, 121 more utility baghouses have been installed, of which 107 are in operation. These 107 baghouses are associated with utility boilers representing 16,607 MW of generating capacity. Another 11 units utilizing baghouses are on order, which will add another 4380 MW and bring the total to 20,987 MW [3].

Utility installations using baghouses range from 6 MW to 820 MW. To give an idea of the number of bags that are involved in utility baghouses, the 6 MW baghouse at Marshall Municipal Utilities in Marshall, Missouri has four compartments with 120 bags per compartment, for a total of 480 bags. The two 820 MW boilers at the Intermountain Power Project's Delta, Utah station each have 48 compartments with 396 bags per compartment, for a total of 19,008 bags per boiler [3]. At a typical cost of 50 dollars per bag, the two 820 MW units will have 1.9 million dollars invested in bags alone.

Filter bags for utility baghouses (nominally 6.1m x 0.2m, or 9.1m x 0.3m) are almost exclusively made from woven fiberglass fabrics. As utility baghouses operate between 80 and 200°C, fiberglass is the only material which can withstand the hot flue gas environment for extended time periods. Since the first utility baghouses were installed, woven fiberglass fabrics have been the only filter media which satisfy this criterion and which are economically practical. The carbon black and the cement industries both have used baghouses for decades with fiberglass bags.

Over 99% of all continuous fiberglass yarns are made from E, or electrical grade glass. This glass is formulated to maintain a high electrical resistivity at elevated temperatures, as its first use was for temperature resistant insulation for fine wires in the 1930's. Today E glass has become the standard for most uses, including filter media. Two other types of fiberglass are made: C glass and S glass. C glass is more resistant to chemical attack than E glass and S glass is a very expensive high strength glass. C glass is about 10% cheaper to produce than E glass [4]. Since E glass has less resistance to acid attack, C glass might represent a better choice for applications where bags could be exposed to an acidic environment.

Fiberglass can deteriorate in an industrial flue gas environment by mechanisms other than acid attack. Unless a fiberglass fabric is coated with some lubricant it will rapidly fail from self-abrasion. Therefore, before these fabrics could be used in the carbon black, cement, or utility industries; temperature and chemical resistant, lubricating coatings had to be developed. The first finishes were combinations of silicone oils and graphite. Later finishes incorporated Teflon®, and in 1974 an all Teflon® finish was patented by the duPont Corporation [5]. This "Teflon® B" finish remains one of the most popular and durable finishes for fiberglass filtration fabrics. One other basic finish has evolved from the original silicone graphite finish. This is the acid resistant finish which was originally developed for carbon black baghouses. This class of finishes combines acid resistant polymers, Teflon®, graphite, and silicone oils into a finish which encapsulates the fiberglass yarns and provides a temperature resistant lubricating coating. Newer finishes based on the acid resistant family of finishes are currently being introduced. Finishes are effective in retarding the failure of fiberglass in flue gas environments. Bag lives of over five years have been recorded at a number of utility baghouses [1].

Eventually bags need to be replaced, usually for at least one of the following reasons: they have failed prematurely due to some mechanical, chemical, or thermal degradation; they have been in service so long that the rate of bag failures has become excessive; or because high pressure drops threaten to limit the unit load. Whatever the reason, utilities with baghouses find that sooner or later they must procure new bags. If the bags that were originally installed were acceptable, then the utility will probably want new bags which are as similar to the old bags as possible. If there were problems with bag failures then a different, better bag would be desired. Regardless, the utility has to decide what properties it wants in new bags, design a bid specification, establish quality control guidelines for bag acceptance, let bids, and procure and install the new bags. In addition, if the new bags are intended to replace bags which have failed prematurely, the utility will want to determine the reason(s) for failure before the new bags are ordered, if possible.

Utilities which have been through the process of buying replacement bags find it very confusing. This is partly because there are no industry wide standards for bid specifications, or published quality

control guidelines. Also, power plant engineers have had little need
to understand textile technology until recently. The confusion is
compounded when bags fail and analyses are required to determine the
cause(s) of failure. The lack of standard test methods to facilitate
such analyses is frustrating.

Through the Electric Power Research Institute (EPRI), an organization funded by member utilities which sponsors research on issues of
interest to the utility community, a program has been initiated to
address the problems discussed above. The scope of this program will
be reviewed below.

TEST METHODS CURRENTLY USED TO CHARACTERIZE FIBERGLASS FILTER FABRICS

Because fiberglass fabrics are used in many industrial applications, standard test procedures have been developed to maintain quality
control in manufacturing fiberglass yarns and fabrics. Some of these
test methods are specific to a particular company while others are ASTM
(American Society of Testing and Materials) or FTM (Federal Test
Method) standards. In some cases, an ASTM or FTM standard which was
developed for another material has been used for fiberglass cloth. In
general, however, the fiberglass yarn producers and weavers have a wide
variety of test methods available which suffice to insure adequate
quality control for the manufacturing of fiberglass yarns and fabrics
[6,7]. One important note is that these test methods are for greige
goods. Greige goods are fabrics that are in the loom state, i.e.
fabrics which have come directly from the loom. These test methods do
not apply to fiberglass fabrics which have been chemically or thermally
cleaned or to fiberglass fabrics which have been finished.

As noted above, fiberglass filtration fabrics must be finished with
a lubricating coating which is impervious to the effects of a hot flue
gas (80 to 200°C). Every fiberglass weaver and filter bag manufacturer
offers at least two such finishes: the "standard" Teflon® B finish and
an "acid resistant" finish [8]. However, not every bag manufacturer
finishes greige fiberglass fabrics. With the exception of DuPont's
Teflon® B finish, fiberglass finishes are proprietary developments and
little is known about the details of their chemistry and composition.

There are few standard test methods for finished fiberglass filtration fabrics. The test methods that are regularly used are those
originally developed for greige goods; and, as mentioned above, some
test methods meant for other materials [7,9]. This is partially
because the users of filtration fabrics have only recently realized the
need for proper standard methods to test finished fiberglass filtration
fabrics. Until the publication of ASTM D 4029-83, Standard Specification for Finished Woven Glass Fabrics, there was no set of standard
test methods that were designed for finished fiberglass fabrics.
Unfortunately, this standard does not specifically address filter media
nor the test methods which have come to be used for this material.

No ASTM or FTM standard test methods have been published for the evaluation of used fiberglass filtration fabrics. However, a number of test methods originally developed for greige fiberglass fabrics and for other materials have been modified for used fabrics [10]. When standard test methods are developed for used fiberglass filtration fabrics, some consideration will have to be given to the ways that these fabrics can fail, as well as to sample preparation. A bag fabric sample from a utility baghouse may differ considerably from a similar sample from a carbon black baghouse. Thus, in some cases, separate test methods will need to be written.

THE NEED FOR NEW TEST METHODS

As noted above, there appears to be a sufficient body of standard test methods to insure adequate quality control for manufacturing fiberglass yarn and cloth. There is one set of standard test procedures which does address the evaluation of finished fiberglass fabrics (ASTM D 4029-83, Standard Specification for Finished Woven Glass Fabrics). However, this standard primarily covers finished fabrics woven from electrical grade (E) glass fiber yarns that are intended for use as a reinforcing material in laminated plastics for structural use. Therefore, the finishes addressed in this ASTM standard differ considerably from any finish developed for a filtration application. There are no standard test methods which are written expressly for the evaluation of finished fiberglass filtration fabrics in either a new or used state. There is also no comprehensive standard specification which is concerned with quality control in manufacturing filter bags, although a number of existing ASTM and FTM procedures can be used for specific tests.

From the utility users perspective, the task of maintaining a baghouse is made more difficult by the absence of comprehensive, standard, test methods for finished fiberglass cloth (new and used) and for new filter bags. When new bags are ordered to meet a specification, specific test methods should exist to determine if the new bags and fiberglass cloth meet the specification. When failed bags are analyzed to determine the cause of failure, test methods should exist to enable any laboratory to obtain the same result (within some established range) from each analysis. This does not imply that each laboratory would draw the same conclusions from the same data, however.

Of course, the lack of codified test procedures does not mean that new filter bags are not tested to insure quality control or that used bag fabric is not tested to estimate remaining service life or to determine the cause of a failure. Many test procedures are used to evaluate the condition of new and used fiberglass filtration fabrics. Most of these procedures have been adapted from standard test methods used for greige fiberglass cloth while others are derived from test procedures developed for other materials. Perhaps the best example of the latter is the "MIT Flex Endurance Test" which is derived from the ASTM D 2176-69, Standard Test Method for Folding Endurance of Paper by the MIT Tester. Though the "MIT Flex" test is widely used, there are well-founded reservations about its interpretation and usefulness, especially for the evaluation of used fiberglass filtration fabrics [10,11].

Because utility baghouse users have no choice, they have had to rely on the judgment of the various independent testing laboratories which routinely test fiberglass fabrics. Since one laboratory may conduct their tests in a manner different from another laboratory, the results of tests made at two different laboratories may or may not agree [13]. This possible disagreement can lead a utility to one testing organization exclusively in order to achieve consistency.

The analysis of used bag fabric to determine the cause of a failure is made extremely difficult because the mechanisms of failure of finished fiberglass cloth (as a filter media) are not well understood, and no body of literature exists which addresses this issue. At best, only general descriptions are available [12,13]. Since the failure mechanisms are not well understood, the various tests that are used to forecast fabric life from a used bag sample should be considered questionable unless they have been proven.

The tests currently used to forecast fabric life are questionable from another standpoint. New and used fiberglass filtration fabrics are quite different in strength and appearance, and used fabric samples will differ in strength depending on where the bag the samples were obtained (fold areas will always be weaker). The test methods used to measure fabric strength that are appropriate for new fabrics may well be inappropriate for used fabrics which have been weakened by exposure to an environment which is known to be corrosive to that material. Also, used fabrics from a utility baghouse will be coated with ash and the presence of ash affects the results of these tests [12]. Vacuuming and washing used fabric samples before for testing may also affect the results of the test. Weakened fibers can be broken by vacuuming. Ashes which have a low pH will, upon wetting, produce a liquid which can corrode fiberglass. The primary mechanism is an electrophilic (H+) attack on the non-silicate structure of the glass [14]. Thus, it is possible that vacuuming and washing a fabric sample may cause further weakening. Therefore, it is conceivable that the tests used for new fabrics may not be capable of providing meaningful data for fabrics which have been in service for some time.

From the above, it is clear that there is a genuine, urgent need for the development of standard test methods which can be used for an unbiased evaluation of new finished fiberglass cloth and fiberglass bags. The same need exists for the development of test methods which will allow the determination of why a bag or fabric has failed. This latter development is contingent on an understanding of the mechanisms of failure of finished fiberglass cloth when it is used as a filter media.

RESEARCH FOR TEST METHOD DEVELOPMENT

Utilities which currently operate baghouses and those that are planning to purchase baghouses are well aware of the concerns which have been discussed above. Recently, EPRI has started joint discussions and research planning with vendors from all phases of the fiberglass industry, including bag manufacturers. In September, 1985, EPRI sponsored a Fabrics and Filterbag Workshop in Atlanta, GA, where issues related to bag fabric testing and characterization were identified and discussed. This meeting was attended by forty representatives of utilities, fiberglass manufacturers, weavers, finishers, and bag fabricators, as well as independent researchers and consultants. The key issues discussed included bag failure analysis, accelerated testing of fabrics, testing of new fabrics, bag standardization, and pilot scale bag testing. Pertinent comments from this meeting for each of these issues are listed below.

Bag Failure Analysis

It was determined that the analysis of mechanical, chemical, and thermal modes of bag failure, and the study of how the modes of failure are related, is a high priority because an understanding of failure mechanisms must precede the development of ways to assess the degree of damage. The primary reasons why filter bags fail is not understood, and work needs to be started to identify and model each failure mode. No single organization was recognized as being capable of singlehandedly performing the research alone; however, several manufacturers indicated their willingness to provide support for various aspects of the work (e.g. scanning electron microscopy, providing the time of knowledgeable scientists for data interpretation, chemical analyses, etc.).

Accelerated Fabric Testing

Once the mechanisms of bag failure are better understood, procedures should be developed to compress into a much shorter time the mechanical, chemical, and thermal degradation that would be expected to occur in fiberglass filter bags over a two to three year period. It was generally conceded that accurate predictions of bag life will be very difficult to achieve, considering the variety of flue gas chemistries, fly ash compositions, and operating conditions encountered in utility baghouses. However, a practical goal is to develop a simple ranking of fabrics and/or filter bags in order of their relative suitability for specific applications.

Testing of New Fabrics

There is a general dissatisfaction with the tests that are presently available for finished fiberglass fabrics and with the consistency of the results of such tests from different testing organizations. Someone, perhaps EPRI, should develop and publish specific

procedures (based on commonly applied test methods) which fabric manufacturers and testing firms would follow when testing NEW fabrics and
filter bags. Until failure mechanisms are better understood, it is not
considered meaningful to specify procedures for monitoring in-service
fabric deterioration or for predicting remaining bag life.

Bag Standardization

Procurement specifications now used provide ample opportunity for
significant variations in fabric and filter bag construction. There
are several areas where specifications can be "standardized" and lead
to more consistent and better quality products. Aside from standard
test methods for finished fiberglass fabrics, test methods would need
to be established for evaluating the quality of a new bag.

Pilot Scale Bag Testing

There is industry-wide support for an experimental program to test
full-scale filter bags in a pilot-scale baghouse. This suggestion is
being implemented. The performance of individual filter bags will be
monitored so that filter bags of differing design can be compared.
Samples of these filter bags will be used in the study of bag failure
mechanisms and for test method development.

Manufacturers Advisory Group

A Manufacturers Advisory Group (MAG) has been formed by EPRI. This
group includes representatives from every part of the filter bag industry, EPRI member utilities, and EPRI contractors. It will be concerned
with all of the topics described above. In part, the MAG will be
involved in guiding the development of standard test methods for new
and used fiberglass fabrics and for new filter bags. It is hoped as
the test methods are developed, a standards organization like the ASTM
would be an avenue to generating generally accepted standard test
procedures.

The MAG will also advise EPRI in the implementation of a pilot-
scale test program, as recommended above. EPRI has dedicated a four
compartment, 2.5 MWe baghouse at its Arapahoe Test Facility in Denver,
Colorado for these tests. Each compartment of 36 bags will be used to
test variations of one parameter of filter bag construction. These
tests will begin in June 1986. The first set of parameters to be
tested are fabric finish, filtration surface texturization, weave, and
yarn construction.

SUMMARY

In the utility industry, baghouses are an accepted control technology for coal-fired boilers because of their inherently high collection

efficiency. Baghouses also offer the capability of flue gas desulfurization with dry scrubbing. Compared to the other technologies available for the control of air pollution from coal-fired boilers, baghouses are a relatively recent development. As with any new technology, there is much to learn in order to optimize the performance of
baghouses. EPRI is sponsoring a large research program devoted to
developing a reliable data base of information for the performance of
baghouses on utility boilers, from which an understanding of their
behavior is emerging [1]. From this EPRI sponsored research, it has
been realized that there is a general lack of knowledge in the area of
bag fabrics, specifically fiberglass fabrics. The lack of standard
test procedures designed for new and used woven fiberglass filter media
is indicative of this lack of knowledge. With the Manufacturers
Advisory Group, EPRI intends to address the issues raised in a Fabrics
Workshop held last September. These issues include: bag failure analysis, accelerated fabric testing, testing of new fabrics, bag standardization, and pilot scale testing. One important conclusion from the
workshop was that an understanding of the mechanisms by which fiberglass fabrics fail is crucial to the development of test methods for
used bag fabrics and criteria for predicting bag life based on tests
made on these fabrics. As these issues are addressed and test methodology is developed, it is hoped that a standards organization such as
the ASTM would become involved in promulgating new standard test procedures designed specifically for woven fiberglass filter media.

ACKNOWLEDGMENT

     This paper is based upon results from research sponsored by the
Electric Power Research Institute, Palo Alto, CA, under contract number
RP-1129-8, R. C. Carr, Program Manager.

REFERENCES

[1] Carr, R. C., and Smith, W. B., "Fabric Filter Technology for
     Utility Coal-Fired Power Plants, Parts I through VI," Journal of
     the Air Pollution Control Association, Vol. 34, Nos. 1-6,
     Jan.-June 1984.
[2] Blythe, G. M. and Rhudy, R. G., "Fabric Filter Interactions in
     Spray-Dryer Based FGD," presented at the Third Conference on
     Fabric Filter Technology for Coal-Fired Power Plants, Scottsdale,
     AZ, 1985; available from the Electric Power Research Institute,
     Palo Alto, CA.
[3] Piulle, W. V., "1985 Update, Operating History and Current Status
     of Fabric Filters in the Utility Industry," presented at the Third
     Conference on Fabric Filter Technology for Coal Fired Power
     Plants, Scottsdale, AZ, 1985; available from the Electric Power
     Research Institute, Palo Alto, CA.

[4]   Loewenstein, K. L., The Manufacturing Technology of Continuous
      Glass Fibers, 2d ed., Elsevier Science Publishers B. V.,
      Amsterdam, 1983, pp. 33-39.
[5]   U. S. Patent No. 3,838,082, September 24, 1974.
[6]   Boudreau, R. J., Hoffman, W.,McCluskey, S. C., "The Manufacture
      and Quality Assurance Testing of Fiberglass Filter Fabrics," in
      Proceedings: First Conference on Fabric Filter Technology for
      Coal-Fired Power Plants, Report CS-2238, Electric Power Research
      Institute, Palo Alto, CA, 1982.
[7]   Knox, C. E., Murray, J., and Schoeck, V., "Technology of Glass
      Filter Fabric Design," in Symposium on the Transfer and
      Utilization of Particulate Control Technology: Volume 2, Fabric
      Filters and Current Trends in Control Equipment, Report
      EPA-600/7-79-044b, United States Environmental Protection Agency,
      Research Triangle Park, NC; available from the National Technical
      Information Service, Springfield, VA, Report No. PB 295227/3BE.
[8]   Grubb, W. T., "Finishing Fiberglass Filtration Fabrics," in
      Proceedings: First Conference on Fabric Filter Technology for
      Coal-Fired Power Plants, Report CS-2238, Electric Power Research
      Institute, Palo Alto, CA, 1982.
[9]   Miller, R. L., Zourides, A. D., and Budrow, W. F., "Extending
      Fiberglass Bag Life," in World Filtration Congress III, Vol. 1,
      The Filtration Society, Croydon, England, 1982, pp. 122-129.
[10] Perkins, R. P. and Westley, M. W., "Routine Testing of Fabrics,"
      presented at the 78th Annual Meeting of the Air Pollution Control
      Association, Detroit, 1985; available as Paper 85-54.3 from The
      Air Pollution Control Association, Pittsburgh.
[11] Westley, M. W. and Lau, K. W., "Processing and End-Use Performance
      of Finished Glass Fabrics for Filtration," in Proceedings: Second
      Conference on Fabric Filter Technology for Coal-Fired Power
      Plants, Report CS-3257, Electric Power Research Institute, Palo
      Alto, CA, 1983.
[12] Budrow, W., "Advancements in Woven and Felted Glass Fabrics for
      the Utility Industry," in 4th International Fabric Alternatives
      Forum Proceedings, published by American Air Filter Co., Inc.,
      Louisville, 1980.
[13] Budrow, W. F., "Durability of Textile Filter Media," Journal of
      the Air Pollution Control Association, Vol. 28, No. 5, May 1978,
      pp. 548-550.
[14] Levine, S. N. and LaCourse, W. C., "Mechanisms of Attack on
      Glasses in Aqueous Media, Interim Scientific Report, June 1 -
      September 20, 1966," State University of New York, Stony Brook;
      available from The National Technical Information Service,
      Springfield, VA, Report AD 641500.

Wayne T. Davis

FABRIC FILTER DESIGN: THE CASE OF A MISSING PARAMETER

REFERENCE: Davis, W.T., "Fabric Filter Design: The Case of a Missing Parameter," Fluid Filtration: Gas, Volume I, ASTM STP 975, R.R. Raber, Ed., American Society for Testing and Materials, Philadelphia, 1986

ABSTRACT: The design of fabric filters for removal of particles from dust-laden gas steams is dependent on the gas and dust characteristics. While the gas characteristics and the mass concentration of the dust are generally specified, the particle size distribution and other dust characteristics described in this paper are frequently unkown. The objective of this paper is to review those dust characteristics which have been found to affect the performance and to propose methods which might be suitable to obtain these missing parameters.

KEYWORDS: particle size distribution, filtration,
           permeability

INTRODUCTION

The design of fabric filters for removal of particles from gas streams is based 1) on the past experience of the designer or vendor in a particular application and 2) on specific information supplied by the purchaser in regard to the process. Frequently the information consists of the total air flow rate, the gas temperature and composition, and the dust loading or concentration. Depending on the experience of the designer, this may or may not be adequate to assure a successful design. This paper provides a brief review of the development of the equations relating pressure drop across filter media to the fundamental characteristics of the particles being filtered. One particular characteristic, K_2 -- the specific resistance coefficient, has been shown to be a key parameter in

Dr. Wayne T. Davis is a professor of Civil Engineering and Assistant Dean of the Graduate School of the University of Tennessee, Knoxville, TN, 37996-2010.

understanding the performance of the fabric filter with regard to pressure drop and filtration efficiency. Techniques for measuring and/or predicting K_2 are summarized. This parameter, which is frequently missing in the design specifications, is fundamental to the design of filter systems.

REVIEW OF FILTRATION FUNDAMENTALS

Resistance of Filters

Davies [1] has defined the following variables as being involved in the measurement or prediction of the resistance to flow through filters:

1. ΔP, the pressure loss across the filter (N/m^2)
2. Q, the flow rate (m^3/sec)
3. A, the superficial facial area of the filter (m^2)
4. h, the thickness of the filter (m)
5. μ_g, the gas viscosity [kg/(m-sec)]
6. ρ_g, the gas density (kg/m^3)
7. λ, the mean free path of gas molecules (m)
8. R, the mean radius of the fibers (m), and
9. c, the packing density of the filter (dimensionless)

The superficial face velocity or filtration velocity (V) incident upon the filter is defined as Q/A and has units of meters per second (m/sec). Historically, this velocity has been measured in feet per minute and has been referred to as the "air/cloth ratio."

Formation of independent groups of these variables and application of Buckingham's Theorem [1] yielded four dimensionless and independent groups:

$$ f \left| \frac{\Delta P A R^2}{\mu_g Q h} , \frac{Q R \rho_g}{A \mu_g} , \frac{\lambda}{R} , c \right| = 0 $$

The second independent variable is the Reynold's number (Re) of the fiber. Davies [1] has shown that for low Reynold's numbers (Re ≤ 0.3) that $\Delta P/Q$ is independent of Reynold's number. It can be shown that this condition would be met for the extreme conditions of industrial fabric filtration. The choice of a velocity of 9 x 10^{-3} m/sec (10 fpm), a fiber diameter of 20 x 10^{-6} m (20 µm), and a kinematic viscosity (μ_g/ρ_g) of 0.15 cm^2/sec, yields a Reynold's number of 0.012 for the flow around the fibers.

The third independent variable is the Knudsen number, λ/R, which is a measure of the aerodynamic slip past the fibers. This occurs at low pressures where the mean free path is significant, or at small fiber diameters in which λ and R are on the same order of magnitude. For atmospheric pressure and fiber diameters greater than 1-2 µm, the reduction in ΔP across a fiber due to slippage is less than 2% and may be assumed to be negligible.

Consideration of the relative independence of $\Delta P/Q$ on Reynold's number and Knudsen number, leads to the conclusion that Darcy's Law must obey the following equation for filters operated under the typical operating conditions of industrial fabric filters [1]:

$$\frac{\Delta P A R^2}{\mu_g Q h} = f(c) \tag{1}$$

For the condition of constant packing density and fabric parameters, equation 1 is a statement of Darcy's Law, i.e.,

$$\Delta P = k\, V \tag{2}$$

for a non-compressible filter material at low air velocities.

Empirical Relationships for Filter Resistance

Several equations have been developed--both experimentally and theoretically--to determine the function of c in Equation 1. One such equation was presented by Carman [2,3,4] in which a study of liquid filtration resulted in the Carman-Kozeny equation

$$\frac{\Delta P A}{\mu_g Q h S^2} = \frac{k}{(1-c)^3} = \frac{k}{(\varepsilon)^3} \tag{3}$$

where k is the Kozeny constant, S is the total surface area of granules per unit volume of bed, and ε is the porosity or void fraction (1-c).

Two special cases of Equation 3 are important in fabric filtration: packed beds of spheres and packed beds of filters. Equations 1 and 3 can be shown to be equivalent as follows. In dealing with beds of spheres, the area of particles per volume of particles, A_p/V_p, is

$$\frac{A_p}{V_p} = \frac{area\ particles}{volume\ particles} = \frac{4\pi R_p^2}{(4/3)\pi R_p^3} = \frac{3}{R_p} \tag{4}$$

where R_p is the radius of the spheres. Since the volume of particles/volume of bed is defined as the packing density, c, then S, the area of particles per volume of bed, can be defined as

$$\left|\frac{area\ particles}{volume\ particles}\right| \left|\frac{volume\ particles}{volume\ bed}\right| = \frac{3}{R_p}(c) = S. \tag{5}$$

Substitution of Equation 5 into Equation 3 yields

$$\frac{\Delta P A R_p^{\ 2}}{9\mu_g Q h} = \frac{k^{'}(c)^2}{(1-c)^3} = \frac{k^{'}(1-\varepsilon)^2}{(\varepsilon)^3} \qquad (6)$$

which is equivalent to Equation 1.

The above equation, which relates the pressure drop across a granular bed to bed characteristics was adapted by Williams, Hatch and Greenburg [3] in 1940 to allow this equation to be used to predict the increase in pressure drop across a filter medium as a dust cake was accumulated during filtration. The dust cake thickness, h, in Equation 6 was expressed as

$$h = \frac{W}{\rho_p(1-\varepsilon)} \qquad (7)$$

where W is the mass of dust on the filter per unit of area, and ρ_p is the particle density. The increase in h with time due to accumulation of filtered dust was incorporated by recognizing that W, frequently called the area cake density, could be calculated as follows:

$$W = C_i V t E_f \qquad (8)$$

where C_i is the dust loading (mass/volume of gas), t is the elapsed filtration time, and E_f is the fractional filtration efficiency (assumed to be 1.0 for most high efficiency filters). Substitution of Equations 7 and 8 into Equation 6 yields the often reported expression describing the pressure drop of a dust cake collected on a filter at a constant loading and velocity:

$$\Delta P = \left[\frac{k^{'}\mu_g}{\rho_p} \left(\frac{A_p}{V_p} \right)^2 \frac{1-\varepsilon}{\varepsilon^3} \right] C_i V^2 t$$

or

$$\Delta P = \left[\frac{k^{'}\mu_g}{\rho_p} \left(\frac{3}{R_p} \right)^2 \frac{1-\varepsilon}{\varepsilon^3} \right] C_i V^2 t \qquad (9)$$

The term in brackets is the specific resistance coefficient of the dust cake and is frequently referred to as K_2:

$$K_2 = \left[\frac{k^{'}\mu_g}{\rho_p} \left(\frac{3}{R_p} \right)^2 \frac{1-\varepsilon}{\varepsilon^3} \right] = \frac{\Delta P}{C_i V^2 t} \qquad (10)$$

K_2 can thus be estimated using the basic characteristics of the dust (R_p, ε, ρ_p) or measured by monitoring the change in pressure drop with time.   By constructing what is referred to as a normalized performance curve, Spaite and Walsh [4] and later Billings [5], Davis et.al. [6,7] and Dennis et al. [8] showed that K_2 could be measured directly as the slope of the curve generated from a graph of $\Delta P/V$ versus $C_i Vt$ (see Figure 1.)

Historically, K_2 has been reported in various units, as follows, leading to some confusion as the move to the International System of units (SI) has been attempted:

$$K_2 = \frac{\Delta P/V}{C_t Vt}$$

$$English \ system \ K_2 = \frac{(inches \ H_2O \ / fpm)}{(lb \ of \ dust/ft^2 \ of \ cloth)}$$

$$modified \ SI: \ K_2 = \frac{(N/m^2)/(m/min)}{(g/m^2)(m/min)(min)} = \frac{N-min}{g-m}$$

$$SI: \ K_2 = \frac{(N/m^2)/(m/s)}{(kg/m^3)(m/s)(s)} = s^{-1}$$

The reluctance to use the SI system has been the fear of losing the physical significance expressed in the English system as compared to (seconds)$^{-1}$ which doesn't relate well to the physical meaning of K_2. For comparison purposes K_2 in the English system can be obtained by multiplying K_2 in (N-min)/(g-m) by 5.97.

Experimental Measurements of K_2

Many researchers have conducted laboratory and pilot-scale fabric filter tests to measure K_2.   Billings et al. [5] reported an extensive field survey of K_2 as a function of the air/cloth ratio (filtration velocity) and particle size.   In this early work, K_2 was determined from the reported values of operating air/cloth ratio (V), dust loading (C_i), filtration time (t), and residual and maximum pressure drops (ΔP_R, ΔP_m).   While this earlier work was quantitative, the wide range of dusts, quality of reported data, configuration of the individual systems (single compartment, multiple compartment, type of cleaning, etc.) lead to considerable scatter. The relationship showed order of magnitude variatgions in K_2 at a given particle size.

In more recent tests, obtained under controlled conditions, the relationship between K_2, particle size and velocity have been shown more clearly.   According to Equation 10, K_2 should be proportional to the inverse of the particle diameter squared.   However, the porosity, ε, has been found to be diameter dependent.   Figure 2 illustrates a typical ε versus diameter relationship (Dallavalle [9]).   Note that ε is a strong function of particle diameter for

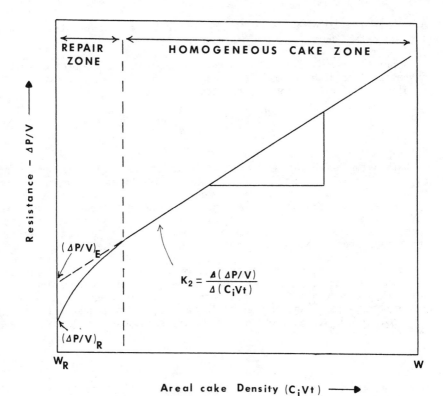

Figure 1. Normalized Performance Curve

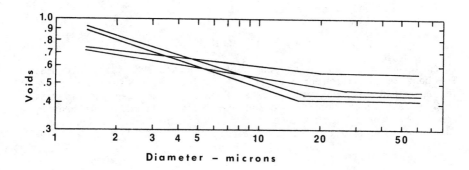

Figure 2. Void Fraction as a Function of Particle Diameter for Powders.

particles less than 10-15 μm, while for particles greater than 20 μm it is constant but dust type dependent.

A brief investigation of this data shows that the term $(1-\varepsilon)/\varepsilon^3$ is extremely sensitive to small changes in ε and is roughly proportional to diameter. In Equation 10, the net result is that

$$K_2 \qquad \frac{1}{R_p^2} \left(\frac{1-\varepsilon}{\varepsilon^3} \right) \qquad \frac{1}{d^2} (d) \qquad d^{-1}$$

Data from Dennis et al [8]. and Davis et al. [6], are shown in Figures 3 and 4, respectively.   The solid lines represent each researcher's best fit to the data where available. The data reported by Dennis et al. [8] were summarized from eight different sources for flyash, mica, and talc at 2-6 fpm: the data by Davis, et al. [6] were on talc dusts at a velocity of 4 fpm. Both sets of data clearly indicate a strong dependence of K_2 on the particle size.

It is evident from these data that velocity also has an effect on K_2, contrary to the K_2 relationship as defined in Equation 10. While this observed effect may be partially attributed to the effect of velocity on ε, and/or Reynold's number, most researchers have reported that K_2 is a function of velocity such that

$$K_2 = k V^x \tag{11}$$

Dennis et al. [8] reported that x had a value of 0.5 for flyash, and varied from 0.5-1.0. Davis et al.[11] in a series of tests on flyash using eleven different filter materials reported an average value of 0.7 for flyash.   The data in Figures 3 and 4, data by Davis et al.[10] and data by Frazier et al. [10] were normalized to a velocity of 3 fpm and replotted in Figure 5 assuming an average value of x of 0.6.   The normalized data show that there is a well defined relationship between K_2 and the particle size.

A best fit equation was determined for the data:

$$K_2 = 118.4 \, MMD^{-1.10}$$

where K_2 is measured in the English system and MMD is in microns. The best fit equation predicts the K_2 value within a factor of 2. The agreement between various sets of data is excellent considering that measurements obtained under carefully controlled laboratory conditions for a constant particle size distribution have shown factor of two variations within a single laboratory [11].

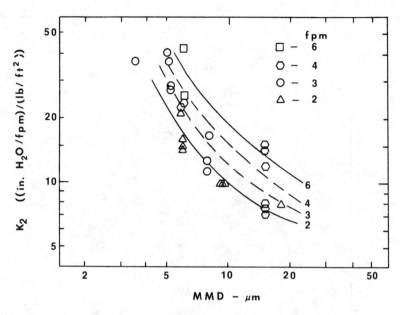

Figure 3. K₂ Versus MMD and Face Velocity [8].

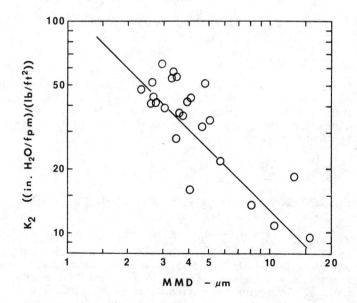

Figure 4. K₂ Versus MMD for Talc at 4 fpm [6].

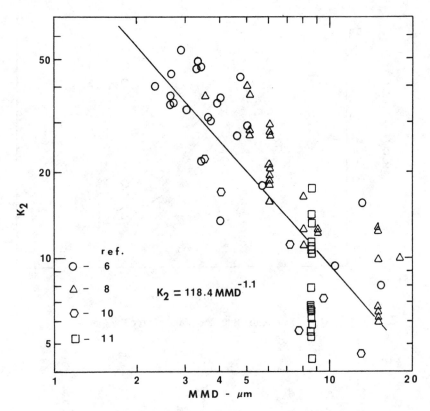

Figure 5. K_2 Normalized to 3 fpm Versus MMD.

USE OF K_2 in Modeling

Many models have been proposed for prediction of the increase in resistance across fabric filter collectors as the dust is accumulated. These have generally been categorized as either linear or non-linear models. The linear model has generally taken the following form:

$$(\frac{\Delta P}{V})_{Total} = (\frac{\Delta P}{V})_E + K_2 C_i Vt \tag{12}$$

or in terms of filter resistance (S = ΔP/V) and W (W = C_iVt)

$$S_T = S_E + K_2 W' \tag{13}$$

The areal cake density, W', is the mass of dust added during the filtration cycle and equals the total areal cake density W, minus the residual areal cake density, W_R, retained in the filter after each cleaning cycle. This linear equation describes the performance curve shown in Figure 1. The projected effective residual resistance, $(\Delta P/V)_E$ or S_E, is the value obtained by linear extrapolation of the homogeneous cake portion of the performance curve to the beginning of the filtration cycle (at the point where W has a value of W_R). If S_E is greater (less) than the actual residual resistance, the linear model will overestimate (underestimate) the filter drag at the beginning of the cycle.

Other models have incorporated a non-linear transition into the linear model to allow the user to predict the non-linear portion of the performance curve. Dennis et.al. [8] developed a model for prediction of the filter drag for flyash filtered on fiberglass fabrics. The basic equation is as follows:

$$S_T = S_R + K_2 W' + (K_R - K_2) W^* (1 - exp(-W'/W^*))$$
(14)

where

$$W^* = (S_E - S_R + K_2 W_R)/(K_R - K_2)$$
(15)

The term K_R is the initial slope of the performance curve at W_R. For large values of W', or small values of W^*, Equation 15 approaches a linear equation.

The significance of either of the modeling approaches is that one of the most critical parameters is the specific resistance coefficient, K_2. Modeling of the fabric filter system requires the ability to predict both the initial filter drag (S_E or S_R) and K_2. However, even if information is not available on S_E or S_R, the prediction or measurement of K_2 allows one to predict the rate of rise of pressure in the collector. This provides insight into the design which is not available unless K_2 is known.

METHODS FOR MEASUREMENT OF K_2

Existing Methods

The previous discussion has identified K_2 as a critical and often missing parameter which is needed in the design of fabric filter collectors. Several options are available for determining the value of K_2. These include the following:

1. Theoretical approach
2. Use of K_2 - MMD empirical data
3. Pilot-Plant measurement

The theoretical approach involves the use of equations such as Equation 10. This requires direct knowledge of the dust properties such as ρ_p, R_p, and ε. The data reported herein shows that a velocity correction is probably necessary, although part of the velocity effect may be accounted for if the effect of velocity on the packing density is considered. A more detailed discussion of the use of the theoretical approach and its accuracy has been presented by Ensor et al. [12]. In some cases it was reported to be within ±±

accuracy. In other cases it was reported to be a factor of 3.3 lower than actual measurements. It was hypothesized that the problem was not with the theory, but with the difficulty in measuring the dust properties.

The second approach is to use existing empirical data to predict the value of K_2. The relationships shown in Figures 3-5 between K_2 and MMD may be used to estimate K_2. By properly correcting for the effects of ρ_p, μ_g, and velocity effects, these data provide a reasonable method of predicting K_2. It should be noted that the data are based on a number of different bench-scale and laboratory pilot scale data. Of significant concern is the large variation in K_2 from test to test. A factor of 2-3 difference in K_2 at the same MMD is shown in these data even when conditions are carefully controlled.

The third approach is to conduct a pilot scale test to determine K_2. This approach has been taken frequently since it also provides information on the other parameters which are needed for design (i.e. $(\Delta P/V)_R$ and C_i). Pilot scale tests can be conducted either by resuspending the dust in a laboratory test collector or by conducting pilot scale tests at the site of existing processes. Several companies provide pilot scale test services for on-site evaluation.

A Proposed Fourth Option: In-situ Measurement

A fourth option is the possibility of the field measurement of K_2 in a manner analogous to the field measurement of electrical resistivity, and the particle size distribution. At the present time industrial source sampling is frequently conducted to determine the design specifications. Standard test procedures, U.S. EPA Methods 1-5 and Method 17, are utilized to extract process emissions from existing sources to characterize particulate emissions [13]. Methods 5 and 17 extract samples isokinetically and collect the mass on pre-weighed filters. The Method 5 sampling train [14] is shown in Figure 6. Method 17 is a modified Method 5 in that the filter is placed upstream of the sampling probe in the stack. A modification of the EPA Method 5 and 17 sampling procedures (by placing a cascade or inertial impactor in the sampling line) allows the operator to determine the particle size distribution. These methods are already employed to obtain two of the parameters, needed to design the fabric filter collector: dust loading, (C_i), and the mass mean diameter (MMD) of the dust.

The design of the Method 5 and Method 17 sampling train is such that the velocity through the 2-4 inch (5-10 cm) diameter filter is in the range of 3 to 30 feet per minute with values typically in the 10 to 20 fpm range. This typical range is higher than the velocity normally chosen for operation of most shaker, shake/deflate, or reverse air cleaned fabric filters. These have been reported by Carr and Smith [15] to operate   in the range of 1-3 fpm on fabric filters applied to coal-fired boilers.

It is proposed that a slight modification of the EPA sampling train could be made by redesigning the filter holder to incorporate sufficient surface area to allow a velocity of 2-5 fpm to be tested.

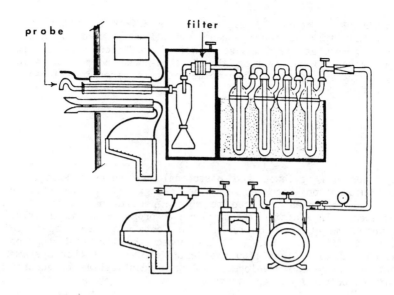

Figure 6. EPA Method 5 Samplng Train (14).

By incorporating pressure taps into the filter holder, it would then be possible to conduct an in-situ measurement of the pressure rise rate at a specified velocity while simultaneously obtaining particulate loading data (C_i). While this would not be a certified EPA Method 5 test, it would provide information on K_2 in a method analogous to the determination of the particle size distribution. By proper choice of a sampling nozzle to maintain near isokinetic sampling, the value of K_2 could be determined at or near the intended design velocity. A velocity correction could then be made using Equation 11. The advantage of this approach over the theoretical or empirical approaches is that it is a direct measure of the rise rate which incorporates the effects of ρ_p, ε, and MMD into the actual measurement. The measurement would also compensate for the effect of temperature (viscosity term in K_2) since the filter and extracted gas can be maintained at the stack temperature. In addition the test would be significantly less complex than placing a pilot plant fabric filter on-site; less information is obtained, however, since no conditioning of the filter would be conducted.

The concerns which would need to be addressed are as follows. First, the above test would have to be operated at a constant velocity in order to interpret the data. This is analogous to inertial impactor testing which is conducted at near-isokinetic conditions at a single point within the stack at a constant velocity. In both cases, fluctuations in the stack velocity would tend to decrease the accuracy of the measurement. Secondly, deposition within the nozzle and heated probe of the sampling train would need to be minimized.   Method 17 minimizes this effect by placing the filter in the stack at the upstream end of the heated extraction probe, rather than at the downstream end.   In addition consideration should be given to eliminating the "button-hook" nozzle to allow sampling directly into the filter holder with no directional changes in the air flow.

CONCLUSION

This paper has reviewed the historical and current status of specific information required in the design of fabric filter collectors.   Techniques utilized to obtain certain parameters have been reviewed.   In addition the author has proposed an alternative method which might be considered for obtaining the specific resistance coefficient, K_2.   It is hoped that this discussion provides additional insight into the design of fabric filter systems and might encourage the development of more quantitative procedures for obtaining critical design parameters.

REFERENCES

[1]   Davies, C.N.   Air Filtration, Academic Press, London and New York, 1973.

[2]   Carmen, P.C., "Fluid Flow Through Granular Beds," Transactions - Institute of Chemical Engineers, 1937.

[3]   Williams, E., T. Hatch, and L. Greenburg.   "Determination of Cloth Area for Industrial Air Filters," Heating, Piping, and Air Conditioning, April 1940.

[4]   Walsh, G.W., and P.W. Spaite.   "An Analysis of Mechanical Shaking in Air Filtration," Journal of the Air Pollution Control Association, 12:2, February 1962.

[5]   Billings, E.   Handbook of Fabric Filter Technology Volume I, Fabric Filter Systems Study, PB 200-648, NTIS, December 1970.

[6]   Davis, W.T. and R.F. Kurzynske.   "The Effect of Cyclonic Precleaners on the Pressure Drop of Fabric Filters," Filtration and Separation, Vol. 16, No.5.

[7]   Davis, W.T., P.J. LaRosa and K.E. Noll.   "The Generation and Evaluation of Fabric Filter Performance Curves from Pilot Plant Data," Filtration and Separation, November/December 1976, pp. 555-560.

[8]   Dennis, R., R.W. Cass, D.W. Cooper, R.R. Hall, V. Hampl, H.A. Klemm, J.E. Longley, and R.W. Stern.   Filtration Model for Coal Flyash With Glass Fabrics, EPA-600/7-77-084, August 1977.

[9] Dallavalle, J.M. Micromeritics, Pitman Publishing Corporation, New York, 1948.

[10] Frazier, W.F., and W.T. Davis.   "Effects of Flyash Size Distribution on the Performance of a Fiberglass Filter," The Symposium on the Transfer and Utilization of Particulate Control Technology: Vol. III Particulate Control Devices, EPA-600/9-82-0050c, July, 1982, pp. 171-180.

[11] Davis, W.T., and W.F. Frazier. "A Laboratory Comparison of the Filtration Performance of Eleven Different Fabric Filter Materials Filtering Resuspended Fly Ash."   Presented and published in the Proceedings of the 75th Annual Meeting of the Air Pollution Control Association, 1982, Paper 82-32.5.

[12] Ensor, D.S., D.W. VanOsdell, A.S. Viner, R.P. Donovan, and L.S. Hovis." Modeling Baghouse Performance," Proceedings: Fifth Symposium on the Transfer and Utilization of Particulate Control Technology, Vol. 3 EPRI CS-4404, February, 1986.

[13] Federal Register, 40 CFR Part 60 Section 60.8.

[14] APTI Course 450 Source Sampling for Particulate Pollutants, EPA 450/2-79-007, December 1979, p. 18.

[15] Carr, R.C., and W.B. Smith." Fabric Filter Technology for Utility Coal-Fired Power Plants," Journal of the Air Pollution Control Association,Vol. 34, No. 1, January 1984, pp. 79-89.

Douglas W. VanOsdell and Robert P. Donovan

ELECTROSTATIC ENHANCEMENT OF FABRIC FILTRATION

REFERENCE: VanOsdell, D. W. and Donovan, R.P., "Electrostatic
Enhancement of Fabric Filtration," Fluid Filtration: Gas, Volume
I, ASTM STP 975, R. R. Raber, Ed., American Society for Testing
and Materials, Philadelphia, 1986

ABSTRACT: Incorporating electrical forces into the design of
fabric filters provides reduced pressure drop and improved
collection efficiency, as compared with conventional practice.
Improvement has been demonstrated for a number of filter and
electrode geometries, and pilot plant operations have been successful at a number of field sites. The mechanisms whereby
electrostatic forces improve fabric filter performance include:
1) reduced dust penetration and/or retention in the depth of the
filter; 2) the formation of a dust cake having increased
porosity; and 3) non-uniform deposition of dust on a filter surface.

KEYWORDS: fabric filtration, fabric filter, electrostatic
enhancement, electrostatic stimulation

Fabric filtration is an accepted technology for the removal of
particles from industrial gas streams and is widely used in a number
of industries. Fabric filters have advantages over competing technologies for many relatively small particle-laden gas streams, for
flammable or explosive gas streams, and efficiently collect valuable
products or noxious materials. For the electric power industry,
fabric filters are especially attractive because they provide a clear
stack. Most fabric filters used on industrial and utility boiler
installations have been highly efficient and fairly successful. However, many utility fabric filters operate at average pressure drops
greater than expected. Most of the time the pressure drop continues
to increase with time during the life of a filter bag. Electrostatic
enhancement (EE) of filtration, with its reduced pressure drop, has
been studied widely for a number of years as a way of minimizing
pressure drop problems. This paper presents descriptions of possible
EE mechanisms and interprets experimental results in terms of those
phenomena.

Mr. VanOsdell and Mr. Donovan are members of the Center for
Aerosol Technology, Research Triangle Institute, P.O. Box 12194,
Research Triangle Park, N.C. 27709-2194.

316

FABRIC FILTRATION BACKGROUND

A useful working definition of fabric filtration is the separation of particles from gases using periodically-cleaned textile filter media. The fabric may be woven, felted, coated with various materials such as the Gore-Tex membrane, heat-treated, or otherwise modified. Its constituent fibers may be glass, stainless steel, polymeric, or any other material or combination of materials suitable for the operating conditions. The most common fabric filter geometry is a hollow tube closed at one end. These "bags" vary in size from 10 cm in diameter and 2 m long to about 30 cm in diameter and 10 m long. For a recent, extensive treatment of fabric filtration for combustion sources, see Donovan [1].

The term "fabric filtration" is something of a misnomer. The actual filtration is performed principally by the dust cake (the deposit of particles previously collected by the filter) and not by the supporting fabric. Only when it is first installed does a fabric work as a filtration medium. However, the flow resistances of both the dust deposit and the dust-laden fabric are significant in an operating fabric filter, since some of the particles do lodge within the fabric, slowly clogging it. The fabric cleaning procedures are designed to efficiently separate the dust cake from the fabric surface with a minimum amount of wear on the fabric. Common cleaning techniques are reversed gas flow, reversed gas pulses, and mechanical shaking. Proper cleaning is necessary for a sucessful fabric filter application, and the cleaning energy requirements vary for different filter materials and gas/particle streams.

A commonly used equation for modeling the operation of fabric filters is:

$$\Delta P = K_1 V + K_2 W V \qquad (1)$$

Where: ΔP = total filter pressure drop, Pa;
         K_1 = residual, or cleaned filter, drag, N s/m^3;
         V = filter face velocity, m/s;
         K_2 = dust cake specific flow resistance, 1/s; and
         W = dust areal density, kg/m^2

This equation approximates the pressure drop behavior of a fabric filter during an operating cycle. Just after cleaning, W is zero, and all of the flow resistance is due to the fabric. As the dust cake builds, the second term of Eq 1 (the dust cake flow resistance) grows in importance. Although Eq 1 indicates a linear relationship between ΔP and W (at constant V), most filters have a nonlinear operating characteristic just after cleaning. Equation 1 is strictly applicable only after the linear operating period begins. All of the parameters are taken as averages obtained from the overall baghouse flows, pressure drop, and dust loads. K_1 is taken as the drag of a used fabric just after cleaning (because the residual drag is caused by both particles that are not removed during cleaning and the drag of the virgin fabric). Equation 1 implicitly assumes that: the dust cake has a uniform flow resistance over a bag's surface, the dust cake loading is uniform, and the face velocity is uniform over the bag surface.

Recent research has shown that the Eq 1 parameters are not actually constant over the surface of a conventional fabric filter. The dust areal density varies dramatically over a filter surface, and K_2 is not the same at all locations on the fabric filter [2]. Because of variations in W and K_2 and a constant filter pressure drop, V also varies over the filter surface. Within these limitations, Eq 1 remains a useful expression, providing a simplified picture of the way a fabric filter operates that is adequate for many purposes.

ELECTRICAL FORCES IN FABRIC FILTERS

Electrical fields and forces occur in conventional filters, as well as in EE fabric filters. Most particles have some charge, and most fabrics are made of dielectric materials that permit some accumulation of charge on the fabric. EE filters deliberately take advantage of electrical effects to improve filter performance by increasing particle removal efficiency and reducing pressure drop across the filter. Filter collection efficiency improves because electrical forces contribute additional collecting forces to the capture of particles by the collecting body (either a fiber or a previously collected particle). Donovan [1] has reviewed both the electrical forces expected to develop in filters and their relative importance. A summary is presented below.

The strongest electrical force in fabric filters is usually the coulombic force, which is proportional to the product of the charge on the particle and the electric field strength. The electric field may be applied externally or caused by accumulation of charges (by charged particle collection) on the filter. In some EE filters, coulombic forces are present because of both external electric fields and charges accumulated on the filter. A second class of electrical forces, gradient forces (dielectrophoretic and image charge), is normally weaker than the coulombic force. Gradient forces arise from gradients in the electric field. In the presence of an electric field (either external or internal to the filter), fibers and particles become electrically polarized. If the polarizing field is non-uniform, a net electric force is created in the direction of increasing field strength. Thus, gradient forces always produce an attractive force between a neutral particle and a charged fiber or particle, and collection is enhanced. The gradient forces depend on the dielectric properties of the particles and/or fibers as well as on charge and field strength. Gradient forces are inversely proportional to the separation distance to the 3rd or 4th power and thus are more sensitive to separation distance than is the coulombic force. Although the gradient forces are not usually as strong as the coulombic force, EE of filtration is improved when they are present.

PERFORMANCE EFFECTS OF ELECTRICAL FORCES

Particle Collection Efficiency

Particle collection efficiency may be enhanced by both coulombic forces and gradient forces. If the collecting body is charged oppositely from the particles in the gas stream, those particles will be attracted toward the collector by coulombic force. Even if the coulombic force has a repulsive effect, the particles may be moved out of the fluid flow path and onto a neutral collecting body. Since the gradient forces are attractive, they enhance collection.

Filter Pressure Drop

If the electrical forces improved only collection efficiency, there would be little interest in EE fabric filters. Conventional fabric filters nearly always meet particle removal standards. However, they often do so at a high cost in filter pressure drop. It is the potential for reduction in pressure drop that leads to the current research interest in EE fabric filters. Examination of Eq 1 shows that fabric filter pressure drop can be reduced by reducing K_1, K_2, or W. Three basic mechanisms have been postulated to explain these reductions [3]. They may operate singly or in combination.

MECHANISMS OF PRESSURE REDUCTION IN EE FABRIC FILTRATION

The first mechanism hypothesized to explain the reduced pressure drop observed to accompany the introduction of electrical forces is enhanced surface deposition. In this mechanism, electrical forces enhance particle collection by fibers on the upstream surface of the filter. Figure 1a is a conceptual drawing of a conventional fabric filter dust deposit, and Figure 1b shows the kind of deposit expected with enhanced surface deposition. All filter fabrics include fibers that protrude into the gas on the upstream surface of the fabric. Some fabric types, such as needled felts and woven materials made of textured yarn, have an especially low fiber density at the upstream face. Dust that would have been collected at the filter surface, becoming part of the dust cake, is collected instead on the surface fibers. Overall, W is not affected, while K_2 is reduced (by increasing the fraction of particles deposited in a low fiber density region, and forming a high porosity deposit). K_1 is reduced also because more of the collected particles are near the surface and hence more easily removed during cleaning. Reduced penetration into the fabric is an advantage for fabric filters for two reasons: flow channels through the filter do not become as clogged; and dust nearer the surface of the fabric is more easily removed during cleaning. It has been shown that less penetration into a filter fabric occurs with EE filtration [4]. Whether the improvement is a permanent or only a temporary advantage has not yet been determined.

A second pressure drop reduction mechanism is based on changes in the character of the dust deposit itself. These changes have been shown very likely to occur in the presence of an electric field. The electric field may cause the dust to form a deposit that is more porous than that found without electric fields. Figure 1c illustrates this mechanism. A porous deposit would be expected to have a K_2 lower than a less porous conventional dust cake. Electrical forces (coulombic and gradient) have been shown to cause the formation of chain-like particle aggregates (dendrites) on a filter fiber [5,6]. High density dendrites produce a less densely-packed dust cake, with a lower K_2 than the conventional dust cake. As before, W does not change. The pressure drop reduction here is primarily a reduction in K_2. Further, it has been postulated that electrical current flow through the dust

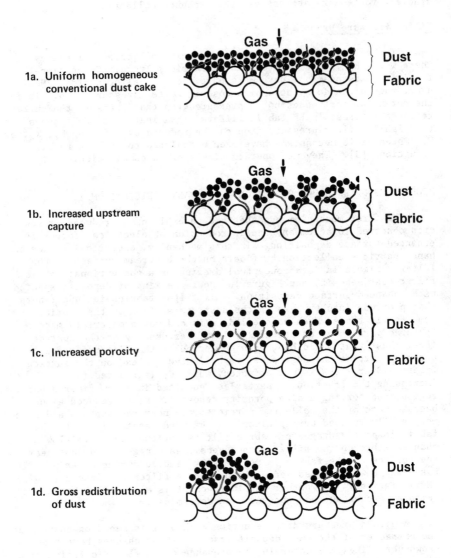

Figure 1. Representations of EE filtration pressure drop reduction mechanisms.

cake strengthens it, making the porous structure less likely to collapse and increase flow resistance [4].

The third proposed mechanism of pressure drop reduction, the
gross redistribution of dust on the filter, is shown in Figure 1d.
Gross changes in the dust distribution appear visually in the form of
regions of light and heavy dust deposition. This pressure drop
reduction mechanism would be due only to coulombic forces. Long-range
coulombic forces cause incoming particles to depart from the flow
streamlines and collect in the vicinity of the high electric field
regions. This action strips the dust load from much of the gas
passing through the filter, in effect creating regions of reduced W
through which most of the gas flows.

ELECTROSTATIC ENHANCEMENT METHODS

There are two approaches which have been used most often in im-
plementing EE in fabric filters. The first is artificial charging of
the particles before they reach the filter (Figure 2a). Conceptually,
all that is required is a corona precharger mounted upstream of the
filter medium. This approach will be referred to as EE by particle
charging (EEPC). The other technique applies an electric field near
the filter surface without a particle charger, and will be referred to
as EE by external field (EEEF). Electric fields oriented parallel
(Figure 2b) to the fabric surface have been used by some researchers,
while others have used fields perpendicular to the fabric surface
(Figure 2c). The results for both parallel and perpendicular fields
have been similar. These two basic approaches--particle charging and
externally applied field--have also been combined and used simultaneously. Examples of each of these approaches will be examined
below, with an emphasis placed on identifying the pressure drop
reduction mechanism.

Electrostatic Enhancement by Particle Charging (EEPC)

For EEPC, another pressure drop reduction mechanism (in addition
to those mentioned above) becomes potentially important. Corona pre-
chargers both charge and precipitate dust. They may precipitate a
significant fraction of the incoming dust before it reaches the fabric
filter. This reduces W in Eq 1. This phenomenon may be initially
mistaken for increased dust cake permeability (lower K_2) if it is
assumed that all the incoming dust deposits on the fabric. For an
operating fabric filter, it is difficult to differentiate between
precollection (reduced W at the fabric) and changes at the filter
(lower K_1 or K_2) because it is difficult to measure the mass of dust
actually collected on the filter. Dust load reductions of 60 to 70
percent in the corona precharger have been reported for a laboratory
filter having a 7.6 cm diameter [7]. For another laboratory fabric
filter (filtering redispersed flyash) most of the pressure drop
reduction observed with the precharger ON was thought to be due to
collection in the charging section [8]. On an even larger scale,
particle removal in a pilot plant precharging section (operating on a
slipstream from an industrial coal boiler) was described as ranging
from 60 to 90 percent [9]. This particular EECP filter, the Apitron,
is described in more detail below.

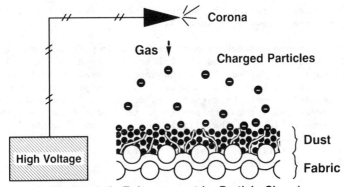

A. Electrostatic Enhancement by Particle Charging

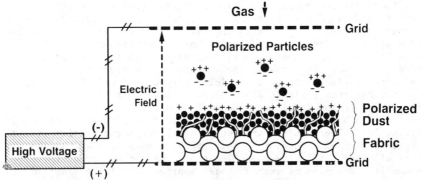

B. Electrostatic Enhancement by External Field Perpendicular to Filter

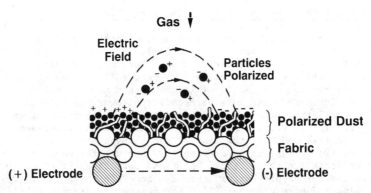

C. Electrostatic Enhancement by External Field Parallel to Filter

Figure 2. Electrostatic enhancement methods.

Once the changes in W are accounted for, it has been found that the presence of charged particles alone is not sufficient to cause a pressure drop reduction unless other conditions are met [7]. A significant filter surface charge was also required in order for the dust cake permeability to be reduced. At a high relative humidity (58 percent), the potential on the filter surface was 60 V with little difference in dust permeability observed between the charged and uncharged particle runs. The potential on the filter surface was maintained by the incoming charged particles. The conductivity of the dust and fabric was such that all the incoming charge was conducted away from the filter at a 60 V potential difference. At a 48 percent humidity, however, the fabric conductivity was lower, and the steady state charge on the filter surface built up to a potential of 940 V. At this potential, the specific dust flow resistance (K_2) was about 40 percent lower than that at the 60 V potential. It was not clear which of the three pressure drop reduction mechanisms might be operating, but it was clear that increased dust charge alone was not sufficient to produce a reduced K_2. Similarly, improvements in particle removal efficiency also depended on filter surface charge.

In general, charging particles is an effective means of charging a fabric. Particle removal has been found to increase whenever an EECP filter is used. For a pilot scale EECP fabric filter, an overall collection efficiency of 99.9998 percent has been reported [9].

The effect of EECP on size dependent filtration efficiency has not been investigated thoroughly. Theoretical calculations for charged fibers and particles suggest that coulombic forces would be dominant at relatively low Stokes numbers, while at higher Stokes numbers the mechanical collection mechanisms would dominate [10]. Experimentally, the fiber-capture efficiency for particles as small as 5 μm has been found to be independent of electrical charge on the particle [10]. However, Smith et al. have reported the collection efficiency to be independent of particle size for a range of 0.5 to 10 μm for an EECP filter (characteristic of precipitation in a precharger) [9].

The Apitron is a commercially available version of an EECP fabric filter. Figure 3 consists of a series of drawings of this filter and shows its cleaning and normal filtration modes [11]. The charging/electrostatic precipitator section is marked. As with electrostatic precipitators, a practical consideration for EECP filters is the need to periodically clean the charging electrodes to maintain efficient particle charging and to avoid back-corona. The Apitron performs both the fabric and the electrode cleaning simultaneously through its induced pulse cleaning. The pulse of high-pressure air injected at the top of the charger's center electrode cleans the electrode with the air blast and also induces a reverse flow through the filter that cleans the filter. For the Apitron operating on a slipstream from an industrial pulverized coal boiler, Smith et al. found that overall pressure drop was reduced by 50 to 60 percent with the charging section ON, as compared to having the charging section OFF [9].

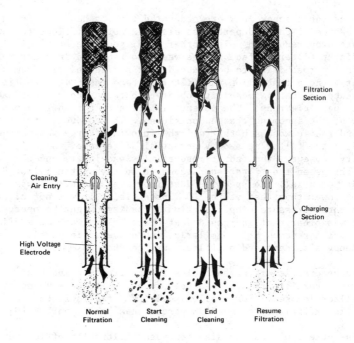

Figure 3. Operating cycle of Apitron® EE baghouse (after Helfritch[11]).

Because of the corona charger, operating costs for the Apitron have at times been quite high. At one site, it was reported that the Apitron charger caused a 60 percent reduction in pressure drop. However, the electrical power cost exceeded the pressure drop savings. Similarly conducted tests at other sites have been more promising, showing that the Apitron can be operated so that the energy savings from the pressure drop reduction exceeds the power consumption in the particle charger [12]. Clearly, careful design of the charging section of an EECP filter is necessary to avoid excessive power consumption.

In summary, EECP apparently offers enhanced performance in two ways, through particle removal in the charging section and through formation of dust cakes having increased permeability. The cause of the increased permeability has not been determined and could be due to any or all of the three basic mechanisms. Increased permeability in the dust cake requires charge buildup on the filter and the consequent presence of an electric field at the filter. For this reason, the electrical properties of the dust and fabric are important, as are the operating conditions. Particularly important is the relative humidity of the gas stream. Efficient operation of the particle charger may be critical to cost-effective EECP operation.

Electrical Enhancement by External Electric Field (EEEF)

Electric fields external to a fabric filter can also be used to achieve enhanced filtration. It is not usually necessary to artificially charge the particles, although charging may increase the degree of enhancement. The obvious advantage of the EEEF approach is that a corona charger, with its potentially high electrical power consumption, is not required. The potential disadvantages are the requirement of some physical means to establish the external electric field and the possibility that the natural particle charge will not be sufficient to obtain the desired enhanced operation.

External fields can be applied in a either parallel or perpendicular fashion to the fabric surface. Havlicek argued that orienting the electric field perpendicular to the fabric surface was the preferred method [13]. He reasoned that an electric field would polarize the dielectric filter materials and cause local electric field gradients. If the field were perpendicular to the fabric, the highest gradients would be located at the upstream and downstream faces of the filter fibers (which correspond to the stagnation regions for gas flow around a fiber). This would maximize the ratio of electrical capture force to particle drag force. Havlicek also supported his contention with data. Fields perpendicular to the fabric surface have been developed by sandwiching a flat fibrous filter between two conductive grids [14, 15] and by suspending an electrode axially in a filter bag [16].

In contrast with Havlicek, Lamb et al. [17] calculated theoretical collection efficiencies for the two orientations and concluded that having the electric field parallel to the fabric surface was the superior orientation. The electric field parallel to the filter was obtained by attaching electrodes to the fabric surface. A number of parallel electrodes were attached with alternate electrodes connected to the positive and negative terminals of a high voltage power supply.

As discussed below, performance improvements of the same magnitude (40 to 90 percent reductions in dust cake pressure drop at constant velocity, or as much as doubling the flow capacity at the same pressure drop) have been reported for both systems. Both have been used with fabric filters. Current experimental evidence does not support the choice of one orientation over the other.

The performance improvement obtained from EEEF may be due to any of the three principal mechanisms cited above. Evidence exists for all three. Increasing the fraction of dust accumulated in the upstream layers of a filter reduces the overall pressure drop for a given mass of dust [18]. Both experimental evidence and a theoretical explanation were provided. The applicability of this evidence to cake filtration is not clear. The theory was tested experimentally with a multilayer filter having an upstream porosity of 0.985 and downstream porosity of 0.77. Applying a 9 kV potential to the electrodes (parallel wires 2 cm apart, alternately charged) caused 80 percent of the dust to collect in the upstream layer and produced a reduction of 90 percent or more in observed pressure drop [18]. Only 40 percent of the dust was collected in the upstream layer without the electric field. This evidence suggests that increased dust deposition in a low density filter region can account for the benefits observed with EEEF. The conclusions could apply to cake, as well as fibrous, filtration.

Chudleigh has also conducted EEEF filter tests [19]. Chudleigh's experimental apparatus had an electric field perpendicular to the fabric. He reported end-of-cycle dust pressure drop reductions of 40 to 80 percent, depending on field strength and polarity [18]. The multi-layer deposition hypothesis was supported in that Chudleigh observed greater reduction in pressure drop for fabrics that had relatively low fiber densities at the filter surface. Chudleigh also observed that the effect of the electric field was greatest during the early part of a filtration cycle (while the dust cake was forming) and became less important after a large deposit had accumulated.

Researchers at the Textile Research Institute (TRI) also conducted extensive research into EEEF filtration with a laboratory scale fabric filter [4,17]. Pressure drop reductions (reported as the ratio of pressure drop increase with and without the electric field) of as much as 90 percent were reported. Figure 4 shows the electrode configuration used for much of the TRI work. The electrode harness was simply tied to the outside of a conventional, pulse-cleaned filter bag and was supported on the inside by an insulated cage. In addition to obtaining the basic performance data, the research showed that filters with a low fiber density at the surface performed better than relatively smooth-surface filters. Also, felted Teflon fabrics performed better than glass fiber bags, presumably because of their lower density of surface fibers [4].

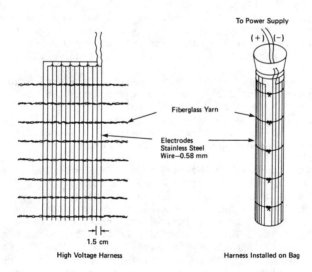

Figure 4. Electrode "harness" for applying external electric field (after Lamb and Costanza[4]).

Results for a pilot scale EEEF filter with electrodes of a similar geometry were reported by VanOsdell et al. [20]. While the operating conditions for the pilot scale unit (operating on a slip-stream from an industrial boiler) were much different from those of the laboratory unit, the fractional improvement obtained with EEEF was much the same. This pilot unit research showed that performance improvement could be obtained through reductions in pressure drop at constant face velocity, or through increased throughput at constant pressure drop. An increase in allowable face velocity from 2 to about 5 cm/s was demonstrated. It was also found that the electrodes could be placed on the clean side of the fabric, a simple change with profound consequences. The vertical electrodes of a pulse-cleaned bag cage thus could be electrically isolated and used as electrodes for introducing the electric field. The operating mode at the pilot unit also emphasized the operating benefits obtained from the reduction in the residual pressure drop of the fabric filter (resulting from EEEF). This reduction of the K_1 term was in addition to the usual reduction of K_2 observed in the laboratory. Economic analysis showed that the most attractive operating strategy was to operate an EE filter at about the same overall pressure drop as a conventional filter but at a higher face velocity. The resulting savings in capital costs were significant.

The pilot plant results suggested that the improved cleaning potential (lower K_1) of an EEEF filter may be more important than a reduction in dust cake drag [20]. Measurements of residual dust mass indicate that the EEEF filter retained less dust than did the conventional filter. Evidence for increased dust cake porosity was found in a bulk density for the EEEF hopper dust that was lower by 15 to 20 percent than that of a conventional filter hopper dust [20]. The particle penetration was about the same for both the EEEF and conventional pilot fabric filters.

A pressure drop advantage of 20 to 50 percent was obtained with a similarly designed EEEF baghouse operating on a slipstream from a pulverized coal electric utility boiler [21]. The electrodes were woven into the woven fiberglass fabric and were operated with the electric field parallel to the fabric.

Another EEEF filter, which has a wire electrode suspended in the center of a bag to obtain an electric field perpendicular to the fabric has also been developed and tested [16]. A small fabric filter operating on re-dispersed flyash was used to test this concept. The wire was charged to a high voltage (20 to 30 kV for a nominal electric field strength of 2 to 3 kV/cm), and was operated as a corona charger. In this configuration, the primary EE operating mechanism appears to be the re-distribution of the dust cake due to coulombic forces. The particles are naturally charged, and are precipitated on the sides of the filter bag. The region of the bag most distant from the inlet remains relatively clean and handles most of the gas flow.

Another pilot unit operation used the same center-wire EEEF configuration [22]. The source of flyash for this pilot unit was a stoker-fired steam boiler. The filter bags were 6.71 m long and 20.3 cm in diameter. The EEEF field was obtained by suspending a cable maintained at 20 to 30 kV in the middle of each filter bag. The electrical current drawn by each cable was approximately 0.5 mA, providing a current density of the same order as is obtained in ESPs. Some of the filter bags were equipped with an auxiliary electrical grounding system of woven-in electrodes (to reduce charge accumulation) and others relied only on the dust cake to provide a current path. Overall, the bags without an auxiliary grounding system appeared to operate at about the same flow resistance as those with auxiliary grounding. The drag of the EEEF filter bags was reduced an average of about 70 percent.

Combinations of EECP and EEEF

Various combinations of EECP and EEEF have been suggested by a number of researchers. Most of these systems would utilize an upstream corona particle charging section, followed by an EEEF filter with the electric field parallel or perpendicular to the fabric [23,24,25]. In the laboratory, the combination systems have always appeared to benefit from having both charging and external field available. The performance, as measured by pressure drop reduction or collection efficiency, was always better with both than with only one.

Masuda has patented a combination system which has an interesting twist [26]. He suggests using a corona precharger to charge the dust, followed by an electric field applied roughly parallel to the fabric. What is unique about Masuda's work is the nature of the electric field. To quote Masuda "... small, light bodies such as dust particles, when electrically charged ..., are repelled and can be electro-dynamically driven in one direction by a series of constantly varying alternating electric fields...." Masuda suggests applying an alternating voltage to parallel electrodes and then controlling the voltage alternation (two or three phase voltage) to keep the particles off the filter and at the same time propel them toward a hopper. A number of different geometries are included in Masuda's patent. The approach is unique but has not been field tested (to the authors' knowledge).

CONCLUSIONS

There is no doubt that EE fabric filtration results in reduced system pressure drop and increased particle removal efficiency. The advantages ascribed to EE fabric filtration have been consistently reported for all configurations.

Three mechanisms have been proposed to account for these advantages. The first entails increased deposition near the filter surface, resulting in a dust cake non-uniform in the direction of flow, and potentially in reduced dust penetration into the fabric. The second suggests an increased porosity of the collected dust deposit, resulting in reduced specific dust cake gas-flow resistance. And the third proposes a non-uniform dust depostion on the filter surface, resulting in thick and thin dust deposit regions. As pointed out in the discussion above, each mechanism is supported by some experimental data. The influences of the different mechanisms are difficult to separate experimentally, because they are all affected by the same parameters. It seems likely that all three operate simultaneously to some degree, with one or the other taking precedence depending on the specific electric field strength, dust and fabric electrical properties, and particle charge.

Economic Viability of EE of Fabric Filtration

The economics of EE filtration may be examined from the standpoint of reduced pressure drop (and energy costs) for a conventionally-sized fabric filter. Alternatively, the EE filter may be used as a way to operate at a higher face velocity, requiring a smaller filtration facility. Overall, pilot plant data suggest that EE filtration has a net cost advantage over conventional filter technology when used to reduce the size and thus the capital cost of the filtration facility. For pulse-cleaned EEEF filtration, the total annual cost savings may be as much as 30 percent for an EE filter operated at twice the face velocity of a conventional fabric filter [20]. A net present value analysis of center-wire, reverse-air, EEEF suggests cost savings in the range of 10 to 30 percent [27]. In all cases, proven long-term commercial operating experience at the conditions of the estimates is lacking. Until the cost advantages are proven in the field, all the economic estimates are speculative.

Outlook

Work on EE fabric filtration continues at the EPA facility in Research Triangle Park, N.C. and at other sites. Successful performance may lead to acceptance by industry. The current field applications are difficult sites for any gas cleaning method. Work continues in the laboratory to determine the operating mechanisms under different conditions and thus improve performance and performance prediction. Chances for commercial application appear best on a small fabric filter currently performing poorly but which could be upgraded through EE. The costs would be modest, and the gains could be impressive. Widespread industrial/utility acceptance will come only after long-term commercial use.

REFERENCES

[1] Donovan, R.P. Fabric Filtration For Combustion Sources, Marcel Decker, Inc., New York, N.Y., 1985.

[2] Carr, R.C. and Smith, W.B., "Fabric Filter Technology for Utility Coal-Fired Power Plants: Part I: Utility Baghouse Design and Operation," JAPCA, Vol. 34, No. 1, January, 1984, pp. 80-89.

[3] VanOsdell, D.W., Donovan, R.P., and Hovis, L.S., "Flow Resistance Reduction Mechanisms for Electrostatically Augmented Filtration," Proceedings: Fifth Symposium on the Transfer and Utilization of Particulate Control Technology, Report No. EPRI CS-4404, Vol. 3, pp 18/1-18/13, February, 1986.

[4] Lamb, G.E.R. and Costanza, P.A., "A Low-Energy Electrified Filter System," Filt. and Sep., Vol. 17, July/August 1980, pp. 319-322.

[5] Oak, M.J. and Saville, D.A., "The Buildup of Dendrite Structures on Fibers in the Presence of Strong Electric Fields," J. Colloid and Interface Sci., Vol. 76, No, 1, July, 1980, pp. 259.

[6] Bhutra, S. and Payatakes, A.C., "Experimental Investigation of Dendritic Deposition of Aerosol Particles," J. Aerosol Sci., Vol. 10, 1979, pp. 445-

[7] Chudleigh, P.W. and Bainbridge, N.W., "Electrostatic Effects if Fabric Filters During Build-up of the Dust Cake," Filt. and Sep., Vol. 17, July/August 1980, pp. 309-311.

[8] Hovis, L.S. et al., "Electrically Charged Flyash Experiments in a Laboratory Shaker Baghouse," Control of Emissions from Coal-Fired Boilers, Vol. I, Report No. EPA-600/9-82-005a, July 1982, pp. 23-24.

[9] Smith, W.B. et al., "Electrostatic Enhancement of Fabric Filter
     Performance," Proceedings: International Conference on
     Electrostatic Precipitation, H.J. White, Ed., APCA, P.O. Box
     2861, Pittsburgh, Pa. 15230, October 1981, pp 84-106.
[10] Loeffler, F., "The Influence of Electrostatic Forces and of the
     Probability of Adhesion for Particle Collection in Fibrous
     Filters," Novel Concepts, Methods, and Advanced Technology in
     Particle-Gas Separation, T. Ariman, Ed., University of Notre
     Dame, IN 46556, 1978, pp. 206-231.
[11] Helfritch, D.J., "Apitron Product Brochure," Dustex Corporation,
     12037 Goodrich Dr., Charlotte, N.C. 28217, 1979.
[12] Felix, L.G. and McCain, J.D., "Apitron Electrostatically
     Augmented Fabric Filter Evaluation," Report No. EPA-600/7-79-070,
     February 1979.
[13] Havlicek, V., "The Improvement of Efficiency of Fibrous
     Dielectric Filters by Application of An External Electric Field,"
     Int. J. Air and Water Poll., Vol 4, No. 314, 1961, pp. 225-236.
[14] Nelson, G.O. et al., "Enhancement of Air Filtration Using
     Electric Fields," Am. Ind. Hyg. Assoc. J., Vol. 39, June 1978,
     pp. 472-479.
[15] Walkenhorst, W. and Zebel, G. "Uber ein neues Schwebstoffilter
     hoher Abscheideleistung und geringen Stromungswiderstandes,"
     Staub-Reinhalt. Luft, Vol. 24, 1964, pp. 444-448.
[16] Hovis, L.S., Daniel, B.E., and Donovan, R.P., "Electrostatic
     Enhancement of Fabric Filtration of Fly Ash and Spray Dryer
     By-Product," Proceedings: Third Conference on Fabric Filter
     Technology For Coal Fired Power Plants, EPRI, 3412 Hillview Av.,
     Palo Alto, CA, November 1985.
[17] Lamb, G.E.R. et al., "Electrical Stimulation of Fabric
     Filtration. Part II: Mechanism of Particle Capture and Trials
     with a Laboratory Baghouse," Tex. Res. J., No. 10, October 1978,
     pp. 566-573.
[18] Morris, B.A., Lamb, G.E.R., and Saville, D.A., "Electrical
     Stimulation of Fabric Filtration. Part V: Model for Pressure
     Drop Reduction," Tex. Res. J., Vol. 54, No. 6, June 1984, pp.
     403-408.
[19] Chudleigh, P.W., "Reduction of Pressure Drop Across A Fabric
     Filter By High Voltage Electrification," Filt. and Sep., Vol.
     20, No. 3, May/June 1983, pp. 213-216.
[20] VanOsdell, D.W. et al., "Electrostatic Augmentation of Fabric
     Filtration: Pulse-Jet Pilot Unit Experience," Report No.
     EPA-600/7-82-062, U.S. EPA, November 1982.
[21] Chambers, R.L., Spivey, J.J., and Harmon, D.L., "ESFF Pilot Plant
     Operation at Harrington Station," Proceedings: Fifth Symposium
     on the Transfer and Utilization of Particulate Control
     Technology, Report No. EPRI CS-4404, Vol. 3, pp 25/1-25/17,
     February, 1986.
[22] Greiner, G.P., Viner, A.S., Hovis, L.S., and Gibbs, R.,
     "Operation of an Electrically Enhanced Pilot Plant on a Stoker
     Fired Boiler," paper presented at the Sixth Symposium on the
     Transfer and Utilization of Particulate Control Technology, New
     Orleans, LA, February 24-28, 1986.
[23] Anderson, W.M. and Dervay, J.R., "Electro-Bag Dust Collector,"
     U.S. Patent No. 3,733,784, issued May 22, 1973.

[24] Penney, G.W. "Electrostatic Dust Filter," U.S. Patent No.
3,966,435, issued June 29, 1976.
[25] Reed, G.A. and Mott, J.A., "Electrostatic Dust Collector," U.S.
Patent No. 3,915,676, issued October 28, 1975.
[26] Masuda, S., "Electrostatic Apparatus for Removing Entrained
Particulate Material From a Gas Stream," U.S. Patent 3,930,815,
issued January 6, 1976.
[27] Viner, A.S. and Locke, B.R., "Cost and Performance Models for
Electrostatically Stimulated Filtration," Report No.
EPA-600/8-84-016, August, 1984.

Duane H. Pontius and Wallace B. Smith

PROGRESS IN FABRIC FILTER TECHNOLOGY FOR UTILITY APPLICATIONS

REFERENCE: Pontius, D. H. and Smith, W. B., "Progress in Fabric Filter Technology for Utility Applications," Fluid Filtration: Gas, Volume I, ASTM STP 975, R. R. Raber, Ed., American Society for Testing and Materials, Philadelphia, 1986.

ABSTRACT: The use of fabric filtration for controlling particulate emissions from power generating plants has proven effective, both technically and economically. In most installations, the collection efficiency of these systems is much better than is required by law. Recent developments have centered on improvements in methods for removing collected ash (dustcake) from filter bags, applications in integrated emissions control systems, and studies leading toward a better understanding of the fundamental processes involved.

KEYWORDS: fabric filtration, electric power generation, air pollution control

INTRODUCTION

Within the past decade or so there has been a rapid growth in the acceptance of fabric filters for control of particulate emissions in electric utility applications. Primarily responsible for this trend is the clear stack condition (no visible smoke emissions) usually seen at coal-fired power plants equipped with well-maintained baghouse systems. In most cases the increasingly strict statutory limits imposed on particulate emissions are easily satisfied by this technology. Particulate mass collection efficiency greater than 99.9% is typical, and values of opacity as low as 0.1% are routinely achieved [1].

Early concerns about reliability, bag life, and excessive pressure drop have, for the most part, been eased. More than 100 baghouses are

Dr. Pontius is Head of the Physics Division, and Dr. Smith is Associate Director of Environmental Sciences Research, Southern Research Institute, P. O. Box 55305, Birmingham, AL 35255-5305.

now in operation at utility power plants. They are installed at plants that burn various types of coal, including eastern, high-sulfur, bituminous coal; western, low-sulfur, subbituminous coal; and lignites [2].

In this paper, the nature of the aerosol emitted by a coal-fired boiler will be described briefly, and some of the general features of fabric filters will be outlined. The mechanics of filtration and methods for removing the collected ash will be described. Finally, the general features of a dustcake will be discussed, and an elementary filtration model will be presented.

Process Description

The exhaust from a coal-fired boiler contains a variety of combustion products, including gaseous components (N_2, CO_2, H_2O, O_2, SO_x, and NO_x) and solids (fly ash). The fly ash is in the form of fine particles suspended in the gas. It consists of noncombustible mineral material that condenses as the exhaust begins to cool from the flame temperature, typically near 1600°C. After having passed through devices that extract usable heat, the flue gas will have been reduced to a temperature of about 150°C. It can then be processed by a fabric filter.

A typical power plant, with an electrical generating capacity of 500 MW, produces effluent gases at a rate of about 700 m^3/s. These gases contain suspended particles in concentrations ranging from ap-

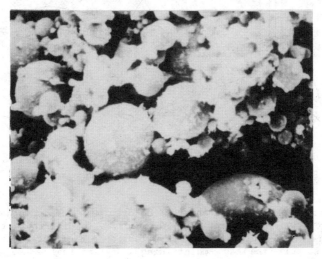

⊢————⊣ 1 μm

FIG 1.--Fly ash particles.

proximately 2 to 20 g/m^3. In most instances, the particles are generally spherical in shape, as can be seen in Figure 1, and may include sizes from less than .01 μm to more than 100 μm in diameter. The distribution of particulate mass as a function of diameter typically peaks between 3 and 20 µm, depending on the coal and the type of boiler.

Baghouse Configuration

The basic component of a baghouse is a filter bag. For power plant applications in the U.S., the bags are made of woven fiberglass. A typical design is cylindrical in shape, nominally 30 cm in diameter and 9 to 10 m long. The usual practice in utility baghouses is to suspend the bags vertically, as indicated in Figure 2. The bottom of each bag

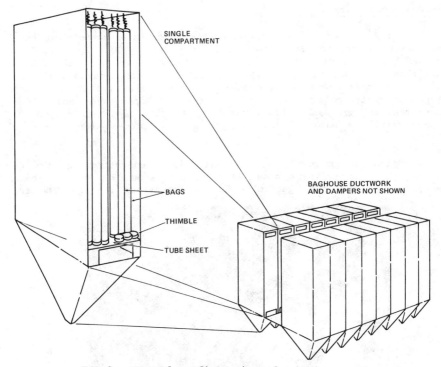

FIG 2.--General configuration of a baghouse.

is clamped onto a "thimble", a short piece of pipe which is welded into a "tube sheet", which is a manifold that serves to distribute the gas among the bags. The top of the bag is attached to a circular metal cap designed to support the bag, maintaining a cylindrical shape at the top. Each bag cap is, in turn, connected to the supporting upper framework by means of a device designed to permit adjusting the longitudinal tension of the bag.

All utility baghouses are divided into compartments, each containing the same number, and approximately the same arrangement of bags. A system of ducting, dampers, and poppet valves (not shown in Figure 2) permits any compartment to be isolated or taken off line. That feature is important not only as a convenience for maintenance, but as an integral part of the routine operational procedure. A filter bag cannot collect fly ash for an indefinite period; it must be cleaned from time to time, just as a used vacuum cleaner bag must be emptied or replaced. Cleaning procedures presently used in utility baghouses require removing compartments from service one at a time.

BAGHOUSE OPERATIONS

Baghouses are operated on cyclic schedules. In a complete cycle, each compartment goes through a cleaning process. Depending upon the operating conditions in a specific installation, the cycle may be less than an hour in duration, or it may last up to about four hours. The shortest cycles occur in those cases where it is necessary to clean continuously--to carry out a sequence of cleaning processes on one compartment after another with no intervening periods during which all compartments are available for filtration.

Filtration

The standard for utility baghouse design is based on bottom entry to the bags. The flue gas is drawn into the volume beneath the tube sheet, from which it flows upward, through the array of thimbles, and into the bags. The gas then flows outward through the fabric at a radial velocity that averages, typically, about 1.0 cm/s. If the particulate mass loading of the flue gas is, for example, 10 g/m^3, then in a three-hour filtration period, an average of about 1 kg of fly ash will be deposited per square meter of fabric.

Flue gas encounters bare fabric only when a bag is absolutely new. A dustcake begins to build up on the fabric as soon as it is put into operation. None of the cleaning processes presently in use will remove all of the accumulated dustcake, nor is it considered particularly desirable to do so. The dustcake has two important effects on the behavior of the system: it increases the aerodynamic drag of the filter, and it improves the particulate collection efficiency. Since a baghouse must normally treat a given total gas volume per unit time, the increasing drag associated with the buildup of the dustcake results in a rising pressure drop across the system. The gas is drawn through the filter by a fan; therefore, as the pressure drop increases, so does the cost of operating the baghouse. (Typical values of pressure drop across a fabric filter - excluding ductwork, dampers, etc. - is in the range of 1 to 2 kPa.) On the other hand, the beneficial effect of the dustcake on collection efficiency requires only a thin layer of ash.

When a filter bag has reached an equilibrium state of operation (a few months after installation), it will support a dustcake of typically 2 to 5 kg/m^2. Since the amount of new ash added in a filtering cycle is of the order of 1 kg/m^2 or less, it must be concluded that a large fraction of the dustcake is relatively permanent. Nevertheless, and in spite of the fact that the new ash accumulated during one filtering cycle is normally a small fraction of the amount in the residual dustcake, the new ash is responsible for large changes in the aerodynamic drag of the filter. Figure 3 illustrates the general behavior of the

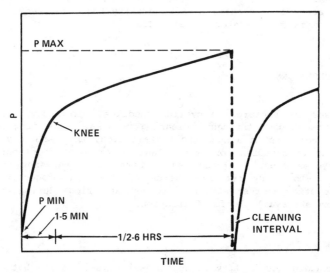

FIG 3.--Buildup of pressure during a filtration cycle.

pressure drop across a single baghouse compartment. The cleaning process results in a sharp reduction in pressure drop, but when the compartment is put back on line the pressure drop increases rapidly for a brief period of time. The rate of increase soon settles down to a more gradual rise which continues until the end of the filtering period.

Cleaning

There are two methods of dustcake removal presently in use for cleaning bags at utility power plants. About 90% of all utility baghouses are cleaned with a reverse-gas procedure, which involves redirecting the gas flow to move backward through the compartment being cleaned. The other method commonly used is to shake the bags mechanically. Sonic power is also employed in some installations as a supplementary technique to help break up the dustcake, but it should not be considered a separate cleaning process because it is not effective used alone. Other approaches, such as pulse-jet cleaning, have been applied in small industrial baghouses but have not yet been exploited in full-scale utility systems.

Reverse Gas: In the reverse-gas cleaning mode, part of the gas from downstream of the baghouse is forced to flow contrary to the normal direction for filtration. The reverse-gas flow rate is of the same order of magnitude as the filtration flow rate; the average gas velocity normal to the fabric surface is only 1 to 2 cm/s. Clearly the cleaning mechanism does not amount to blowing the fly ash off of the fabric.

When the gas flow is reversed, a bag tends to collapse. Bags designed to be cleaned by reverse gas have metal rings attached at intervals of 1 to 2 meters from bottom to top. The rings prevent the bags from collapsing completely when the gas flow is reversed. Instead, the bags flex along several longitudinal folds so that a cross section exhibits a lobed appearance, as illustrated in Figure 4. The dustcake

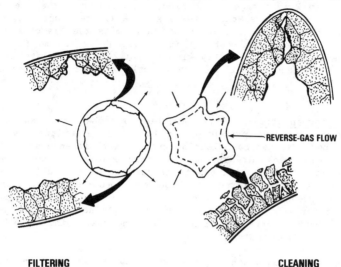

FILTERING   CLEANING

FIG 4.--Bag collapse during reverse-gas cleaning cycle.

attached to the fabric is inelastic, so the flexing causes cracks to appear throughout its structure. Along the deep fold lines the flexing action is especially severe. The residual dustcake is thinnest on these folds.

Reverse gas is operated for, typically, 30 s. When the operation is completed, the compartment is put back into normal operation in parallel with the other compartments.

Shaking: A fairly obvious way to remove dustcake from a filter bag is to shake it with some sort of mechanical device. That approach is used in a few utility baghouses. Shaking is not as widely applied as reverse gas because it had been thought that such a vigorous cleaning method would reduce the useful life of the bags. Experience in the field has shown, however, that bags cleaned by shaking can last well over three years. Usually the shaking is done with a motor-driven cam or crank, coupled to the upper bag supports. A frequency of about 4 Hz is typical, and the amplitude of the oscillations ranges up to approximately 2 cm measured at the bag cap. Shaking is usually carried on for 10 to 15 s.

A filter bag in its normal inflated state is relatively rigid. It cannot be shaken effectively in this state, so the shaking procedure is preceded by a deflation of the bags. This operation is carried out by using a small quantity of reverse gas to reduce the pressure differential across the bag, and thus to relax the tension in the fabric. The reverse gas flow does not usually continue through the shaking period. The filter bags in a baghouse cleaned by shaking are not equipped with anti-collapse rings. They are otherwise similar in size and configuration to the bags in systems where reverse-gas is used.

The principal advantage of shaking, compared with the use of reverse gas, is that it generally results in a lighter dustcake. Ultimately, the effect is reduced drag, which permits the use of a smaller total area of fabric (fewer bags) for a given gas volume flow rate. There is usually a significantly greater rate of particulate emission from a shaker baghouse, but there is little risk of even approaching regulatory limits.

Sonic Power Augmentation: There have been a number of cases in which problems with excessive pressure drop have been alleviated by accompanying the usual mechanical cleaning process with intense sound in the compartment being cleaned. The use of sonic energy as a supplementary cleaning method has been studied at various pilot- and full-scale baghouses. The results achieved in those test programs have shown that reductions in pressure drop by about 20 to 50% are possible, depending on the characteristics of the fly ash and the flue gas. The greater improvements have occurred at plants where the coal being burned has a relatively low sulfur content.

Sonic power augmentation requires sound to be generated in baghouse compartments. The source must therefore be capable of withstanding temperatures up to about 150°C and a strongly acidic atmosphere. Pneumatic, diaphragm-operated horns have proved suitable for this harsh environment. These horns have only one moving part--a vibrating diaphragm which controls the escape of compressed air from a small plenum chamber.

Both the intensity and the frequency of the sound must be in appropriate ranges if the use of horns is to succeed in improving dustcake removal. Again depending on the characteristics of the ash and the gas, an effective horn will produce a sound pressure greater than 80-160 Pa (132-138 dB), averaged throughout a compartment. Just as importantly, the sound must be strongly concentrated in the range of frequencies below about 250-300 Hz. It is not sufficient that the fundamental frequency of the horn be in that range; the entire sonic spectrum must be taken into account [3].

DUSTCAKE CHARACTERISTICS

Dustcakes are porous, disordered structures. Intuition may tend to suggest that particles in the gas would lodge tightly into the pores of the dustcake to form a densely packed aggregate, but in fact, the porosity of dustcakes ranges, in general, between 60 and 80%. The porosity of fly ash packed tightly by hand can be as low as about 40%. The principal reason for the looseness of dustcakes derives from the fact that the forces binding individual particles together are large in

comparison with the weights of the particles. Furthermore, the particles become attached to the structure one at a time by forces that
involve inertia, diffusion, and electrostatic effects.

General Description

The macroscopic features of a dustcake include a profusion of fine
cracks and a broad variability in thickness. These characteristics are
visible in Figure 5, a macrophotograph of a cross section of a dustcake
taken from an operational baghouse configured for reverse-gas cleaning.

1 cm

FIG 5.--Dustcake cross-section.

The sample was cut from a used bag during an equipment outage. The
condition of the dustcake is as it would be immediately after cleaning
under normal operating conditions. Thus, only the residual, or relatively permanent dustcake remains. The sample was prepared by infusing
a fluid epoxy into its porous structure. (When this technique is used,
both the viscosity and the surface tension of the epoxy must be as low
as possible to avoid damaging the specimen.) After the epoxy had hard-
ened, the sample was sliced and polished to reveal the surface shown in
the photograph [4].

The mechanics of filtration tend toward the development of a uniform dustcake. A thin region poses a low aerodynamic resistance to the
passage of flue gas, and hence it accepts more than its normal share of
the total flow. Since the fly ash is more or less evenly distributed
throughout the flue gas, the dustcake in such a region will grow at a
more rapid rate than would a thicker part of the dustcake. This self-
regulating effect has been demonstrated in laboratory experiments [5].
Unquestionably, then, the network of cracks and the irregular shape of
the dustcake shown in Figure 4 must result from the cleaning process.

The cracks in the dustcake suggest an explanation for the general
shape of the curve shown in Figure 3. Immediately after the cleaning
operation, the cracks offer a network of open passages through the

dustcake. These passages are much larger than the average pore size, and therefore behave as something like a shortcut for the incoming flue gas. Thus, a disproportionately large fraction of the gas streams comparatively freely through the cracks to the fabric. The cracks, however, occupy only about 3 to 5% of the dustcake volume, so in taking on a large fraction of the gas flow, they will fill up rapidly with ash. The rapid rise in pressure drop shown in Figure 3 can thus be interpreted as a result of an accumulation of fly ash in the cracks that were created by the cleaning process. When the cracks are filled, they are indistinguishable from the dustcake at large. They are no longer paths of low resistance, so the gas flow through the dustcake becomes more uniformly distributed. The result is a more gradual increase in pressure drop.

Elementary Model Development

The behavior of a dustcake may be simulated mathematically by treating the system as a parallel combination of independent paths conducting the flow of gas. It will be assumed that all new fly ash deposited by the filtering process becomes an integral part of the dustcake, with the same permeability as the residual matter. To set up the mathematical model, divide the surface of the filter into elements of area, each having thickness y. Then let a_i be the sum of the areas of all elements having thickness y_i. Assuming Darcy's equation applies [6], the amount of gas q_i flowing through an area a_i per unit time is proportional to the product of the permability K of the dustcake and the pressure drop P across the filter, and inversely proportional to the viscosity μ of the gas.

$$q_i = \frac{KPa_i}{\mu y_i} \tag{1}$$

The rate of change of the depth of the dustcake over the region that has depth y_i is proportional to the gas flow rate per unit area, so

$$\frac{dy_i}{dt} = \frac{Lq_i}{a_i} \ , \tag{2}$$

where L is a constant for the operating condition. Specifically, L is the particulate mass per unit volume of flue gas divided by the bulk density of the dustcake. The total amount of gas Q flowing through the filter is the sum of the components passing through the various elements of area.

$$Q = \frac{KP}{\mu} \sum_i \frac{a_i}{y_i} \tag{3}$$

In principle it is possible to express a_i as a continuous variable a(y), in which case the sum would translate into an integral over the surface of the filter. The form used for Eq 3, however, is readily incorporated into a computer simulation.

There is a technical difficulty in applying Eq 3 to an example that includes cracks in the dustcake. Although it seems appropriate to assign a thickness of zero to the area occupied by cracks, that action would drive Q to infinity. In fact, it is not realistic to set the thickness at zero, because, even in the absence of a dustcake, the filter material offers some resistance to gas flow. The problem can be resolved by adding a constant c to y_i in the denominator of Eq 3. The value of c is the thickness of a dustcake whose aerodynamic resistance is equivalent to that of the fabric. Equation 3 thus becomes

$$Q = \frac{KP}{\mu} \sum_i \frac{a_i}{(y_i + c)} \; . \qquad (4)$$

The externally controllable variables in Eq 4 are the pressure drop across the filter and the total gas flow rate. In a multi-compartment baghouse, the overall constraint is a fixed value of total gas flow rate, but the distribution of gas flow among compartments varies continuously. Fluctuations occur as compartments are taken off line for cleaning and then returned to service. Furthermore, complex flow interactions result from the constant modulation of aerodynamic drag that comes from the buildup of dustcake. This effect is different for every compartment, since each one is in its own phase of the filtration cycle. The filtration model under development here applies to one compartment at a time. It requires that the operating conditions or constraints be provided from separate computations or empirical tabulations.

The simplest solution of Eq 4 is for the case of constant pressure drop. Under that condition the differential equation, Eq 2, is readily integrated to yield

$$y_i(t) = \left[y_i^2(0) + \frac{KLPt}{\mu} \right]^{1/2} , \qquad (5)$$

which may be substituted into Eq 4 to obtain the total gas flow rate as a function of time.

A more realistic illustration of the application of this model is the solution in terms of a constant value of the total gas flow rate through the filter. The calculation of P as a function of the time for this constraint is also based on Eq 3, as solved for P:

$$P = \frac{Q\mu}{K} \left[\sum_i \frac{a_i}{y_j} \right]^{-1} \qquad (6)$$

Because P and the individual q_i are variables, a continuous solution cannot be derived as it was for the case of constant pressure drop. But a simple iterative procedure can be used. First the initial values of the y_i are set, and P is calculated by Eq 6. Then, appropriate increments in all y_j are computed by the discrete analog of Eq 2, which takes the form

$$\delta y_j = \frac{Lq_j}{a_j} \delta t , \qquad (7)$$

corresponding to an increment of time δt. The new values $y_j \rightarrow (y_j + \delta y_j)$

are then inserted into Eq 6 to update the value of P. Continuing this procedure results in a determination of P as a function of the time.

The iterative solution can accommodate temporal variations in the pressure drop across a compartment. Since it is necessary to maintain a fixed operating condition for only one increment of time per calculation, the value of Q (as determined from a separate computation or tabulation) can be adjusted for each iteration.

Example

A computer program has been designed to carry out the iterative procedure described in the foregoing description of the dustcake model, simulating operation of a filter at a constant rate of total gas flow. A sample result is shown in Figure (1). A value of gas flow rate was chosen so as to result in a component of velocity equal to 1.0 cm/s in the direction normal to the filter surface. The particulate mass loading was taken to be 5.75 g/m^3. Cracks were estimated to occupy 5% of the dustcake. Unevenness of the dustcake thickness was described in terms of the following table of values:

TABLE 1--Dustcake configuration

Dustcake thickness, cm	Fraction of area
0.1	0.2
0.2	0.4
0.4	0.2
0.5	0.1
0.7	0.1

As can be seen in Fig. 6, maintaining a constant rate of gas flow through the filter results in a pressure drop that rises quickly at the

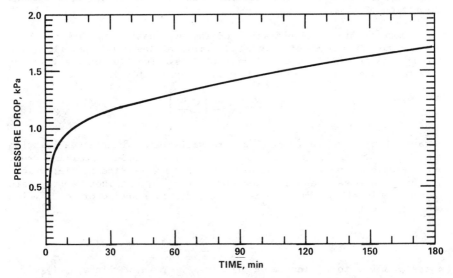

FIG 6.--Computer simulation of filtration cycle.

start of the filtration process. The slope of the curve rapidly decreases until a nearly linear rise in P is established. These features
are in general agreement with the behavior of filters studied in the
laboratory and in pilot scale facilities. In full scale systems, however, neither the pressure drop nor the total flow rate in a compartment is held constant. A complete model of a typical working baghouse
would require additional computations to determine the distribution of
gas flow among the compartments. The flow rate per compartment would
then be adjusted for each time increment.

SUMMARY

The fabric filter technology as applied to coal-fired power plants
has been characterized by steady improvements in performance and reliability. Particulate collection efficiency is typically better than
99.9%, and there are usually no visible emissions from the smoke stack.
All utility baghouses are divided into compartments which are operated
in phased cycles. About 90% of all utility baghouses are cleaned by
reverse gas, and the remainder are cleaned by shaking. Sonic augmentation may be useful in cases where dustcake removal is inadequate.

Dustcakes are generally porous structures, but during the period of
the first few minutes after cleaning, the flue gas is conducted
principally through cracks in the dustcake. As filtration continues,
the gas flow becomes more uniformly distributed over the surface of the
filter. This behavior has been verified in terms of an elementary
mathematical model.

ACKNOWLEDGMENT

This paper is based upon results from research sponsored by the
Electric Power Research Institute, Palo Alto, CA, under project number
RP-1129-8, R. C. Carr, Program Manager.

REFERENCES

[1] Carr, R. C., and Smith, W. B., "Fabric Filter Technology for Util-
     ity Coal-Fired Power Plants, Parts I through VI," Journal of the
     Air Pollution Control Association, Vol. 34, Nos. 1-6, Jan.-June
     1984.
[2] Felix, L. G., Merritt, R. L., and Carr, R. C., "Performance Evalu-
     ation of Several Full-Scale Utility Baghouses" Proceedings: Second
     Conference on Fabric Filter Technology for Coal-Fired Power
     Plants, CS-3527, Electric Power Research Institute, Palo Alto, CA,
     Nov. 1983.

[3]   Pontius, D. H., and Smith, W. B., "Characterization of Sonic Devices Used for Cleaning Fabric Filters," Journal of the Air Pollution Control Association, Vol. 35, No. 1, December 1985.

[4]   Felix, L. G., and Smith, W. B., "Preservation of Fabric Filter Dust Cake Samples," Journal of the Air Pollution Control Association, Vol. 33, No. 11, November 1983, p. 1092.

[5]   Kistler, W. G., Steele, W. J., Pontius, D. H., and Albano, R. K., "Laboratory Measurements of the Spatial Distribution of Gas Velocities through Seasoned Baghouse Fabrics throughout a Simulated Filtration Cycle," Journal of the Air Pollution Control Association, Vol. 36, No. 1, January 1986, pp. 34-41.

[6]   Davies, C. N., Air Filtration, Academic Press, New York, 1973, p. 30.

Ronald C. Scripsick

NEW FILTER EFFICIENCY TESTS BEING DEVELOPED FOR THE DOE

REFERENCE: Scripsick, R.C., "New Filter Efficiency
Tests Being Developed for the Department of Energy,"
Fluid Filtration: Gas, Volume I, ASTM 975, R.R.
Raber, ED., American Society for Testing and
Materials, Philadelphia, 1986

ABSTRACT: The United States Department of Energy (USDOE)
is considering adoption of new test specifications and new
test techniques for quality assurance testing of nuclear
grade high efficiency particulate air (HEPA) filters.
Proposed new filter test specifications require penetration
measurements to be made at a particle diameter in the range
from 0.1 µm to 0.2 µm rather than at a particle
diameter of 0.3 µm which is currently required. The new
specifications are aimed at making penetration measurements
at or near the size of maximum penetration for the current
generation of nuclear grade HEPA filters. The new test
specifications provide a high estimate of filter
penetration relative to the predicted penetration of field
challenges in terms of aerosol size and aerosol density.
This conservative estimate is required to assure that USDOE
air cleaning systems adequately protect public and worker
health and the environment in the event of unplanned
release of hazardous materials.
      Operational evaluation of a high flow alternative filter
test system (HFATS) developed for the USDOE by Los Alamos
National Laboratory is currently being completed. Interim
results of this evaluation and results of laboratory and
full-scale evaluations demonstrate that the HFATS is
capable of providing measurements of filter penetration at
specific sizes over the diameter range from ~0.1 µm to
~0.4 µm within the operational requirements of the
USDOE Filter Test Facilities.

Mr. Scripsick is a research scientist with the Industrial Hygiene
Group, Health, Safety and Environment Division of Los Alamos National
Laboratory, P. O. Box 1663, Mailstop K486, Los Alamos, NM 87545.

KEYWORDS: filter efficiency, filter tests, penetration
measurement system, filter efficiency standards, HEPA
filters

INTRODUCTION

The United States Department of Energy (USDOE) uses nuclear grade
high efficiency particulate air (HEPA) filters in critical air
handling systems for protection of public and worker health and the
environment. The filters are used to decontaminate ventilation
airstreams of hazardous particulate radionuclides and other hazardous
particulate materials. The performance of these filters is directly
related to the quality of the air that is introduced in the workplace
or released to the environment. The lowest allowable airborne
concentrations (AACs) of these hazardous materials can be on the order
of 10^{-13} g/m^3 for workplace releases and on the order of
10^{-14} g/m^3 for environmental releases [1,2]. The highest
filter challenge concentrations of these materials can be in the range
from 100 mg/m^3 to 10 g/m^3 during severe upset
conditions [1,3]. Consequently, the air cleaning systems must have a
design capacity capable of decontaminating airstreams by as much as a
factor of 10^{12} to 10^{15}. To achieve these capacities system
designers often use HEPA filter banks in tandem. For example, to
obtain a decontamination factor on the order of 10^{14} requires four
tandem banks of HEPA filters with each bank having a collection
efficiency $\geq$99.97 per cent.

To assure that filters placed in air cleaning systems meet design
specifications, every nuclear grade HEPA filter purchased for use in
USDOE facilities must pass a quality assurance (QA) test at a USDOE
Filter Test Facility (FTF) before it is forwarded to the purchaser.
This QA testing includes filter efficiency measurements made at rated
flow and also at 20 per cent of rated flow for filters with rated flow
of 125 cubic feet per min (CFM, ~3.5 m^3/min) and higher. The
current test specifications call for measurement of filter efficiency
at 0.3 µm using a thermally generated di-(2-ethylhexyl) phthalate
(DEHP) or di-(2-ethylhexyl) sebacate (DEHS) aerosol, an Owl polarized
light aerosol size analyzer and a scattered-light photometer (SLP).
New standards for USDOE filter testing are in the final stages of
being adopted [4-7]. Specifications of the filter efficiency test
proposed for these standards are discussed in this report. In
addition a new filter efficiency test system is described that was
developed for the USDOE by Los Alamos National Laboratory.

PROPOSED FILTER EFFICIENCY TEST SPECIFICATIONS

Aerosol Size

The major change proposed in the new filter efficiency test
specifications is to require penetration measurements to be made at a

particle size in the diameter range from 0.1 μm to 0.2 μm rather than at a particle diameter of 0.3 μm which is currently required [5]. Examination of the predicted rated flow performance of HEPA filters available when the currently designated test systems were developed indicates that the size of maximum penetration was near 0.3 μm, which corresponds to the particle size at which the current systems were designed to operate [8]. This result suggests that the designers of the current systems intended to measure penetration near the aerosol size of maximum penetration. The merit of such a test of filter performance is that the measurement result is a conservative or worst case estimate of filter performance in terms of challenge aerosol size. For a given air flow, no challenge aerosol consisting of particles similar in shape to and of the same or greater density (ρ, specific gravity) as the test aerosol particles should have a greater penetration than the test penetration. This conclusion follows directly from theoretical and experimental evidence that shows HEPA filter penetration as a unimodal function with respect to aerosol size [8-13].

Theoretical evaluation of the performance of the current generation of nuclear grade HEPA filters operating at flows between rated flow and 20 per cent rated flow predicts the maximum penetration aerosol diameter to be in the range from ~0.1 μm to ~0.2 μm [8]. This predicted shift from a maximum penetration diameter of ~0.3 μm is predominately a result of the median fiber diameter used in the media being reduced from 1 - 2 μm to median diameters in the range from 0.3 μm to 0.5 μm [8]. Experimental measurements of the maximum penetration diameter at rated flow for the media used in modern HEPA filters and the HEPA filters themselves show the diameter to be in the range from ~0.13 μm to ~0.17 μm [11,13, and 14]. Variations in the size of maximum penetration arising from differences in the fiber composition and fiber volume fraction of HEPA filter media, as well as other factors, make specification of a precise size of maximum penetration impossible. In terms of the USDOE test, the predicted maximum penetration aerosol diameter at rated flow is different than the diameter at 20 per cent rated flow. Therefore, a maximum penetration test should specify measurement of penetration at a particle size within a range of particle sizes. The theoretical and experimental evidence cited above indicates that this diameter range should be 0.1 μm to 0.2 μm for the current generation of nuclear grade HEPA filters.

Penetration Rejection Criterion

The penetration rejection criterion of the proposed new filter efficiency test specifications is 0.03 per cent penetration at a particle size in the diameter range from 0.1 μm to 0.2 μm. This criterion is in general more stringent for intact filters than the currently designated penetration rejection criteria because measurements made with the current system are approximately equivalent to the penetration near 0.3 μm which is distinctly below the penetration at the size of maximum penetration. However, penetration measurements on filters with pinholes or other defects can be

independent of aerosol size (see the "Damaged Filter Tests" section)
with penetration at a diameter of 0.3 μm being approximately equal
to the penetration at a particle size in the diameter range from
0.1 μm to 0.2 μm. In this case the current and proposed criterion
are equally stringent. Therefore, the 0.03 per cent penetration limit
is necessary to guarantee that in all cases the proposed criterion is
at least as stringent as the current criterion.

Examination of the penetration of over 800 nuclear grade HEPA
filters at specific aerosol diameters over the range from ~0.1 μm
to ~0.4 μm shows that all but a small fraction of the filters
could meet the above criterion [14]. However, the examination also
showed that the fraction of filters failing to meet the proposed
criterion was larger than the fraction of filters failing to meet the
current penetration rejection criterion [14].

Test Material Density

The new filter test specifies that penetration measurements are to
be made with an aerosol material with a density near
$\rho = 1$ g/cm^3. A theoretical study by Tillery shows the magnitude
of maximum HEPA penetration decreasing as the ρ of the challenge
aerosol material increases [9]. Experimental support for this finding
was observed by Tillery in a study where the maximum HEPA filter
penetration of plutonium dioxide aerosol particles
($\rho = $ ~10 g/cm^3) was observed to be significantly lower than
the penetration measured using the current approved DEHP/DEHS
($\rho = 0.983/0.915$ g/cm^3) test method [15,16]. These findings
indicate that, for a given aerosol size, test measurements made with a
material of low ρ relative to the ρ of materials encountered in
the field provide a worst case estimate of filter penetration. For
the USDOE, much of the particulate airborne contaminants of concern
are composed of actinide compounds which rarely, if ever, have
densities less than 1 g/cm^3. The actinide compounds also include
materials with AACs that are among the lowest of all AACs, so that
conservative filter efficiency estimates may be necessary to insure
that exhaust air is sufficiently decontaminated [1,2]. Certain
particulate materials found in USDOE facilities, like lithium
compounds, may have densities less than 1 g/cm^3. In general,
these materials have higher AACs than the actinides so that
conservative estimates of filter efficiency may not be as critical as
they are for filters used in actinide air cleaning systems [1,2].

NEW FILTER EFFICIENCY TEST SYSTEM

The method for QA penetration testing of size 5 HEPA filters
(rated flow of 1000 CFM [~28 m^3/min]) at USDOE FTFs comes
largely from military standard MIL-STD-282 [17]. Since the adoption
of MIL-STD-282, there have been many advances in aerosol technology
which have potential for beneficial application to QA penetration
testing in the areas of reproducibility, accuracy, ease of operation,

and development of more detailed and meaningful filter performance data.

An investigation of alternative filter penetration test methods was undertaken at Los Alamos National Laboratory starting in 1982 with funding from the USDOE Airborne Waste Management Project Office. Commercially available aerosol instrumentation and technology was reviewed with regard to needs identified in the current high flow (500 CFM [~14 m^3/min] and greater rated flow) test system (Q107 test system). Consideration was given to developing a test system that would meet the current test specification and at the same time be capable of measuring penetration at the size of maximum penetration which is required in the proposed new test specifications. Once the most promising alternative test system components were selected, a laboratory evaluation of the components was carried out to determine the best match to meeting the identified needs of the Q107. From this evaluation, a set of components comprising the High Flow Alternative Filter Test System (HFATS) were selected. Laboratory-scale and full-scale systems were evaluated and refined at Los Alamos. In late 1984 and early 1985, full-scale proof-testing and a public demonstration of the prototype system were conducted at the Oak Ridge Filter Test Facility (ORFTF), Oak Ridge Gaseous Diffusion Plant at Oak Ridge, Tennessee. A summary report on the development of the HFATS was published in 1985 [18]. A final report is in preparation [19]. An operational evaluation of the HFATS was initiated in late 1985 at the ORFTF. Data collection for this evaluation was completed in early 1986. A report on the results of the operational evaluation is to be published in the proceedings of the 19th DOE/NRC Nuclear Air Cleaning Conference to be held August 17-21, 1986, in Seattle, Washington [14].

HFATS Description

The HFATS takes advantage of commercially available aerosol technology. The system, which is shown diagrammatically in Fig. 1, uses a modified Laskin nozzle aerosol generator/aerosol neutralizer system to provide the filter challenge [19,20]. The neutralizer is used to standardize electrical charge on the aerosol challenge. The challenge has a measured geometric mean diameter of ~0.2 μm and a geometric standard deviation of ~1.6 [19]. The aerosol concentration produced by the system in an ~28 m^3/min flow test stream has been measured to be ~5 mg/m^3 which is approximately one-tenth the concentration produced by the currently used thermal generation system [19]. Operation of the generation system over a period of months under actual test conditions indicates that the lower challenge concentration may significantly reduce potential operator exposure to test aerosol materials [14]. In addition, the new generation system is easier to operate than the current thermal generator, operates well below the flash point of DEHP, and is expected to produce no decomposition materials.

A laser aerosol spectrometer (LAS, Model LAS-X, Particle Measuring Systems, Inc., Boulder, Colorado) interfaced with a microcomputer (Model HP-85B, Hewlett-Packard Co., Corvallis, Oregon) and an aerosol diluter has shown the greatest potential for fulfilling the aerosol

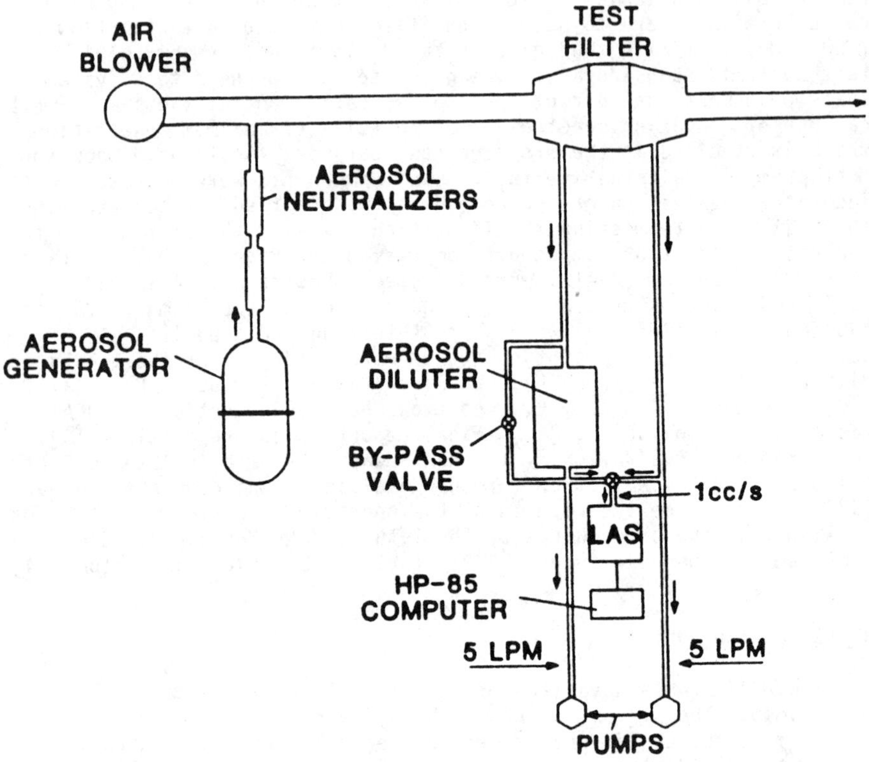

Figure 1.   A schematic diagram of the HFATS showing the major
components of the system which includes the aerosol
generator, the aerosol neutralizers, the HFATS aerosol
diluter, the laser aerosol spectrometer, and the HP-85
computer.

monitoring needs of the HFATS [18,19]. Laboratory evaluation
indicated that the LAS can accurately measure aerosol diameter size
from ~0.12 µm diameter to over 0.4 µm and can accurately measure
aerosol size distributions with concentrations up to
~3 x 10^9 particles/cm³. Because filter challenge
concentrations over 10^5 particles/cm³ are required for filter
testing it was necessary to use a diluter in conjunction with the
LAS. The LAS/microcomputer/diluter aerosol monitoring system combines
the function of the Owl and SLP. It is capable of measuring
penetration at a specific aerosol size or over a range of sizes.

All evaluations of the alternative test system conducted thus far
have been performed using DEHP. The system was designed so that
operation with a variety of test materials is possible. During the
course of the investigation, alternatives to DEHP have been identified
in terms of certain toxicological and physical criteria [18,19].
Enough information has been obtained so that alternatives to DEHP
could be put into use at the FTFs with limited testing should USDOE
decide to eliminate the use of DEHP.

HFATS Performance

Damaged Filter Tests: A series of damaged filter tests were
performed to evaluate the response and sensitivity of the HFATS to
filters with 2 mm holes. After one hole was placed at the edge of the
filter, penetration measurements were made in four different positions
on the test chuck, each time rotating the filter in the plane of the
chuck by 90°. At airflow rates of ~28 m³/min and 200 CFM
(~5.7 m³/min) a significant difference in the measured
penetration was observed relative to the intact filter but no
significant difference was observed in the penetration measurements
made in the four positions. These results indicate that the
penetration measurements using the alternative test system were not
sensitive to the position of leaks.

A series of penetration measurements were also made at both
airflow rates on the filter with 0, 1, 3, 5, and 9 2-mm holes. The
penetration values were calculated by dividing the aerosol
concentration at a specific size measured downstream of the filter by
the upsteam concentration measured at the same aerosol size. Figure 2
shows, as expected, that for both filter flow rates, penetration
increased with increased damage to the filters. For every level of
damage, penetration measurements were greater for the lower airflow
rate than for the high airflow rate. This finding agrees with filter
"pinhole leak" theory which states that the fraction of flow passing
though holes in a filter increases as total flow through the filter
decreases over the regime where flow through the holes remains
turbulent [21]. For the tests on the intact filter, this trend was
reversed with the 5.7 m³/min test showing a lower penetration
relative to the ~28 m³/min test.

Penetration dependence on aerosol size is evident for the intact
filter operating at ~28 m³/min (Figure 2) with maximum
penetration at an aerosol diameter of ~0.15 µm, and penetration

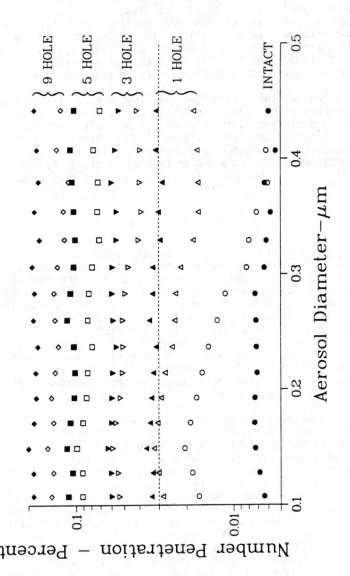

Figure 2. Result of damage filter tests. Solid symbols indicate tests conducted at 5.7 m³/min. Open symbols indicate tests conducted at 28 m³/min.

for 0.3 μm diameter particles approximately half the maximum. For damaged filters operated at ~28 m^3/min, the "shape" of the penetration curve flattens as damage to the filter progresses, with penetration virtually independent of aerosol size for the measurements of the filter with five and nine holes. Independence of penetration as a function of aerosol size is a result of the increased fraction of the downstream aerosol that is associated with leaks. In contrast, intact filter penetration, as shown in Figure 2, is a function of aerosol size, which is a result of different upstream and downstream aerosol size distributions. As the aerosol penetrating through the holes dominates the aerosol penetrating the filter, any dependency of penetration on size is eliminated.

Results of this series of studies on damaged filters demonstrates that the HFATS can be used to measure penetration of size 5 filters operated at airflows of ~28 m^3/min and ~5.7 m^3/min and is sensitive to leaks in these filters.

Intact Filter Tests: Because the HFATS is capable of measuring penetration as a function of size in the aerosol diameter range of ~0.1 μm to ~0.4 μm, the measurements made with the HFATS provide information on the performance of the HEPA filters purchased by DOE and DOE contractors that the Q107 system is not capable of providing. Some typical penetration measurement results on individual filters are shown in Figures 3, 4, and 5. The measurements were made on size 5 filters operating at ~28 m^3/min and ~5.7 m^3/min.

The data presented in Figures 3, 4, and 5 are typical of the penetration measurement results routinely observed for the filters tested in HFATS operational evaluation where over 800 nuclear grade HEPA filters were tested [14]. For the full rated flow, maximum penetration is normally observed in the vicinity of 0.15 μm diameter which agrees with theoretical and experimental evaluations of nuclear grade HEPA filter media and constructed filters [8,11, and 13]. For 20 per cent of rated flow, penetration is observed to be largely independent of aerosol size.

This latter observation does not agree with the theoretical and experimental findings for nuclear grade HEPA filter media [10,12]. The fact that the penetration is independent of aerosol size for constructed filters at 20 per cent of rated flow suggests the possibility that penetration through filter defects created during construction is dominating the overall filter penetration. From pinhole leak theory, if indeed the defect penetration is dominating overall penetration, this dominance would become more pronounced as flow is decreased [21]. Some limited testing of size 5 filters at 80 CFM (~2.25 m^3/min) was performed by the OR FTF staff to evaluate this hypothesis. These data showed a marginally higher penetration at the lower flow relative to the penetration at ~5.7 m^3/min which supports the defect penetration hypothesis. Additional low flow studies conducted by Los Alamos staff in early February 1986 indicate that the higher penetration measurements associated with the ~2.25 m^3/min flows may be the result of background aerosols and not filter penetration. Additional studies will be required to make definitive conclusions.

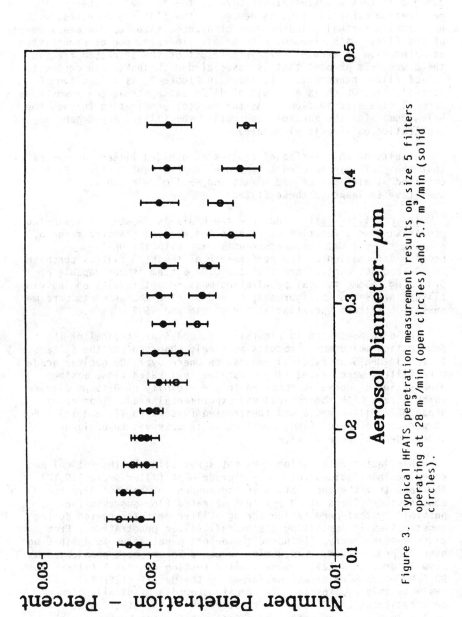

Figure 3.   Typical HFATS penetration measurement results on size 5 filters operating at 28 m³/min (open circles) and 5.7 m³/min (solid circles).

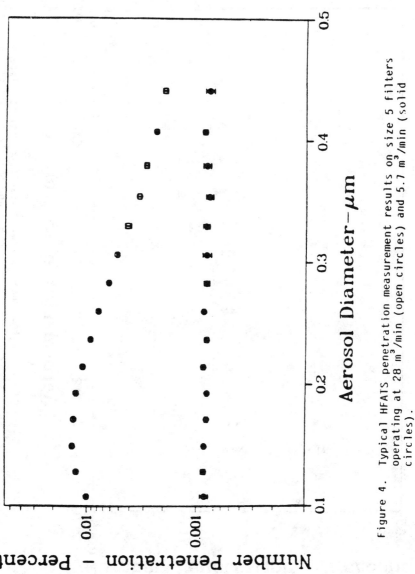

Figure 4. Typical HFATS penetration measurement results on size 5 filters operating at 28 m³/min (open circles) and 5.7 m³/min (solid circles).

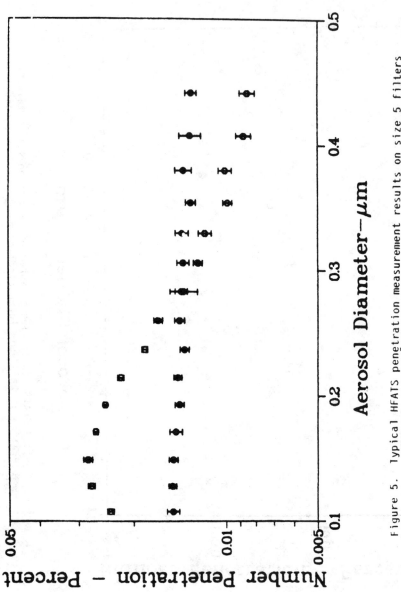

Figure 5.   Typical HFATS penetration measurement results on size 5 filters operating at 28 m³/min (open circles) and 5.7 m³/min (solid circles).

The three filter penetration curves shown in Figures 3, 4, and 5 demonstrate typical measurements on size 5 HEPA filters in the HFATS operational evaluation. These curves also represent three classes of HEPA filter performance that have been observed in the operational evaluation which are: (1) 20 per cent flow penetration equal to or greater than the rated flow penetration over the HFATS size range (see Figure 3), (2) 20 per cent flow penetration below the rated flow penetration over the HFATS size range (see Figure 4), and (3) 20 per cent flow penetration curve intersecting the rated flow penetration curve in the HFATS size range. These curves demonstrate that, depending on the filter and the aerosol size, the penetration at rated flow may be greater than, equal to, or less than the penetration at 20 per cent rated flow.

HFATS Measurement Uncertainty: Two key design criteria for the HFATS were (1) the system was capable of making filter penetration measurements within the individual filter test time requirements of the USDOE FTFs and (2) the precision of the system measurements were within acceptable limits. The allowable time for HFATS measurements was based on the existing time required for testing of filters on the Q107. This time was estimated to be in the range of 3 to 5 min for tests at 100 per cent flow and 20 per cent flow.

No guidance was found to independently suggest acceptable limits for the precision of test system measurements. A precision of ~10 per cent coefficient of variation was adopted for penetration values of 0.03 per cent or larger.

A theoretical model was developed to estimate the uncertainty of alternative test system penetration measurements. The model is based on standard propagation of error techniques neglecting covariance terms and uses Poisson statistics to estimate uncertainties in the upstream and downstream LAS concentration measurements. The model equation is as follows:

$$CV_p = [(P\ N\ T_d)^{-1} + (D/(N\ T_u)) + CV_D^2]^{1/2}\ , \qquad (1)$$

where
    CV_p = coefficient of variation for penetration,
    P = aerosol number penetration,
    N = undiluted upstream count rate, counts/s,
    T_d = downstream counting time in seconds,
    D = dilution ratio,
    T_u = upstream counting time in seconds, and
    CV_D = coefficient of variation for dilution ratio.

A plot of Eq. 1 over the penetration range specified by NE-F-3-43 is shown in Figure 6 for $N = 1.4 \times 10^4$ counts/s (undiluted count rate at 0.3 μm)), $T_d = 60$ s, $D = 220$, $T_u = 10$ s, and $CV_D = 0.05$, which are the selected operating conditions for the alternative test system [5]. The value for N made at ~0.3 μm

aerosol diameter on the HFATS challenge aerosol. The value of CVp
was measured in a HFATS diluter evaluation reported elsewhere
[18,19]. Under these operating conditions, the total time required
to measure the penetration of a filter at two airflow rates would be
less than 5 min, which meets the first design criterion. The plot in
Figure 6 shows that for penetration measurements greater than 0.02
per cent, the CVp is less than 0.1, which satisfies the second
design criterion. The higher CVp values for penetrations below
0.02 per cent are acceptable because the precision of penetration
measurements in this range is not as critical as the precision of the
measurements above 0.02 per cent.

In order to examine the accuracy of the model, the predictions of
the model were compared to estimates of uncertainty made from testing
of five size 5 filters at rated flow. Six penetration measurements
were made on each filter using the alternative test system operating
with a 10 s upstream count and a 60 s downstream count. The average
penetration and the coefficient of variation associated with each
filter were calculated and the results plotted in Figure 6. This
comparison is limited in that only five filters were studied.
However, it is encouraging to note that four of the five measured
data points were at or below the predicted coefficient of variation.
The single CVp measurement in excess of this prediction is for a
measured penetration of <0.005 per cent, so that the relatively high
CVp (~20 per cent) still represents a relatively small variation
in penetration (<$\pm$ 0.001 per cent).

DISCUSSION AND CONCLUSION

Conservative Quality of Test Efficiency Measurements

From the discussion above it is apparent that the current and
proposed filter tests were designed to provide conservative estimates
of filter efficiency. Estimates of filter penetration from
experimental studies of flat sheet HEPA filter media indicate that
penetration at a particle size of ~0.5 µm can be an order of
magnitude lower than the penetration at the size of maximum
penetration [11]. This finding indicates that penetration
measurements at the size of maximum penetration are a very
conservative estimate of filter penetration and that depending on the
challenge aerosol size can over-estimate field penetration by an
order of magnitude or more. This conclusion coupled with the
conservative penetration estimate related to the relatively low
density of the DEHP/DEHS test aerosol serve to give the USDOE filter
test the appearance of being ultra-conservative.

Certain factors serve to mitigate this conservative quality of
the test. HFATS penetration measurements on HEPA filters at 20 per
cent rated flow show an independence to aerosol size so that the
penetration measured at 0.17 µm may be the same penetration that
would be measured at larger sizes. This finding suggests that the

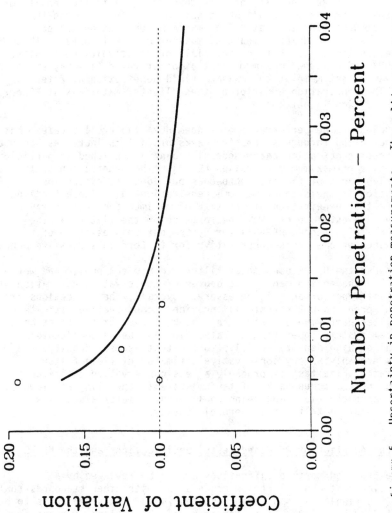

Figure 6.   Uncertainty in penetration measurements. The solid curve is the theoretical prediction of uncertainty. The data points (0) are uncertainty estimates based on actual penetration measurements.

20 per cent rated flow measurements may not be as conservative an estimate of field penetration as the rated flow penetration measurements appear to be. The extent of the aerosol size independence of penetration should be investigated.

As is pointed out in the text, not all materials encountered by USDOE filters have a density greater than or equal to the density of DEHP/DEHS. For the materials which have a density less than the density of DEHP/DEHS the rated flow maximum penetration may be greater than the DEHP/DEHS measured maximum penetration [9]. This observation indicates that for materials of relatively low density, the DEHP/DEHS measured maximum penetration at rated flow may not be a conservative estimate of the maximum field penetration at rated flow. The penetration behavior of these lighter materials at 20 per cent rated flow is not known.

Studies of the performance of tandem HEPA filters indicates that the penetration through successive banks of filters increases because the aerosol penetrating each successive bank is enriched in particles of the size of maximum penetration [9,22]. The penetration of the successive banks of filters approaches but does not exceed the penetration at the size of maximum penetration [9]. These findings indicate that penetration at the size of maximum penetration may be a conservative estimate of field performance of the filters in the first bank of a tandem HEPA filter system but this estimate of performance becomes less conservative for filters in successive banks.

Because the USDOE uses these filters to protect public and worker health and the environment, some degree of conservative estimation of field filter performance is necessary. Overall, the indications are that the proposed USDOE test will provide a conservative estimate of HEPA filter performance. For single bank systems or the first bank of tandem systems, operating at rated flow and being challenged with material of high density relative to the test aerosol density, the test is probably a very conservative estimator of field filter performance. The test is probably a less conservative estimator for filters in the backup banks of tandem systems, operating at flows other than rated flow, and being challenged by materials of low density relative to the test aerosol density.

Plans for Adoption of Proposed Tests Specifications and the HFATS

The proposed test specifications are to be reviewed by a USDOE-selected technical review group along with other standards that are in the final stages of being adopted [4-7]. This review is to be completed by Fall 1986. The standards will be revised as necessary to accommodate the comments of the technical review group. These standards will be identified as mandatory standards in USDOE order 6430 [23].

The USDOE is to be petitioned to consider adoption of the HFATS in mid-1986 according to procedures specified in NE-F-3-43 [5]. Upon acceptance of the petition, Los Alamos is to submit evidence supporting adoption of the HFATS to USDOE. The USDOE with input from the technical review group will decide whether to adopt the HFATS as an improved test method. This decision could be made by the end of 1986. If adopted, the HFATS would be ready for placement in the FTFs by the end of 1987.

In 1987, the US Army Product Assurance Directorate (PAD) plans to initiate evaluation of the HFATS for use in the PAD QA HEPA filter testing program. Implementation of the HFATS at PAD is scheduled to be completed by the end of 1987.

Also in 1987, the USDOE and PAD plan a cooperative effort to adapt the HFATS technology to a Low Flow Alternative Filter Test System (LFATS) for testing size 3 and smaller HEPA filters (rated flows of 125 CFM [~3.5 m^3/min] and lower). The adaptation is scheduled to be completed by the end of 1987. The USDOE plans to perform an operational evaluation on the LFATS in 1988.

ACKNOWLEDGMENTS

The proposed test specifications were co-authored by Ronld C. Scripsick and Marvin I. Tillery of Los Alamos, James F. Bresson of USDOE/Interim Waste Operations, Sidney C. Soderholm of the University of Rochester, and Werner Bergman of Lawrence Livermore National Laboratory. Lloyd Wheat of Los Alamos as well as Soderholm and Tillery participated in the development of the HFATS. Harry J. Ettinger was the program manager at Los Alamos and funding for these activities came from Bresson and Tom Thomas of the USDOE Airborne Waste Management Program.

REFERENCES

[1] Hinds, W. C., Aerosol Technology, John Wiley and Sons, New York, 1982, pg. 10.
[2] Sowby, F. D., "Limits for Intakes of Radionuclides by Workers: A Report of Committee 2 of the International Commission on Radiological Protection," Annals of the ICRP, Vol. 3, No. 1-4, 1979.
[3] Selby, J. M., Watson, E. C., Corley, J. P., et al., "Considerations of Effluents from Mixed Oxide Fuel Fabrication Plants," Pacific Northwest Laboratories Report BNWL-1697, June 1975.
[4] Filter Test Facility Standards Writing Group, "Operating Policy of DOE Filter Test Program," U.S. Department of Energy, DOE Nuclear Standards number NE-F-3-42, February 1986.
[5] Filter Test Facility Standards Writing Group, "Quality Assurance Testing of HEPA Filters and Respirator Canisters," U.S. Department of Energy, DOE Nuclear Standards number NE-F-3-43, February 1986.

[6]   Filter Test Facility Standards Writing Group, "DOE Filter Test
      Facilities Quality Program Plan," U.S. Department of Energy, DOE
      Nuclear Standards number NE-F-3-44, February 1986.
[7]   Filter Test Facility Standards Writing Group, "Specifications
      for HEPA Filters Used by DOE Contractors," U.S. Department of
      Energy, DOE Nuclear Standards number NE-F-3-45T, February 1986.
[8]   Lee, K. W. and Liu, B. Y. H., "On the Minimum Efficiency and the
      Most Penetrating Particle Size for Fibrous Filters," J. Air
      Pollution Control Association, Vol. 30, 1980, pp. 377-381.
[9]   Tillery, M. I., "Determination of Protection Factors for Tandem
      HEPA Filters" in the Proceedings of the 27th Annual Technical
      Meeting of the Institute of Environmental Sciences, 1981 ,
      pp. 47 - 51.
[10] Lee, K. W. and Liu, B. Y. H., "Theoretical Study of Aerosol
      Filtration by Fibrous Filter," Aerosol Science Technology,
      Vol. 1, 1982, pp. 147 - 162.
[11] Liu, B. Y. H.; Rubow, K. L.; and Pui, D. Y. H., "Performance of
      HEPA and ULPA Filters," in the Proceedings of the 31st Annual
      Technical Meeting of the Institute of Environmental Sciences,
      1985, pp. 25 - 28.
[12] Lee, K. W. and Liu, B. Y. H., "Experimental Study of Aerosol
      Filtration by Fibrous Filter," Aerosol Science Technology,
      Vol. 1, 1982, pp. 35 - 46.
[13] Bergman, W.; Bierman, A.; Kuhl, W.; Lum, B.; Bogandoff, A.;
      Hebard, H.; Hall, M.; Banks, D.; Muzumder, M.; and Johnson, J.,
      "Electric Air Filtration Theory, Laboratory Studies, Hardware
      Developments and Field Evaluations," Report UCID-10052 DE84
      004668, Larwence Livermore National Laboratory, Livermore,
      California, 1984.
[14] Scripsick, R. C.; Smitherman, R. L.; and McNabb, S. A.,
      "Operational Evaluation of the High Flow Alternative Filter Test
      System," in preparation to be published in the Proceedings of
      the 19th DOE/NRC Nuclear Air Cleaning Conference, 1986 (to be
      published).
[15] Tillery, M. I.; Gonzales, M.; Elder, J.; Royer, G.; and
      Ettinger, H. J., "Measurements of the Penetration of High
      Efficiency Air Filters Using Plutonium Oxide Aerosols," in
      AEROSOLS: Science, Technology and Industrial Applications of
      Airborne Particles, Liu, B. Y. H.; Lui, D. Y. H.; and
      Fissan, H. J., Eds., Elsevier, New York, 1984, pp. 625 - 627.
[16] Hinds, W., Macher, J. and First M. W., "Size Distributions of
      Aerosols Produced From Substitute Materials by the Laskin Cold
      DOP Aerosol Generator," in the proceedings 16th DOE Nuclear Air
      Cleaning Conference, Report CONF-801038, U.S. Department of
      Energy, Vol. 1, 1981, pp. 125-138.
[17] "Filter Units, Protective Clothing, Gas Mask Components and
      Related Products, Performance Test Methods," Military Standard
      MIL-STD-282, May 28, 1956.
[18] Scripsick, R. C. and Soderholm, S. C., "Summary Report:
      Evaluation of Methods, Instrumentation and Materials Pertinent
      to Quality Assurance Filter Penetration Testing," Los Alamos
      report LA-UR-85-2104, June 1985.

[19] Scripsick, R. C. and Soderholm, S. C., "Final Report: Evaluation of Methods, Instrumentation and Materials Pertinent to Quality Assurance Filter Penetration Testing," being prepared as a Los Alamos report.

[20] Echols, W. and Young, J., "Studies of Portable Air-Operated Aerosol Generators," NRL report No. 5929, Naval Research Laboratory, 1963.

[21] Thomas, J. W. and Crane, J. D., "Aerosol Penetration Through 9 mil HV-70 Filter Paper With and Without Pinholes," in the Proceedings of the 8th AEC Air Cleaning Conference held at Oak Ridge National Laboratory, TID-7677, October 22-25, 1963.

[22] Schuster, B. G. and Osetek, D. J., "A New Method for In-Place Testing of Tandem HEPA Filter Installations," American Industrial Hygiene Association Journal, Vol. 40, 1979, pp. 979 - 985.

[23] USDOE, "General Design Criteria Manual," USDOE Order 6430, Issued June 10, 1981 (in revision).

Douglas E. Fain

STANDARDS FOR PRESSURE DROP TESTING OF FILTERS AS APPLIED TO
HEPA FILTERS

REFERENCE: Fain, D. E., "Standards for Pressure Drop
Testing of Filters as Applied to HEPA Filters,"
Fluid Filtration: Gas, Volume I, ASTM STP 975, R.R.
Raber, Ed., American Society for Testing and
Materials, Philadelphia, 1986

ABSTRACT: A study has been made of the pressure drop testing performed by three filter test facilities for the
Department of Energy. The testing is performed for both
pressure drop and filter efficiency to assure that the
filters meet specifications for use in the nuclear industry.
It is shown that serious errors in pressure drop testing
can occur if mass flow rates are used without correcting
the pressure drop to a standard condition. Several alternatives are given for testing methods, including use of
volume flow rates or laminar flow meters as testing standards. Similar standards should be considered for general
filter pressure drop testing.

KEYWORDS: gas filters, filter pressure drop, filter testing standards, filter physical properties

INTRODUCTION

     The Department of Energy operates Filter Test Facilities at Oak
Ridge, TN, Rocky Flats, Co, and Hanford, WA. These three facilities
are operated primarily to test pressure drop and penetration efficiency
for HEPA filters used in the nuclear industry. It is common practice
to use pressure drop as one figure of merit for the performance of
filters. Usually, not very high accuracy is needed for the pressure
drop measurement and since in many cases the filters are operated at
ambient atmospheric pressure, often little attention is given to how
the pressure drop measurement is acquired. Some quality assurance
efforts have been made to determine if the three facilities were
measuring the same results on the same filters. A comparison of the
results from these tests has been less than satisfactory. Therefore
some work has been performed to improve the quality of the pressure
drop testing capabilities for the three facilities. The purpose of

     D. E. Fain is Section Head for the Process Physics Section of the
Materials and Chemistry Department at the Oak Ridge Gaseous Diffusion
Plant operated by Martin Marietta Energy Systems, P.O. Box P, MS 271,
Oak Ridge, TN 37831.

this paper is to present some of the results of this work and indicate how they may apply to filter pressure drop testing in general.

There are several aspects of filter pressure drop testing that will be addressed. Major attention must be given to the purpose of the testing. Testing may be performed primarily as a quality assurance measure to ensure that the filters meet the manufacturer's specifications, to evaluate the performance characteristics of filters, or to measure some physical properties of the filters. The purpose of the testing should govern what type of tests will be performed and the standards that are needed for those tests.

The first standard needed for pressure drop testing is a known measure of the flow rate. This sounds simple, but good flow standards are hard to acquire and a particular flow device may be more convenient for certain measures of flow rate. Attention must be given to both the type of flow standard (hardware) and to the measurement of the parameters required to determine the flow rate. It will be necessary to have precise and accurate pressure measurements for both ambient pressure and pressure drop across the filter. It also maybe necessary to have precise and accurate temperature measurements. Attention should be given to how these measurements are acquired. The cost of the testing is almost always a major concern. The speed, improved accuracy, data management, and general efficiency of modern data acquisition and analysis techniques can have significant economic impacts and should be carefully considered for use in acquiring and managing the data.

Depending on the objective, some of the types of tests that could be performed are:

Quality Assurance
    Pressure drop for a single constant mass flow rate
    Pressure drop for a single constant volume flow rate
    Mass flow rate for a fixed pressure drop
    Volume flow rate for a fixed pressure drop

Measure performance characteristics
    Pressure drop measurements over the range of normal atmospheric
    ambient pressures
        with a range of mass or volume flow rates

Measure filter physical characteristics
    Pressure drop over large range in ambient pressure
        with a constant mass or volume flow rate

SUMMARY

The major point to be made from this study is that pressure drop testing alone can be very misleading unless considerable care is taken to ensure that the pressure drop measurement can be compared with an expected value or standard. When mass flow is used for the tests and no correction is made for ambient measurement conditions, variations in pressure drop of as much as 50% could be observed at different testing locations. Such an error will not occur in the case of performance or characterization testing, because a detailed analytical evaluation of the measurements is required to achieve the desired results. When pressure drop only is desired, a simple solution is to use a specified volume flow rate for testing.   Specified volume flow rates will generally result in an accuracy of 5% or better regardless of where the filter is tested, so long as it is tested at ambient atmospheric pressure. A better solution is to correct the measured pressured drop to some specified conditions.   In this case the correction will be smaller if volume flow rates are used for the test measurements, but either volume or mass flow rates can be used for the tests.   The correction can be made just for variations in ambient pressure and temperature or can also include corrections for known filter performance characteristics.   A precision of 1% or better can be achieved with reasonable care and quality control.   Another solution is to use a viscous laminar flow meter for metering the gas flow in the test.   In this case only the pressure drop for the standard laminar flow meter and the pressure drop for the filter need to be measured. Precision and accuracy of 1% or better can be achieved with reasonable care and quality control. Detail for these methods, the needed care and possible exceptions are given below.

ELEMENTS OF FILTER TESTING

Significant Measurable Parameters

Since the purpose is to get information about the filter and not the flow measurement, the flow characteristics of the filter are of particular importance.   Generally, the pressure drop information is needed for quality control, to compare with other filter's performance, measure physical characteristics expected from a manufacturing process, or for information to develop or improve a manufacturing process. The more that is known about the filter's physical characteristics, the less quantity and variety of pressure drop data will be needed to get the desired information. Conversely, the less that is known about the filter's physical characteristics, the more quantity and variety of pressure drop data will be needed to get the desired information.   If little is known about the filter's physical characteristics, then the range of predictable performance (for example for designing filter systems) and the range for comparison of performance with standards or with other filters is restricted to the range for which experimental data have been acquired.

Filter pressure drop performance can be assessed and can be predictable in terms of several different types of flow that may be associated with the filter or the construction of the filter.

1. Inertial Flow (Bernoulli Flow)

This flow is characterized by a simple exchange of potential energy to kinetic energy when the gas velocity is increased. For the application to filter pressure drop measurements, the pressure drop is given by

$$\Delta P_i = K_a \rho V^2 \tag{1}$$

or
$$\Delta P_i = K_i F^2 / \rho \tag{2}$$

2. Viscous Laminar Flow (Streamline or Poiseuille-Hagen Flow)

This flow is characterized by a parabolic velocity profile over the cross section perpendicular to the flow direction. The forces required to produce flow are determined by the shear stress in the gas. The pressure drop is given by

$$\Delta P_m = \frac{\eta\ TF}{K_m M \bar{P}} \tag{3}$$

3. Slip Flow (Maxwell Slip Flow)

This flow is an addition to the laminar flow and is caused by slip at the flow path wall. A natural assumption for boundary conditions for laminar viscous flow is a velocity of zero at the flow path wall. Actually the velocity at the wall is not zero but is finite and determined by diffusion at the wall. When the smallest dimension of the flow path perpendicular to the flow velocity is very large in comparison to the mean free path of the gas molecules, this diffusional slip velocity is small in comparison to the velocity at the center of the flow channel and can be ignored. However, in many filters the effective pore size of the filters is such that this slip velocity should not be ignored. With the slip flow included, the flow rate is

$$F = \frac{K_m M \bar{P}}{\eta T}\ \Delta P_s + K_s \left(\frac{M}{T}\right)^{1/2} \Delta P_s \tag{4a}$$

the pressure drop is

$$\Delta P_s = F / \left[\frac{K_m M \bar{P}}{\eta T} + K_s \left(\frac{M}{T}\right)^{1/2}\right] \tag{4b}$$

and the specific flow is

$$F / \Delta P_s = \frac{K_m M \bar{P}}{\eta T} + K_s \left(\frac{M}{T}\right)^{1/2} \tag{4c}$$

4. Turbulent Flow

Turbulent flow generally will not become a factor until the Reynolds number is greater than about 2000. The Reynolds number should be calculated for a given filter for reference. However, because of the nature of filters it is very unlikely that a filter will be operated under conditions of turbulent flow. Therefore, turbulent flow will be ignored in this paper.

It should be noted that the pressure drop is affected differently by variations in the flow, ambient pressure, and temperature for each of the above types of flow. When all three types occur simultaneously, non-linear behavior occurs which might even appear to be random error in the measurement. A rather comprehensive set of flow measurements must be made in order to determine the relative magnitude of the effect on pressure drop of these various types of flows. These will be discussed in some detail below.

Noting the variables in the above equations, the important measurable parameters are the pressure drop across the filter, the entrance or exit (ambient) pressure, the temperature of the filter (hopefully the ambient temperature), and the flow rate. The flow rate can be expressed as a mass flow rate or a volume flow rate. This will be discussed in more detail below.

Significant Physical Characteristics of Filters

The constants in the above flow equations will contain the physical properties of the filter. Therefore variations in the physical properties of the filter will be observed by measured changes in these constants. The physical properties exhibited by these constants include such parameters as the area open to flow (perhaps indicated by a void fraction), a pore or flow channel diameter characteristic for the filter, the filter thickness, the specific surface area, and perhaps others depending on the type of model used to describe the filter. Since these physical parameters are model dependent, they are not very precise and may be interdependent. But for a given filter these are constant and variations in these constants imply variations in the physical properties of the filters. The details of the relation between these constants and the physical properties will probably be of interest only when developing new filters or investigating the physical changes that might result from using a filter under various conditions that might cause physical changes in the filter. A discussion of the relation between physical properties and these constants is beyond the scope of this paper.

Relationship Between Parameters and Performance

In a filter application, the predominate flow mode should be viscous laminar flow, with various amounts of slip and inertial flow. The fractional amount of slip flow increases as the filter pore size decreases. The inertial flow occurs at the entrance and exit to the filter and produces an additive pressure drop in addition to the pressure drop from the laminar and slip flow. Inertial flow occurs when the gas velocity changes, as for example when the area open to flow changes. When the gas velocity changes, a pressure change occurs to produce the acceleration or deceleration. The pressure change is usually small and in most systems can be ignored relative to other pressure drops. However, filters are usually operated at low pressure drops and therefore these inertial pressure drops frequently are not ignorable. In addition, the manufacturing practices usually involve methods of putting large amounts of the filter material into a small space to reduce the pressure drop across the filter. The small space between folds in the paper needed to produce large increases in frontal area causes a larger acceleration of the gas at the entrance and the exit.

It would be helpful to know the relative amounts of the different types of flow for sizing and for the general design of specific applications of filters. It is perhaps even more important to know the partition of the different types of flows when the filters are to be tested at different locations. Since the different types of flow have different dependencies on ambient pressure and temperature, tests performed at different locations may produce different results which would be interpreted as deviations from expected performance or deviations from the manufacturer's specifications.

An estimate of the relative amount of the different types of flow can be determined from appropriate flow measurements. Analytically, the total pressure drop can be expressed as a sum of the inertial pressure drop and the laminar and slip pressure drop,

$$\Delta P = \Delta P_i + \Delta P_s \qquad (5)$$

These equations can be put in more useable form by appropriate algebraic manipulation with Eqs. 2 and 4b

$$\frac{M\bar{P}\,\Delta P}{\eta T F} = \frac{1}{K_m\left(1 + \frac{K_s\eta}{K_m\bar{P}}\,(\frac{T}{M})^{1/2}\right)} + K_i\frac{F}{\eta} \qquad (6)$$

Using this form for a correlation equation and acquiring appropriate sets of data under conditions that will emphasize particular modes, the relative magnitude of the various flow modes can be established through regression analysis. It is helpful to define a quantity called the Y-Cluster by

$$Y = \frac{M\bar{P}\,\Delta P}{\eta T F} \qquad (7)$$

The Y-Cluster is actually the ratio of all the variables in the equation for viscous laminar flow divided by the measured mass flow rate. If the flow were totally viscous laminar flow then Y would be a constant. Therefore deviations from a constant represent variations from viscous laminar flow.

If data are acquired at a constant pressure and temperature and a varying mass flow rate, a linear regression analysis of the Y-Cluster with the ratio of the flow rate to viscosity will provide an estimate of the inertial flow constant. The viscous laminar and slip constants can be estimated by acquiring data at a constant temperature and flow rate with a varying average pressure. Using the estimated inertial flow constant, the pressure drop from the inertial flow can be calculated and subtracted from the measured total pressure drop to give the slip flow pressure drop, ΔP_s. Then, a linear regression analysis of the ratio of the flow rate to the slip flow pressure drop variable with the average pressure will determine an estimate of the viscous laminar and slip constants, as in Eq. 4c.

Homogeneous and Layered Filters

The above analysis does assume that the filter is a homogeneous structure. That means that the properties such as pore size and void fraction are uniform in all directions. Some filters are purposely made non-uniform. For example, some filters are made in layers with the layers having different pore sizes or different void fractions. This type of structure makes the analysis much more complicated, but the analysis can still be performed with good accuracy using non-linear regression techniques.

FLOW CHARACTERISTICS OF A HEPA FILTER

In order to better understand the testing requirements needed to test HEPA filters at the three different DOE Filter Test Facilities, the flow properties of one of the filters were measured. Pressure drop measurements were made over a broad range of flows, but at essentially constant ambient pressure. In Fig. 1, the pressure drop is plotted against the flow rate. Since the measurements were made at constant average pressure, this plot should be a straight line if the flow were pure viscous laminar flow. The curve implies the presence of inertial

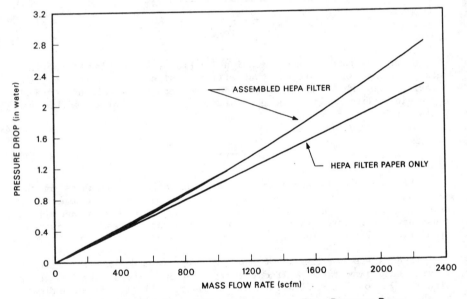

Figure 1 - Effect of Filter Assembly on Filter Pressure Drop

pressure drop. The straight line shows in Fig. 1 shows the results based on measurements made on the filter paper along. In Fig. 2, the Y-Cluster is plotted against the ratio of the flow to the viscosity for the same set of data. As can be seen, this results in a linear correlation from which the inertial flow constant can be estimated from the slope of the straight line. The particular configured test setup did not allow the acquisition of data over a range of average pressure (a different configuration could have allowed a range of pressure from about 12 to 20 psia). This data indicates that the inertial pressure

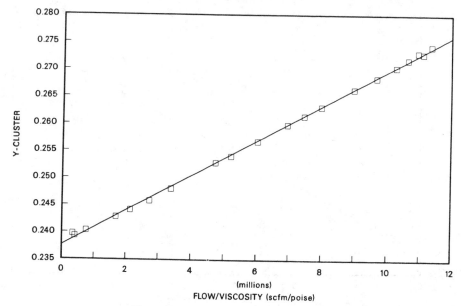

Figure 2 - Y-Cluster for Assembled HEPA Filter

drop contribution varies from zero at zero flow rate to about 15% at 2000 scfm.

In order to determine the physical characteristics of the filter paper alone, it was much more convenient to mount and test small pieces of the filter paper in a laboratory testing system. The flow rates used for these small pieces were in the standard cc/min range. The paper folds present in the finished filter assembly that are the major source of the inertial term are not present to complicate the measurements. Pressure drop data were acquired with three different flow rates over a pressure range of 1 to 25 psia. Two different pieces of paper were tested and one was tested with the flow in both directions. In Fig. 3, the specific flow (ratio of flow to pressure drop) is plotted against the average pressure. As can be seen, an excellent linear correlation is obtained that extrapolates to a finite value at zero pressure as indicated by Eq. 4c. The intercept represents the slip flow contribution and the increase in specific flow with average pressure is the contribution from viscous laminar flow. A similar plot can not be made for the above data for the assembled HEPA filter because all of the data were taken at constant average pressure. However, using the paper only data, the Y-Cluster plotted against ratio of flow rate to viscosity is shown in Fig. 4 and the Y-Cluster plotted against average pressure is shown in Fig. 5. These plots show the complications that result when all three types of flow are present and why it is necessary to separate them in order to understand the filter performance. This complication results because the slip flow becomes a larger fraction of the total flow as the average pressure decreases. For this filter, slip flow is about 18% of the total flow at atmospheric pressure. In testing these filters for quality assurance or for performance, the effect that testing conditions have on the relative changes in the three types of flow and the subsequent change in measured pressure drop may need to be taken into account.

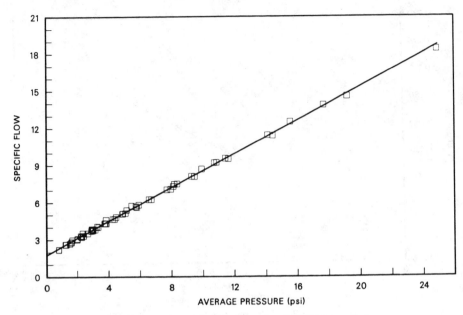

Figure 3 - Specific Flow for HEPA Paper Only

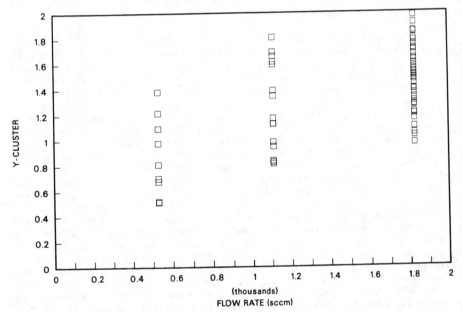

Figure 4 - Y-Cluster for HEPA Paper Only

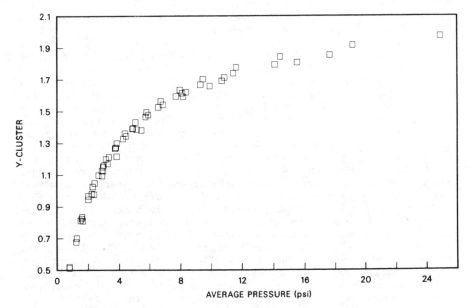

Figure 5 - Y-Cluster for HEPA Paper Only

FLOW STANDARDS FOR TESTING

    In order to perform gas flow tests on filters, some device that
can produce a known reproducible flow rate is necessary. The quality
of the flow standard needed for testing is dependent on the type of
testing that will be done and whether the test results will be compared
with results from other tests from other locations. If the measurement
is needed for manufacturing quality assurance, then the device need only
produce the same flow rate at every occasion needed for testing. If
filter characterization is the intent of the measurements, then the
device must be capable of producing a range of flow rates that have
approximately the same relative accuracy over the whole range. If
measurements are to be compared with results from other locations or
represent results with true accuracy, then the flow device should be
calibrated with the same standard or with standards traceable to the
same standard or the National Bureau of Standards.

    There are many different types of devices for producing known flow
rates. The device used by the DOE FTFs is an orifice plate. The orifice
plate is a flat plate with an appropriate number of equally sized
symmetrically arranged holes drilled in the plate. This device must be
calibrated unless an accuracy of only 10% is acceptable. If the ambient
pressure and temperature are measured, the reproducibility of the device
is as good as the reproducibility of the pressure, pressure drop and
temperature measurements. It is relatively easy to get these measurements to accuracies of better than 1%. Two sets of orifice plates were
calibrated for each of the FTF's, one for flow rates of about 25 to
300 scfm and another from about 200 to 2500 scfm. The smaller plates
were calibrated to a precision of about 1%, but the precision on the
larger plates was about 0.25%. The standard used for the flow measurements was a viscous laminar flow meter with a special ORGDP designed
housing. The accuracy of the standard used was ±0.5% traceable to the
Bureau of Standards. The poorer precision with the smaller orifice

plates was primarily related to the unsteady (pulsating) nature of the pressure drop across the orfice plates.

Usually, the type of device is not important except in terms of available instrumentation, accuracy needed for the tests, ability to calibrate the device, and the repeatibility and stability of the device. However, there are some special considerations in the case of filter testing, especially if the testing is primarily quality assurance. Since the flow through filters is mostly viscous laminar flow, there is a significant advantage in using a flow meter based on the same principle. If the functional dependence of the flow on pressure and temperature are the same for both the flow meter and the filter (and they are both at the same pressure and temperature), then the only measurements needed are the pressure drop across the flow meter and the pressure drop across the filter. A reproducible volume flow rate can be set on the flow meter by setting a specified pressure drop. The pressure drop across the filter will then be comparable to the pressure drop measured at any other measurement conditions (a small correction may be necessary depending on the relative amounts of inertial and slip flow in the filter). The use of any other flow device will require the measurement of more parameters and more corrections. A properly maintained and calibrated viscous laminar flow meter should be seriously considered for use as the flow standard for pressure drop testing of filters.

PRESSURE AND TEMPERATURE STANDARDS

Primary devices for measuring pressure and temperature are certainly more available than flow meter standards. However, these measurements should not be ignored. If an inclined gauge is used, care should be taken to assure that the gauge is at the correct angle and that it contains clean fluid with the correct density. No matter what type of instrument is used, care should always be given to the setting of instrument zeros and to proper reading techniques. Good practices can contribute substantially to the precision and accuracy of the measurements. Modern pressure and temperature transducers can also contribute to ease of acquisition and precision if proper attention is given to calibration of the transducers. Relatively inexpensive desktop computers or personal computers can now be used to acquire data and process it to produce accurately corrected reports of the test information in whatever form the manufacturer or customer may need.

SELECTION OF THE TESTING METHOD

The criteria for selecting a test method has been discussed above. However, there are a few other aspects of the testing method that should be mentioned. If quality assurance is the primary reason for the testing then it is likely that a simple measurement of pressure drop at one or two flow rates will be sufficient.

Pressure Drop Only

The most common method for single point pressure drop measurements is to set a specified mass flow rate and measure the pressure drop across the filter. This is a useful and comparable measurement only if the filter is at the same ambient pressure and temperature each time a measurement is made. This may not be a problem for a manufacturing or testing facility with a test unit in an air conditioned constant temperature area and the local atmospheric pressure does not vary much. However, if the same filter is tested for pressure drop at another facility with different ambient pressure and temperature the measured pressure drop can be significantly different. The results will be interpreted as a deviation from specification unless the proper correction is made to the measurement. As an example, suppose a filter is measured for pressure drop with a specified mass flow at a sea level location on a cold day when the atmospheric pressure is 14.7 psi and the temperature is 50°F. The pressure drop for this filter is measured to be 1.0 inches of water. The filter is shipped to Denver and tested for pressure drop with the same specified mass flow on a hot summer day when the atmospheric pressure is 12.0 psi and the temperature is 100°F. If the effect of inertial and slip flow are small, then the pressure drop at Denver on that particular day would be measured to be about 1.5 inches of water or 50% higher than that measured at the sea level location. Therefore, it is clear that if constant mass flow is selected as the testing method, a correction to a common standard condition should be applied to the pressure drop measurement before it can be expected to be compared with measurements made at other measurement conditions.

Volume flow is simply the ratio of mass flow to the gas density. The volume flow rate is also a measure of the gas velocity averaged over a cross section normal to the flow path. With viscous laminar flow, the volume flow is independent of the gas density. Volume flow is simply proportional to the ratio of the pressure drop to the gas viscosity. The equation for pressure drop for viscous laminar flow using volume flow is

$$\Delta P_v = K_v \eta Q \tag{8}$$

In this case the pressure drop is directly proportional to the product of the volume flow and the gas viscosity. The gas viscosity is independent of pressure but is temperature dependent. Even so, the dependence on ambient measurement conditions is considerably less than when the mass flow rate is used as the specified flow condition. In the example given above, if a volume flow specified measurement condition is used for the test and set at a value to produce 1.0 inches of water pressure drop, then the pressure drop measured on the same day in Denver, as in the example above, would be 1.14 inches of water or 14% higher than that measured at the sea level location. Therefore for pressure drop only measurements, it is perferrable to use a specific volume flow rate for the measurement.

In Fig. 6, the volume and mass flow rates needed to produce one inch of water pressure drop for three different temperatures for a HEPA filter are plotted as a function of ambient pressure. These flows were calculated using a model with constants inferred from the data obtained with the HEPA filter. The variation is volume flow is considerably less than the variation in mass flow.

Filter Characterization or Performance

The testing method chosen for characterization and performance information should definitely include pressure drop measurements at a constant ambient pressure over a range of flow rates. Some pressure

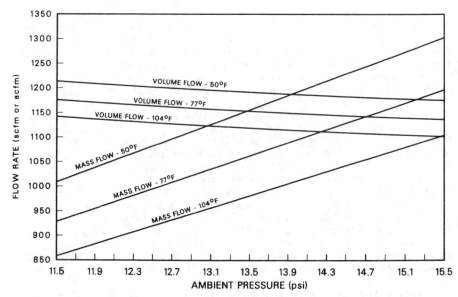

Figure 6 - Mass and Volume Flow Rates for Constant Pressure Drop

drop data taken with a broad range of average pressure would also be important to separate the effects of inertial and slip flow. Whether mass flow rates or volume flow rates are used is relatively unimportant in this case, because acquisition of pressures, temperatures, pressure drops, and flow rates are all required information. A mathematical analysis of the data is required to infer the information needed. The analysis can be performed with either a measured mass flow or with a measured volume flow.

Viscous Laminar Flow Meter for a Standard

A considerable amount of work has been performed at ORGDP using viscous laminar flow meters as standards. The experience has been quite good. Using appropriate correlation techniques [2] similar to those mentioned here, this type of flow meter can be calibrated and used over a broad range of pressures and flow rates. Calibration precision of 0.1% has been routinely achieved. No significant changes in the calibration have been observed from accumulation of dust or other contaminants. If the test gas is particularly dirty a filter can be used upstream of the flow meter it can be cleaned at appropriate intervals, or it can be or recalibrated at appropriate intervals.

The viscous laminar flow meter provides a significant advantage for testing for pressure drop with filters which exhibit primarily viscous laminar flow. Since both the flow standard and the filter have the same dependence on pressure and temperature, the pressure and temperature are not needed to interpret the filter pressure drop and therefore do not need to be a measured parameter of the test. The flow rate for the test would be obtained by setting the pressure drop across the viscous laminar flow meter at a specified value. It would then be necessary to measure only the pressure drop across the filter. That measured pressure drop would be comparable to the pressure drop measured on that filter at any other ambient operating conditions.

QUALITY ASSURANCE MEASURES

First priority on quality assurance is the flow standard. No matter what flow standard is chosen to set the flow rates for the tests, it should be calibrated by a good calibration laboratory. Once the standard has been calibrated, care should be taken to use the flow standard properly and acquire all the appropriate parameters to assure a correct setting of the flow rates. The same practices used in calibrating the flow meter should be used in making the flow measurements, i.e., the same pressure taps should be used in both cases.

Good measurement practices are essential to assure the acquisition of quality data. Pressure tap lines must be leak free and measuring devices properly zeroed. Instrument checks should be made on a routine basis. These practices should be documented according to a quality assurance plan to track the performance progress.

While the flow standard can be used continuously in the test system to monitor and assure that the flow remains constant, frequently the standard is removed from the test system and a secondary device is used to monitor and maintain the flow rate constant. It is good practice to check the monitor at the end of the testing period or at some regular interval to assure that the monitor itself is stable and not changing with time.

It is also good practice to implement a quality control program. Select two or three production filters and keep them in a safe protected area. Test these filters at regular intervals (but different frequency for at least one of them). A control chart (a plot of test result vs time) can be used to estimate the precision of the test and to indicate when the test precision changes or when an erroneous bias creeps into the test results (trouble with the system in either case) and maintenance is needed to correct the change.

RECOMMENDATIONS AND CONCLUSIONS

The problem of defining standards is an issue only for quality control or quality assurance where the testing consists of one or two point pressure drop measurements which will be the sole measure of figure of merit. The standards required for performance and characterization measurements are the usual well calibrated flow meters and pressure and temperature measuring devices. Analysis and correlation of data are an important part of the performance or characterization measurements so modern data acquisition equipment and small computers can be cost effective and produce other information that might otherwise be overlooked.

Standards are needed for tests involving one and two point pressure drop measurements. The character and variation of character among filters significantly complicates the task of setting standards. In fact, unless the filter flow character involves only one type of flow, then absolute standards can not be set. Depending on the accuracy of the quality control needed, it may be plausible to ignore the contributions from inertial and slip flow and set standards based on viscous laminar flow alone. This possibility needs to be examined for each type of filter by obtaining appropriate filter characterization flow measurements. These measurements can be used to estimate the limits of uncertainty for the given filter type to determine if ignoring all but viscous laminar flow to establish standards will provide adequate accuracy. If more accurate results are needed, these data can be used to develop charts, graphs, or analytical functions to make appropriate corrections to some standard operating condition.

These are some clear and straight forward alternatives.

1. Use a viscous laminar flow meter for the flow standard.

The best results will usually be achieved with the method that requires the acquisition of the smallest amount of data and corrections to that data. With this flow standard only the two pressure drops and the temperature are needed and in some cases the temperature may not be needed. In addition, because filters generally will be expected to exhibit mostly viscous laminar flow, the use of this flow standard will require the smallest correction to the measured pressure drops. Since the effects of inertial and slip flow tend to offset each other and high accuracy is usually not required, in most cases a correction to the measured pressure drops will not be necessary.

2. Use a flow meter that will provide a volume flow for the flow standard.

In most cases this alternative will be adequate. For best results the pressure drop should be corrected to some standard temperature such as 25°C. In many cases the correction will be small enough to be ignored. The correction needed is relatively simple. The measured pressure drop will be multiplied by the ratio of the viscosity of the test gas at the standard temperature to the viscosity of the gas at the test ambient temperature. For more accuracy, specific filter characteristic correction for the contribution from inertial and slip flow may also be made.

3. Use a flow device that will provide a mass flow for the flow standard.

In most cases the measured pressure drop should be corrected to specified standard conditions, such as a standard atmosphere (14.696 psia) and standard temperature (25°C). This alternative will usually give the largest correction. The measured pressure drop is corrected by multiplying by the ratio of the ratio of gas density to viscosity at test ambient conditions to the ratio of gas density to viscosity at the standard conditions. A characteristic correction for the contribution from inertial and slip flow may also be made.

Depending on the accuracy of the quality control that is needed, it may be plausible to ignore the contributions from inertial and slip flow. This needs to be examined for each type of filter by performing appropriate performance and characterization measurements.

Data indicate that there is no reason why very accurate, reproducible, and comparable pressure drop measurement cannot be achieved with proper data acquisition and analysis. It is recommended that serious consideration be given to the use of a small desktop or personal computer to acquire and process the data. The appropriate corrections can be incorporated in the software. In addition, the data is assessable to make production reports, reports to customers, quality control charts, and system performance and problem analysis.

Definition of Symbols used in equations (1) through (8).

ΔP_i = pressure drop from gas acceleration
ΔP_m = pressure drop from pure viscous laminar mass flow
ΔP_v = pressure drop from pure viscous laminar volume flow
ΔP_s = pressure drop from combined viscous laminar and slip flow
ΔP = total measured pressure drop
K_a = constant relative to acceleration using velocity
K_i = constant relative to acceleration using mass flow
K_m = constant relative to viscous laminar mass flow
K_v = constant relative to viscous laminar volume flow
K_s = constant relative to slip mass flow
ρ = gas density
V = gas average velocity
F = gas mass flow rate
Q = gas volume flow rate
η = gas viscosity
P = gas average ambient pressure
T = gas temperature
M = gas molecular weight

$Y = \dfrac{M\ \overline{P}\ \Delta P}{\eta\ T\ F}$ a correlation variable also referred to as the Y-Cluster

REFERENCES

[1] Fain, D. E., Selby, T. W., "Calculation and Use of Filter Test Facility Orifice Plates," K/PS-5046, Martin Marietta Energy Systems, Inc., Oak Ridge, TN, 1984.

[2] Fain, D. E., "Calibration of a Laminar Flowmeter," in Flow: Its Measurement and Control in Science and Industry, Volume 2, 1981, Instrument Society of America, Research Traingle Park, NC, 1981.

Applications and Testing:
Protection of Processes

J. Gordon King

AIR CLEANLINESS REQUIREMENTS FOR CLEAN ROOMS

REFERENCE: King, J.G., "Air Cleanliness Requirements for Clean Rooms", Fluid Filtration: Gas, Volume I, ASTM STP 975, R.R. Raber, Ed., American Society for Testing and Materials, Philadelphia, 1986

ABSTRACT: The cleanlines of air required for clean rooms is defined in terms of nature, quantity, size distribution, test methods and removal mechanisms of contaminants. Test instrumentation is described and some practical limitations are given. Included is information regarding some of the changes recommended by the Institute of Environmental Sciences for the latest revision of Federal Standard 209. Some projections are made for what future clean room air cleanliness may be required. Present knowledge indicates that future cleanliness needs beyond the present limits may be restricted to very small, confined zones, where exposure to contamination is transient and of short duration.

KEYWORDS: Air Cleanliness, Cleanrooms, Federal Standard 209, Test Methods-Particles.

FORWORD

Of all the imperatives which determine manufacturing success, none is more pervasive than control of the physical environment. In that niche of manufacturing with which this paper is concerned, the highly technical arenas of electronics, pharmaceuticals, chemical synthesis, genetic engineering and many others, one common factor of paramount importance is the requirement to protect the product from contamination during critical phases of processing. Since air is the environment to which most product, tooling, processes and people are continually exposed, special efforts are needed to make that air clean; hence, clean rooms.

The degree of cleanliness required is best determined by examination of the product in its most contamination-sensitive stages, and by the process equipment required to produce it. Perhaps the greatest user of clean rooms today is the microelectronics industry in the manufacturing of integrated circuit chips. It will be obvious that most of this discussion will be based upon that

J. Gordon King is President of King Consulting, Inc. and affiliated with Henningson, Durham and Richardson, Architects and Engineers, 12700 Hillcrest Road, Dallas, Texas 75230.

industry. A rule of thumb in chip production is that particles one-tenth the size of the minimum critical feature will cause "killer" defects.

DEFINITION - AIR CLEANLINESS REQUIREMENTS

A set of criteria defining the allowable limits of specified contaminants in a controlled environment. For our purposes here, and generally accepted primary criteria for clean rooms, air is routinely examined for particles (whether liquid or solid, viable or non-viable) and moisture content (because of process dependency). Other contaminants such as vapors or gases are monitored, or not, based upon life-safety or specific process considerations.

STANDARDS AND SPECIFICATIONS

Some of the industries using clean rooms are so dynamic in change to an ever-decreasing critical-feature size and increasing complexity that the Standards-Specifications documents cannot keep pace with the need. A case in point is the Federal Standard 209B "Clean Room and Work Station Requirements, Controlled Environment" (3,6). Although a four year long updating effort is soon to be culminated, the highest class to be defined is a Class 1 (Fig. 1). This classification is based upon one (1) particle 0.5 micrometers and larger in diameter per cubic foot of air. To provide acceptable numbers of particle counts for statistical confidence of meeting the class criteria, particles may be counted at 0.1, 0.2 or 0.3 micrometers and larger. These larger number limits will be given in the new version.

In the semiconductor Wafer Fabs, critical feature size is already below 1 micrometer and predicted to be below 0.5 micrometers some time next year. From the "rule-of-thumb" we need to be able to control the particles to less than 0.05 micrometers. This means that the Federal Standard 209C will be outdated in less than a year!

Measurement Methods

Our ability to define, measure and use clean rooms has been in continual "leap-frog" contest with circuit designers and process engineers. An enhanced ability in one area presents either an improved capability or increased demand in another. Economic consequences of these improvements are so enormous that large pressures are brought to bear on forcing acceleration of the (need/sensing/control/economic impact) cycle.

Airborne particle measurement has led the way in pushing clean room technology into the realm of submicrometer control. As the ability to measure the particles went from 5.0 (microscopic) to 0.5 (light scattering) to 0.1 (laser) micrometers; so the ability to filter followed, yields jumped, processing improved, etc. Many intertwined cycles have occured. The next jump in ability to measure particles (down to 0.005 micrometers) is already available (albeit

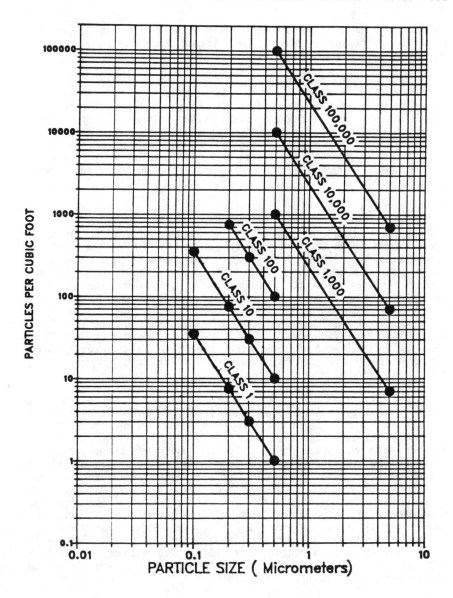

FIGURE 1 – Class Limits In Particles/Cubic Foot of Size
Equal To Or Greater Than Particle Sizes Shown

Note: The class limit particle concentration shown in
Table 1 and Figure 1 are defined for class purposes
only and do not necessarily represent the size
distribution to be found in any particular situation.

cumbersome) through the condensation nucleus counter in combination with a diffusion battery, aerodynamic particle sizer or a differential mobility analyzer.

Whether or not the sub 0.1 micrometer sized particles measurement is of great importance is debatable (5). It remains for the future accumulation of data to tell us in a statistically acceptable way whether the inferred elevated filtration efficiency on very small particles can be relied upon. It is quite certain that the small particles can be "yield killers" if they are on the Wafer. If they arrive through the air as a transfer medium, we have one set of problems. If not, then processing must take the blame and provide the cure.

AIR FILTRATION

Air filtration has two primary characteristics to be discussed here. First, High Efficiency, Particulate, Air (HEPA) and Ultra Low Penetration, Air (ULPA) filters are both depth filters composed of fiberglass mats with an acrylic binder. Even though they are very highly efficient (up to 99.9995% on a particular size range of particles), they are still percentage type devices passing some particles. Therefore, the ultimate contamination load of the air in the clean room that passes through the filters is a function of the quantity of contamination in the air as it approaches the final (HEPA, ULPA) filters. Assuming that the sealing system really seals and that all leaks have been plugged it is reasonably easy to meet any given particle content criteria.

Secondly, reasonbly good control of processes and people in a clean room will result in a very small per-cubic-foot loading of the recirculated air. The recirculated air, representing from 92-98% of the air passing through the final filters, should become cleaner and cleaner with each pass, and it does. Neglecting those internally generated particles from chemical combinations, oxidation, material sloughing from filters, frames, lights, finishes, people, etc., the source of the particles we find must be the make-up air. Unfortunately the make-up air has usually not been filtered to the same degree as the recirculated air, so the final filters are challenged by a much higher load of contamination than necessary. Therefore, in the clean room air the majority of the particles in the most penetrating size range (approximately 0.2 micrometers for HEPA and 0.15 for ULPA) come straight in through the make-up air system and through the final filters.

INTERNAL SOURCES

Particles generated within the clean room tend to be very small (less than 0.01 micrometers) or very large (greater than 100 micrometers). Examples are ammonium chloride particles and skin flakes. Both may be reduced to a manageable level by good manufacturing practices, but cannot be affected by filtration systems.

FUTURE OF CLEAN ROOM AIR CLEANLINESS

How far it is necessary to go in air cleanliness must of necessity be addressed to the function of the clean room. For microelectronics (by far the greatest user at present), the need would appear to be related not only to critical dimension limitations but to type of photolithography and other critical process attributes. In microelectronics, there are rules of thumb to guide us in design of clean facilities. The following table is an illustration of such.

TABLE 2 - Design Rules for VLSI Circuits

Smallest Critical Feature-μM	4	2.5	1.5	0.9	0.5	0.25
Memory Level	16K	64K	256K	1M	4M	16M
Critical Particle Diameter-μM	0.4	0.3	0.17	0.09	0.05	0.03
Transistors in a 100μM Circle	20	50	150	400	1500	4500

The critical particle diameter represents the size of particle which can cause a "killing" defect. This in itself is an important parameter, but when it is considered that yield is a combination of functions of particle size, quantity of particles per square cm and numbers of critical levels, then the extreme importance of cleanliness at every step becomes evident.

The "Transistors in a 100μM Circle" attribute was included for visualization purposes. It represents the cross section of an average human hair.

A method has been described (5) for predicting semiconductor yields using the Bose-Einstein model. As expected, yield may be expected to go down as complexity increases according to the relationship:

$$Y = (1 + AD)^{-n},$$

where: Y = yield
A = chip area - sq. cm.,
D = defect density in defects/sq.cm.
/level and
$\underline{N}$ = number of critical levels

For a typical 64K DRAM, the chip area is 0.25 sq. cm., the typical defect density is 0.5/sq. cm./level and 5 critical levels. A yield of 55% can be expected. However, with 256K DRAMS, the critical feature size is reduced by three (resulting in an increase in defect density of a factor of 9). In addition, one may expect an increase in number of critical levels to 9 or more. The net result is that one must be able to reduce contaminant level by orders of magnitude in order to even hope for yields in excess of 1%.

When the next round of critical-dimension shrinking occurs, one of the candidates for photolithography is x-ray exposure, providing of course that acceptable x-ray photoresists are developed, particulate contamination becomes less of a problem because the particles are transparent to the x-rays. The most probable of the methods is a continuation in the use of "steppers", the latest of which can take us down to a critical feature size of 0.25 micrometers. This, of course, makes the wafer vulnerable to 0.025 micrometer sized particles.

Pharmaceutical/medical/food   From what we know now about sizes of viable organisms, it would appear that our present technology is good enough - all we have to do is apply it. Known organisms, detrimental to our physiology and support systems, appear to be in the 0.01 to 100 micrometer size range. Certainly we have control systems capable within that regime.

Conclusions:

1.   We are constantly approaching the limits of our knowledge. From where we stand at present it would seem that we are almost in control to those limits.
2.   It is predictable that two things will happen. First, the limits of our knowledge will expand, showing us things that are not under control. And finally, there will be a continued effort to control those things which are not controlled.

REFERENCES

(1)   Ensor, D.S. and R.P. Donovan "The Application of Condensation Nuclei Monitors to Clean Rooms", Proceedings of the 30th. IES Annual Technical Meeting, Orlando, 1984.

(2)   Locke, B.R., R.P. Donovan and D.S. Ensor, "Assessment of the Diffusion Battery for determining Low Concentration Submicron Aerosol Distributions in Microelectronic Clean Rooms", Proceedings of the 31st. IES Annual Technical Meeting, Las Vegas, 1985.

(3)   "Federal Standard Clean Room and Work Station Requirements, Controlled Environment", Fed. Std. No. 209B, April 24, 1973.

(4)   CuCu, D. and H.J. Lippold, "Electrostatic HEPA Filter for Particles 0.1 Micrometer", Proceedings of the 30th. IES Annual Technical Meeting, Orlando, 1984.

(5)   Lieberman, A., "Air Cleanliness for State-of-the-Art Microelectronics Manufacturing," Proceedings of the 1st. Microcontamination Conference, San Jose, 1985.

(6)   Cooper, D.W., "Rationale for Proposed Revisions to Federal Standard 209B (cleanrooms)", Journal of Environmental Sciences, Vol. XXVIII, Number Two, Mar./Apr. 1986.

James L. Flannery and James P. Walcroft

AIR CLEANLINESS VALIDATION FOR CLEANROOMS

REFERENCE: Flannery, J. L. and Walcroft, J. P., "Air Cleanliness Validation for Cleanrooms," Fluid Filtration: Gas, Volume I, ASTM STP 975, R. R. Raber, Ed., American Society for Testing and Materials, Philadelphia, 1986.

ABSTRACT: Cleanroom air cleanliness validation is becoming increasingly important with the further miniaturization and susceptibility of products to airborne contamination. The rational for validation discussed, both for particle counting and for other testing. The documents covering cleanroom air validation are described, as is the general topic of "room classification." Topics covered include validation methodology, testing modes, and room types. Factors affecting air cleanliness are discussed as well as other testing procedures to characterize cleanroom performance.

KEYWORDS: verification, certification, testing, air cleanliness, cleanroom, classification

INTRODUCTION

The procedure of sampling a specific volume of air using standardized methods, state-of-the art instrumentation, and basic principles of physics to identify air cleanliness levels has become a science of its own. This presentation will address air cleanliness levels and environments within enclosures commonly called "Cleanrooms." The data and information presented herein will identify various factors which affect air cleanliness levels as well as define specific validation procedures. In doing so, we shall also discuss current trends, standardizing procedures, and practical application of cleanroom standards and testing.

In today's marketplace validation of cleanrooms is vital, to insure that products are manufactured in environments free of airborne particulate contamination.

Mr. Flannery is the National Operations Manager of ENV Services, Inc. and Mr. Walcroft is the Systems Operations Manager of ENV Services, Inc., 216 Goddard Blvd., King of Prussia, PA 19406.

Technology continues to grow toward further miniaturization of products. With components and physical geometries becoming smaller and smaller, validation of cleanroom facilities becomes even more important than ever before.

BACKGROUND HISTORY

In the late 50's and early 60's, airborne particulate contamination in the aerospace, microelectronics, and the pharmaceutical industries presented special manufacturing problems. The U.S. Air Force, NASA, and others greatly involved in the emerging space program which dominated the 60's required subminiature components to be manufactured and assembled in "dust free" environments. Air filters which had been developed by the U.S. Army Chemical Corps, Naval Research Laboratory and the Atomic Energy Commission were used to provide a suitable manufacturing atmosphere. The supply air coming into manufacturing areas was passed through filtering systems, creating a relatively "clean" environment.

As technology advanced, requirements dictated stricter control. In 1961-62 Willis Whitfield, Sandia Corporation, Albuquerque, NM, designed prototypes of the "laminar airflow clean room," using HEPA filters.[1] (A HEPA filter is defined as having a minimum filtering efficiency of 99.97% for .3 micrometer particles of thermally generated dioctylphthalate.) The basic principle consisted of air being forced through a wall of HEPA filters and into a confined work area at a uniform velocity with minimum turbulence. These concepts provided removal of particulate contaminants, reduced cross contamination between adjacent work zones, and minimized the effect of personnel generated contamination. In general, the elements of filtration, dilution and control of the positive pressure airflow proved very effective in meeting stringent requirements.

As the technology gained acceptance and was applied as an effective tool to control airborne particulates, the need for testing and performance verification became obvious. A decade after Whitfield's prototypes the technology was also being applied for biomedical and health care uses. Surgery isolation, safety containment facilities and the like used HEPA filtration to successfully protect people from microbial airborne contamination. These types of applications were the origins of the "certification" industry as we know it today. For product manufacturing, yield rates and rejected product were one issue, but where personnel health and safety was involved performance validations became essential. Currently, with advances in both biomedical and microelectronics industries air cleanliness validation remains an extremely important concern.

WHAT IS VALIDATION

Commonly referred to as certification, validation as applied to cleanrooms is the verification that the design performance criteria of environmental conditions specified are being met at a particular point in time. The dictionary states that it is to confirm, support, or corroborate by objective truth on a sound or authoritative basis.

Validation in the context of air cleanliness within cleanrooms is usually considered to be "room classification" validation. The room "classification" or "class" refers to the level of airborne particulate within the room, as measured by a specific particle counting procedure. For example, a "Class 10" cleanroom would, by definition, contain less than 10 particles at .5 micrometers or larger per cubic foot of air sampled. The clean room is designed to meet a given cleanliness class in order to minimize particle contamination problems with the product. Since different products have differing levels of vulnerability to particle contamination, there are distinct classification requirements necessary for the particular application. Verification of the proper air cleanliness level determines whether or not the design and operational objectives of air cleanliness have been achieved.

Cleanroom characterization and function, however, is not accurately described in terms of room class air cleanliness levels alone. In order to understand the operating parameters and the factors that affect particle levels and room integrity, other validation procedures are necessary as well. These other factors impact upon the success of the clean room to remove particles quickly, to avoid cross contamination, to stop bypass of contaminated air into the room at every point, to maintain outward flow of air, etc. Specific tests (to be described later)are performed for the purpose of qualifying these dynamics and system components. When an accurate analysis of all factors contributing to the clean room performance is made, then 1) corrective action can be taken for any deficient part of the system and 2) the effect of the cleanroom in its contribution toward particle load can be put into perspective.

There are several prerequisites necessary prior to starting a validation procedure. First, the performance characteristics and expectations of a cleanroom must be clearly defined. It is of paramount importance that all parties involved (contractors, users, certifiers, etc.) know in advance and be in agreement with how the system should function. The acceptance criteria of performance testing must also be established. Once these issues are resolved standardized testing procedures and methods must be detailed. Finally, the instrumentation to be used for performance evaluation shall be capable of meeting the parameters of the testing procedures. As you can see, there are many issues which require definition and

agreement prior to validation testing. In addition, with
agreement on the above issues, it is also very important
that the test results data are collected, recorded, and
presented in a form which is understandable and "user
friendly."

A recent development which has proved very helpful in
the validation process is the review and up-dating of
Standards and recommended practices for testing
Cleanrooms. The IES-RP-CC-006-84-T [2] hereafter referred
to as RP-6 is one such document. The recommended practice
establishes standard tests, methods, procedures, and
instrumentation requirements to evaluate and characterize
the performance of the cleanroom system. These factors are
addressed in the RP and used in preparing detailed
specifications for procurement and for assuring operational
compliance. It is invaluable and serves as one reference
assisting planners, designers, manufacturers, buyers, and
certifiers of cleanrooms. Although the document is
relatively new (November, 1984) it is much referenced and
has received industry-wide acceptance.

It should be noted, however, that the RP-6 document does
not establish acceptance levels. This and other
requirements in the verification of cleanroom operation are
left largely to the "agreement between buyer and seller,"
commonly specified in the RP-6 document. RP-6 is meant to
serve as a compendium of testing methods for characterizing
cleanroom performance, not as a specification.

In the case of validation of existing cleanrooms, the
acceptance criteria are already established. The basis for
these acceptance levels generally reflects the minimum
acceptable environmental condition necessary for successful
yield or product sterility. The ultimate concern is to
keep the product as free as possible from airborne
contamination.

In summary, validation (or certification) uses standard
methods and instrumentation along with defined performance
criteria and acceptance levels to evaluate and verify the
performance characteristics of a particular Cleanroom.
Often, validation is viewed by many to be considerably more
than just verification of performance. At times, however,
greater significance is attributed to the "certification of
validation" than the data which support validation. The
established criteria may be met, yet there may be important
information found during the course of the test that is
being overlooked by the recipient of the test report.
Compliance with a specification does not mean that other
factors effecting performance should be ignored. Also,
there are those who associate validation with a period of
time, typically one year, assuming performance levels are
guaranteed. These are just some of the misunderstandings
commonly found in relation to "certification" or
"validation." The test data can only indicate whether or
not performance compliance has been achieved at a
particular point in time.

AIR CLEANLINESS DEFINITION AND CLASSIFICATION

Federal Standard 209B has been the major document covering definitions of air cleanliness classes in cleanrooms. This document has recently undergone extensive revision and will soon emerge as Federal Standard 209C. The revision will better meet the needs of modern day clean room environments requiring ultra clean conditions.

Although the revised standard references the same size particle (0.5 micrometers) as the former version, one major difference is the defined extrapolation to smaller particles. Thus, particles at 0.1, 0.2, and 0.3 micrometers can be utilized to measure room levels to a referenced standard, (although room class is still described at the 0.5 micrometer level). This is of particular usefulness when taking particle counts in rooms that are cleaner than class 100. In these environments, the "larger" particles of 0.5 microns are less numerous and results are thereby affected by greater statistical error. Measuring the more numerous, smaller particles relieves this problem to a great extent. It is acknowledged that greater statistical validity results from measurement of a greater number of particles.

Figure 1 represents the particle concentration limit for given classes. The circles on the lines showing the classes are the points at which measurement can be made to determine the air cleanliness classification.

A provision in the proposed standard allows for measurements at any one of the points listed to classify a room or zone. Classification of a room is, as stated, based upon the 0.5 micrometer level. The RP-6 document, designed to compliment the Federal Standard, also bases particle levels on the 0.5 micrometer level (or as agreed between buyer and seller). However, there seems to be a current trend to classify rooms at levels other than 0.5 micrometers. If this is a buyer/seller and certifier understanding, this is not a problem. However, since these other class categories may cause confusion, it is better, if possible, to stay with the prescribed definitions.

Another factor to be considered in the validation of air cleanliness is the actual particle size distributions occurring in a given situation. The proposed revised standard indicates that the distributions it uses are for "class purposes only," so the issue remains, at least theoretically. It is likely that there could be at least minor differences between the particle size distributions in the chart and the actual distributions in a particular situation. This is one reason why taking samples at several different size ranges may be informative. Results from one study utilizing a laser particle counter document a steeper slope than expected from Federal Standard 209B on particles from .5 to .1 micrometers in clean rooms with a

high degree of cleanlines.[4]   More research is needed to
determine general validity of the distribution curve.

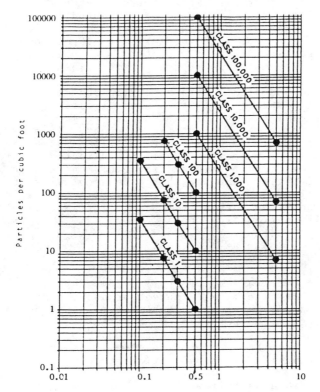

Figure 1. Class limits in particles per cubic foot of size
equal to or greater than particles sizes shown.[3]

INSTRUMENTATION AND METHODOLOGY

The common method of validation or certifying of air
cleanliness utilizes airborne particle counters.   White
light and laser particle counters are available, and while
both use the same principles of counting particles from
measuring their light scattering effects, currently only
the laser counter has the ability to measure at the 0.1 and
0.2 micrometer levels.   At levels above 5.0 micrometers,
manual counting and sizing methods utilizing membrane
filters and microscopic examination or image analysis are
employed (if these size categories are deemed necessary to
measure).   The size ranges tested with the airborne
particle counters, particularly from 0.1 to 1.0
micrometers, seem to constitute the particle size range
most able to penetrate the HEPA filter.

Other issues related to particle sampling would include
sample volumes and grid pattern size of tests. Generally,
as particle levels increase, the number of samples needed
decreases.   The size of the test grid, as well, is

proportionally larger with greater particle amounts and
higher classes. A "sample" of air represents the
appropriate volume (as described in proposed Federal
Standard 209C) for the size range being tested and the room
class. For example, in a class 10 cleanroom, the following
samples would be appropriate volumes: at 0.5 micrometers,
2 cubic feet of air; at 0.1 micrometers, .1 cubic feet of
air. The air is sampled as it reaches the clean zone.

It should be noted that RP-6 specifies the methodologies
to test the <u>room</u> rather than the dynamic process within the
room. Operational testing is listed, but <u>room</u>
characterization is the goal of the document. This is an
intentional emphases of the document as illustrated in
Section 8.2.2.2: "Test points should not directly sample
contamination released from operating equipment and/or
people activity; i.e., the air should be sampled as it
approaches the work activity area. This will generally be
immediately above the work activity level or immediately
upstream in laminar airflow rooms."

ROOM TYPES AND TESTING MODES

There are a wide variety of applications for
cleanrooms. There are many different products manufactured
within them. Needless to say, they come in many shapes and
sizes. Conceptually, there are three distinctly different
types of cleanroom design. Depending upon product,
processes, and acceptable cleanliness levels each are
effective in reducing airborne particulate contaminants.

They are defined as follows:[1]
Laminar Airflow Cleanroom; a cleanroom in which the
filtered air entering the room makes a single pass through
the work area in a parallel flow pattern, with a minimum of
turbulent flow areas. Laminar airflow rooms have HEPA
filtered coverage of at least 80% of the ceiling (vertical
flow) or one wall (horizontal flow), producing a uniform
and parallel airflow. (Net area under filter medium versus
total filter area = 0.80).

Turbulent Airflow Cleanroom; a cleanroom in which the
filtered air enters the room in a non-uniform velocity or
turbulent flow. Such rooms exhibit non-uniform or random
airflow patterns throughout the enclosure.

Mixed Airflow Cleanroom; a hybrid cleanroom consisting
of a combination of Laminar Airflow and Turbulent Airflow
Cleanrooms.

Although there are functional differences between the
three types of cleanrooms they all are effective in
delivering clean air. The laminar airflow cleanroom
provides the tightest control and efficiency in removing

---

[1]Definitions are from IES-RP-CC-006-84-Y, November 1986,
Section 3.2.

particles generated by processes and personnel within the
room. The mixed airflow concept provides similar control
in specified zones within the room, while the turbulent
airflow cleanroom is least effective in controlling
contaminants generated within the room.

Regardless of the concept employed, the performance
expectations of a cleanroom must be well defined. Specific
tests are recommended for each concept for performance
validation. These test procedures are identified in Figure
2, and described in the following sections.

Tests	Laminar Airflow	Turbulent Airflow	Mixed Airflow
Velocity/Uniformity	1	–	0
Filter Leak	1	1	1
Parallelism	1.2	–	0 (1,2 only)
Recovery	1,2	1,2,3	1,2
Particle Count	1,2,3	1,2,3	1,2,3
Particle Fallout	0	0	0
Induction	1,2	1,2	1,2
Pressurization	1,2,3	1,2,3	1,2,3
Air Supply Capacity	1	1	1
Lighting Level	1	1	1
Noise Level	1,2	1,2	1,2
Temperature	1,2,3	1,2,3	1,2,3
Humidity	0	0	0
Vibration	0	0	0

Key:
0   Test optional, depending on process requirements
1   Test suited to as-built phase
2   Test suited to at-rest phase
3   Test suited to operating phase

Figure 2. Recommended Tests by Cleanroom Type[2]

The certification of cleanrooms is necessary in an
"as-built" phase to establish the performance of a new
cleanroom. Testing thereafter should serve to verify the
room performance and to test operational conditions. This
is usually best shown by testing in both an "at-rest" and
"operational" mode. The operational mode is the most
revealing of an active cleanroom/process system. It is
also the testing mode that demands the most attention as to
issues of timing, test location and placement, and testing
parameters. Naturally, air cleanliness is compromised
during dynamic operation. "At rest" testing is performed
with room equipment operating, but void of personnel.

FACTORS AFFECTING AIR CLEANLINESS LEVELS

As mentioned earlier, the air cleanliness within the
cleanroom is affected by many variables. Such factors as
routine housekeeping, room condition and activity level can
affect results, as well as the length of time since the

last period of activity or production within the room.
However, the greatest factor in general room air
cleanliness is the performance of the cleanroom itself.
This is why particle counts alone do not tell the whole
story. Even with standard testing procedures,
particularly in static testing modes, cleanrooms can have
particle levels that are not truly representative of the
room. Therefore, further analysis of other room dynamics
is essential to qualifying the air cleanliness
characterization.

The proposed revisions to Federal Standard 209B
(fourteenth draft) states the following: (Section 5.1.2.2)
"Environmental and use parameters. The applicable
environmental and use parameters of the cleanroom or clean
zone shall be recorded. These conditions or measurements
may include (but are not limited to) air velocity, air
volume change rate, room air pressure, makeup air volume,
unidirectional airflow parallelism, temperature, humidity,
vibration, equipment, and personnel activity." RP-6 lists
14 different tests (some are optional or dependent upon
room types) so that cleanroom performance can be adequately
evaluated. Following are tests that must be performed to
complement particle counting, if actual room performance is
to be comprehensively evaluated:

HEPA Filter and Room Integrity

One of the most critical factors related to the success
of the cleanroom system is HEPA filter integrity. Filter
leak testing can reveal major problems that directly affect
air cleanliness levels. Induction leakage, (intrusion of
unfiltered air through joints, cracks, doorways, etc.)
particularly from positive pressure filter plenums, can be
a major source of particulate. Testing per the procedures
described in RP-6 is essential in new systems. The choice
of method given in RP-6 is dependent upon the configuration
of the cleanroom and customer requirements.

The methods of HEPA filter leak testing are described in
Section 5.1 of RP-6. The first method described utilizes an
aerosol photometer with an upstream challenge. The benefit
of using this method is that there is an advantage in leak
discrimination and definition. In scanning filters and
structural components for leakage, the analog signal from
the photometer often gives the technician more reliable
information than the counter method. This method requires
use of a relatively large particulate challenge, which
usually sets ambient background levels as an insignificant
complicating factor.

However, the second leak testing method described in
RP-6, the particle counter method, must be used in
particular circumstances. It is usually necessary to use
this method in large plenum systems, where the amount of
challenge particulate necessary for the photometer method
is impractical. Also, with the photometer method, the time
necessary to scan a large area would mean lengthy exposure

to the high concentration levels of particulate generated. While particles from DOP and similar substances tend to conform to the filter fibers, and thereby have little effect on filter blockage, it would be prudent to avoid lengthy exposures to high concentration levels. However, the artificial challenge is still necessary to bring the upstream level up to 10^6 particles per cubic foot, which is necessary for determining leakage corresponding to the .01% level.

Using the particle counter to scan HEPA filters is also necessary when use of an artificial challenge is not favored by the user. Leak testing of a system with only an ambient challenge presents a special problem, since leak discrimination is made considerably more difficult when there is a low ambient count, as is often the case.

Filter and induction leak testing usually reveals problems that can be readily corrected, and this greatly improves the cleanroom environment.

Air Flow and Pressurization

Air velocities and volume of air from HEPA filters is an important performance factor for both laminar and turbulent flow cleanrooms. In vertical laminar flow rooms, the effectiveness of air from the ceiling is dependent on both velocity and uniformity of airflow. Detailed measurements of air flow, normally utilizing a hot-wire anemometer within 12" of the filter face, can provide important information on how successful the "air piston" is at washing away particulate.

In turbulent flow rooms, or "dilution controlled systems," [5] the amount of clean air entering the room is critical, since this provides the dilution to the contaminated air. Thus greater air volume will provide greater dilution, and a cleaner room. In this case, the specific effect of the face velocity is less important than the overall amount of air changes per unit time.

The amount of make-up air affects room pressurization and introduces a possibly higher level of contaminated air to the system (depending upon the level of filtering to the make-up air supply). The amount of make-up air entering the system should be measured so that a baseline of information (relating most directly to room pressurization) can be obtained.

Room pressurization is highly critical to the integrity of the cleanroom, since air must always escape outward from the room, not inward. Measurement of each room to surrounding rooms or adjacent areas provide knowledge of air direction and force.

Other Factors

The recovery rate is the measure of time necessary for a room to recover from an induced room particulate challenge. This test sheds light on the length of time necessary for "clean down" to acceptable levels. It is particularly useful in the non-laminar or turbulent flow room, since general room dilutions can be clearly profiled. Neutral challenges can now be utilized, such as polystyrene latex spheres, so that the room surfaces are not exposed to an unacceptable oil-based substance.

Parallelism testing provides a visual account of laminarity in vertical laminar flow rooms. This test, utilizing neutral density helium bubbles or other techniques, actually shows the amount of lateral shift of air at particular points. Usually corrections can be made to increase parallelism so that cross contamination can be minimized.

Tests of other environmental conditions, such as temperature, humidity, sound, light, and vibration round out the performance evaluation so that room dynamics can be understood and dealt with accordingly.

Summary

The objective of air cleanliness validation within a clean room is to determine the level of airborne particulate concentration at a particular time by a specific particle counting procedure. The classification of a cleanroom is based upon the number of particles at a specific size (usually at the 0.5 micrometer level) as determined from the airborne samples taken.

Factors related to cleanroom construction, design, and maintenance, in addition to the manufacturing activity, seriously impact upon the level of air cleanliness within the cleanroom. These factors must be profiled by means of a total cleanroom validation procedure, if the total function and character of the room is to be understood. Both Federal Standard 209B and RP-6 serve to standardize and define the terms and procedures necessary to perform a total validation procedure. In this way, the air cleanliness data is made meaningful and more useful, since the affect of the cleanroom itself is revealed.

The trend in microelectronic manufacturing seems to be toward further miniaturization and product vulnerability. Validation procedures once found to be adequate must now be upgraded and given the attention appropriate in this progressing industry. The validation procedure is cost effective and necessary, and becoming more so as technology advances.

REFERENCES

[1]   Whitfield, W. J. 1962. "A New Approach to Clean
      Room Design", Sandia Corporation, Albuquerque, NM.
      Technical Report No. SC-4673(RR).

[2]   Institute of Environmental Sciences, November, 1984.
      "Testing Clean Rooms," Recommended Practice
      IES-RP-CC-006-84T.

[3]   Institute of Environmental Sciences, April, 1986.
      "Airborne Particulate Cleanliness Classes for
      Cleanrooms and Clean Zones," Proposed Federal Standard
      209C.

[4]   Lieberman, A., "Particle Control and Air Cleanliness
      for State-of-the-Art Microelectronics Manufacturing,"
      Microcontamination, February 1986, pg. 29.

[5]   Stokes, K. H., "Clean Room Classification," The
      Journal of Environmental Sciences, Nov/Dec 1985.,
      pg. 37.

Mauro A. Accomazzo and Donald C. Grant

MECHANISMS AND DEVICES FOR FILTRATION OF CRITICAL PROCESS GASES

REFERENCE: Accomazzo, M. A., Grant, D. C.,
"Mechanisms and Devices for Filtration of Crit-
ical Process Gases," Fluid Filtration: Gas,
Volume I, ASTM STP 975, R. R. Raber, Ed.,
American Society for Testing and Materials,
Philadelphia, 1986

ABSTRACT: The particle capture mechanisms in
aerosol filtration result in a most penetrating
particle size for process gases used by the microelectronics industry. The most penetrating particle size for fibrous filter media and microporous membrane filter media have been measured to
be typically 0.15um and 0.05um, respectively. Filter performance ratings should be based on the retention of particles at the most penetrating particle size. The retention ratings for HEPA fiberglass filters have been determined to be $\geq$99.99%
and for typical membrane filters $\geq$99.9999999%.

The shedding of particles from the downstream side
of various types of Point-of-Use filter devices
was evaluated, using laser and condensation nucleus type particle counters. All types of filter
devices performed similarly during steady flow
operation but performance differences were observed when either pulsed flow or mechanical
vibrations occurred. In general, the membrane
filter devices shed less particles than the fiberglass filter devices, and the membrane filter
devices that used stacked disc molded supports
were cleaner than the pleated filter devices.

KEYWORDS: fibrous filter, membrane filter,
aerosol filtration, filtration mechanisms, particulate shedding, particulate penetration

Dr. Accomazzo is Director of Applications Engineering
and Mr. Grant is Manager of the Particle Technology
Laboratory, in the Research and Development Department of
Millipore Corporation, 80 Ashby Road, Bedford, MA 01730.

INTRODUCTION

It is generally recognized that the feature sizes of integrated circuits are becoming smaller and smaller. Today, 256 Kilobit and 1 Megabit VLSI (very large scale integration) memory devices are commercially available, and the critical dimensions for these devices are in the 1-2um range. Future devices (4 Mbit and 16 Mbit) will use line widths below one micrometer.

This reduction in feature size has been made possible by refinements in microlithography and by replacing wet etching with gas-based etching technologies[1]. In addition, many other processes (e.g., oxidation, diffusion, nitridation, epi, and chemical vapor depositions) also utilize gases in critical steps[2].

An entirely new grade of specialty gases, made specifically for the microelectronics industry, is now available from commercial gas suppliers. These suppliers along with users are establishing purity standards through ASTM and SEMI (Semiconductor Equipment and Materials Institute) committees. At present, these standards apply only to gaseous impurities and do not consider particulate impurities which can also cause device failure[3].

Typical VLSI gaseous impurity concentrations are currently stated in ppm (parts per million, vol/vol) but Japanese suppliers are beginning to provide gases with impurities at the ppb (parts per billion) level. A simple calculation shows that a particulate impurity concentration of 1 ppm consists of 2×10^{12} particles/liter of a 0.1um spherical particle[4]. Even at 1 ppb, the number of 0.1um particles would be 2×10^{9} particles/liter.

These numbers are obviously many orders of magnitude greater than the number of particles found in the clean room environment (Class 100 is 3.5 particles/liter >0.5um), where these devices are manufactured. And yet, particles of this size (0.1um and less) must be removed from gases if one expects to obtain reasonable yields of the next generation of devices with feature sizes of one micrometer or less. Fortunately, these particles can be removed effectively from gas streams by membrane filtration.

Aerosol science has made great progress over the past two decades and its application to gas filtration has resulted in excellent correlation between theory and experiment. A review of the literature on aerosol filtration can be found in Rubow's Ph.D. thesis[5]. In addition, other useful references are listed in back of this paper[6-10].

This paper will discuss the mechanisms of particulate removal by filter media and review the performance data

of several types of filter devices that are currently
available for removing particulates from critical process
gases used by the microelectronics industry. Since
particles can be generated from elements in the gas
delivery systems (i.e., regulators, valves, flow meters,
flow controllers, fittings and piping), the discussion
will be limited to POU (point-of-use) filters.

MECHANISMS OF AEROSOL FILTRATION

Particles are filtered from gas streams by simply making
contact with the filter matrix. It can be assumed here,
and has been demonstrated, that once particles contact the
filter structure they will adhere to it[10]. Particles are
brought into contact with the filter surface by five
possible mechanisms: gravitational settling, electrostatic
deposition, impaction, interception, and diffusion.

Gravitational settling is simply the removal of particles from the flow field (Fig. 1) caused by their large
size (or mass) and low velocities through the filter
matrix. This mechanism plays only a small role in the
filtration of gases utilized by the semiconductor industry.

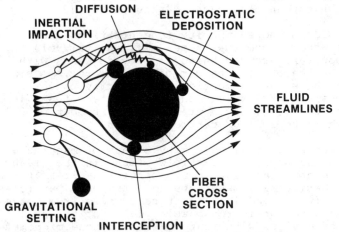

Fig. 1--Schematic of various particle capture
mechanisms.

The mechanism of electrostatic deposition results from
the attractive forces between charged particles and an
oppositely-charged filter matrix. However, aerosols are
often both positively and negatively charged and filter
matrices can lose their surface charge with time. Hence,
this mechanism should not be relied upon for high
efficiency filtration.

The impaction mechanism occurs when a particle leaves
the flow field as it is diverted around the filter matrix

(see Fig. 1) and impacts with the filter structure. This mechanism is favored by high fluid velocities and larger particle sizes. However, flow velocities are usually low due to pressure drop considerations and hence the impaction mechanism has minimal effect in removing submicron size particles.

Particles are removed by the interception mechanism if the location of the stream line and the particle size result in contact between the filter matrix and the particle (Fig. 1). For submicron aerosols this capture mechanism plays a major role in filter removal efficiency. As one would expect, for a given filter matrix, removal efficiency increases as the particle size increases. Removal efficiency is also independent of velocity.

The diffusion mechanism is extremely effective for the removal of very small particles. These particles move by "Brownian motion" in which their motion is affected by collisions with the molecules of the gas. Hence, these particles do not follow the fluid stream lines through the filter matrix but instead travel a zigzag path which increases their probability of contacting the filter matrix. The diffusion mechanism becomes more efficient as the particles get smaller due to increased particle diffusion velocities.

As discussed above, filters designed to remove submicron particles from gases rely primarily on the interception and diffusion capture mechanisms. Two types of filter matrices are utilized: fibrous and membrane. However, before discussing the performance properties of these filters, we need to discuss the concept of the most penetrating particle size.

Most Penetrating Particle Size

The particle removal effectiveness of very high efficiency aerosol filters is often expressed in terms of filter penetration where:

$$\text{Fractional Penetration} = \frac{\text{Downstream Particle Concentration}}{\text{Upstream Particle Concentration}}$$

Particle removal efficiency is related to penetration by the following equations:

$$\text{Efficiency} = \frac{(\text{Upstream Conc.-Downstream Conc.})}{\text{Upstream Concentration}}$$

$$\text{Penetration} = 1 - \text{Efficiency}$$

or as percentage

$$\% \text{ Penetration} = 100 - \% \text{ Removal Efficiency}$$

A typical plot of fractional penetration as a function
of particle size and flow rate is shown in Fig. 2. Note
that for a given flow condition, there exists a particle
size that exhibits maximum penetration. This maximum
occurs because the two major particle removal mechanisms
(interception and diffusion) are influenced by particle
size in opposite ways. The interception mechanism results in less penetration as particle size is increased.
However, the diffusion mechanism becomes more effective
as particle size is decreased.

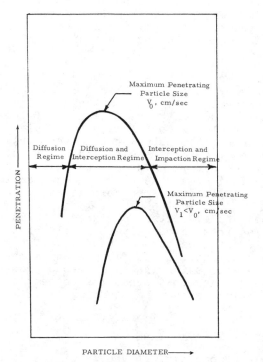

Fig. 2--Schematic presentation of fraction penetration
vs. particle size and flow velocity.

Increasing the flow rate increases penetration through
the filter as one might expect. However, the most
penetrating particle size decreases with increasing flow
rate. This is because the interception mechanism is
independent of flow rate but at higher flows, the residence
time in the filter is decreased. Hence, the diffusion
mechanism has less time to capture particles in the filter
matrix resulting in greater penetration and a shifting of
the maximum penetrating particle size to smaller particles.

The determination of the most penetrating particle size
is of practical importance because it is the most stringent
test for a filter. Studies conducted at the University of
Minnesota have shown that the most penetrating particle

size for fiberglass and membrane filters is typically
0.15um and 0.05um, respectively[5,11,12]. These
results are summarized below.

FILTER MEDIA PERFORMANCE

Fibrous Filters

Fibrous filters consist of a randomly-laid bed of
small diameter (0.1-1um) fibers usually held together
with a bonding agent (Fig. 3). Typical fiberglass filters can be 1000um thick and yet have a pressure drop of
1-5 centimeters of water (.01-.05 psi), at a flow rate of
32 liters/minute per 100 square centimeters of filter
area (5 cm/sec face velocity).

Fig. 3--SEM of fiberglass filter (Millipore AP-15)
at 800x magnification.

Filter efficiencies are determined by an aerosol
challenge test conducted at a specified flow rate and
with particles sized for maximum penetration. Typically,
fiberglass filter media used in HEPA (high efficiency particulate air) filters will remove 0.3um dioctylphthalate
(DOP) particles with efficiencies in excess of 99.97%
($\leq$0.03% penetration) at flow rates of 32 liters/minute
per 100 cm^2 of area. At higher flow rates one would
expect lower retention efficiencies as well as a decrease
in the particle size for maximum penetration.

Recent studies[11] conducted at the Particle Technology
Laboratory at the University of Minnesota have shown that
the most penetrating particle size for today's HEPA and
ULPA (ultrahigh efficiency particulate air) filter media

is 0.15um (Figs. 4 and 5). Fractional penetrations of the most penetrating DOP particles were 10^{-4} and 10^{-5} (3.5 cm/sec face velocity) for the HEPA and ULPA filters, respectively.

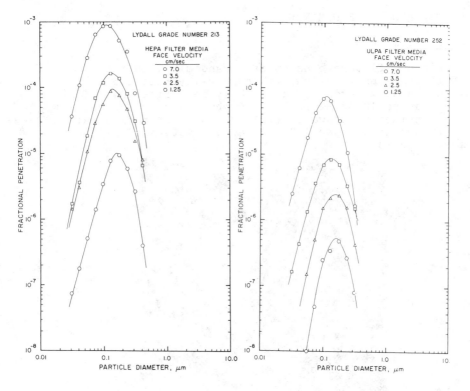

Fig. 4--Fractional pene-
tration of DOP aerosol
through HEPA filter
media[11].

Fig. 5--Fractional pene-
tration of DOP aerosol
through ULPA filter
media[11].

Because the spaces between the fibers can be quite large (2-20um), submicron particles are captured within the depth of the filter structure. Fig. 6 shows an SEM of a fiberglass filter (AP-15, Millipore) that has been challenged with a 0.05um NaCl aerosol at a 10 cm/sec face velocity and clearly shows particles adhering to fibers within the structure. Hence, the filtration efficiency is also a function of filter thickness.

Major advantages of fibrous-type filters are low pressure drop, high particle loading capacity, good aerosol retention efficiencies, and relatively low costs.

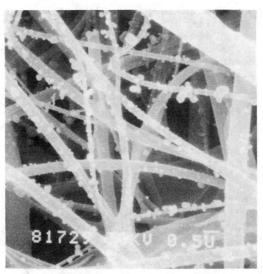

Fig. 6--SEM of fiberglass filter challenged with
0.05um NaCl aerosol (10 cm/sec). Filter loading
is 10^8 particles/cm^2 (magnification-8,000x).

Membrane Filters

Membrane filters consist of a continuous thin sheet
of plastic (100-150um thick) with a sponge-like
structure (Fig. 7).

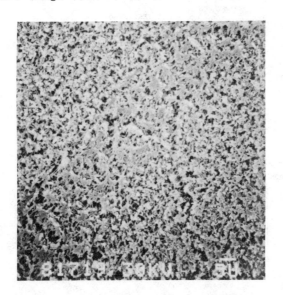

Fig. 7--SEM of PVDF membrane filter at 800x
magnification.

The structure, which is typically 80% void volume, acts as a screen filter for particles greater than its pore size and as a depth filter for particles that can penetrate into the matrix. Because of their very fine structure, membranes are very efficient at removing submicron particles.

The pressure drop across membrane filters is significantly greater than fibrous filters and dependent on pore size. For a 0.8um pore size membrane, the pressure drop at a flow rate of 60 liters/minute/100 cm^2 area (10 cm/sec velocity) is approximately 3 kPa (0.4 psi) whereas the pressure drop across a sterilizing-grade membrane (0.2um) would be about 35 kPa (5 psi) at the same flow rate.

Filtration efficiencies are again determined by aerosol penetration studies conducted at various flow rates with either liquid DOP or solid sodium chloride particles typically ranging in size from 0.03 to 1.0um. Rubow's thesis studied the effect of particle size and flow rate on the DOP penetration through membranes ranging in pore size from 0.8um to 8.0um[5]. Penetration at the most penetrating particle size was less than 10^{-5} and 10^{-2} for the 0.8um and 8.0um membranes, respectively, (face velocity of 10cm/sec). Figure 8 shows that the most penetrating particle size for a Millipore type AA (0.8um) membrane filter is between 0.03um and 0.08um for face velocities between 250 and 40 cm/sec, respectively.

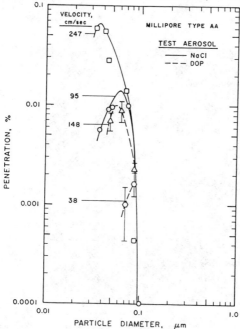

Fig. 8--Penetration of DOP and NaCl aerosols through 0.8um membrane filter[12,13].

Figure 9 shows an SEM of the surface of a 0.2um polyvinylidene fluoride (PVDF)membrane (Durapore[R]-Millipore) that has been challenged with the same quantity of 0.05um NaCl aerosol as the fibrous filter shown in Fig. 6 (9 cm/sec). Note that by comparing Figs. 6 and 9, one can see that a larger fraction of the aerosol challenge is visible on the membrane surface.

Fig. 9--SEM of membrane filter challenged with 0.05um NaCl aerosol (9 cm/sec). Filter loading is 10^8 particles/cm^2 (magnification-8,000x).

Studies conducted by Accomazzo, Rubow and Liu[12,13] have shown that the above sterilizing grade PVDF membrane and devices containing it have retention efficiencies in excess of 99.9999997% (≤ 3 x 10^{-9} penetration) when challenged with 0.05um NaCl aerosol at velocities between 10 and 100 cm/sec (5-50 psi pressure drop). The predicted aerosol penetration through this PVDF filter is less than 10^{-15} (Rubow's model). Hence, the 10^{-9} penetration claim for sterilizing grade membrane filters is very conservative and limited by the sensitivity of the aerosol detection system.

The above filtration efficiency results make membrane filters ideal candidates for filtration devices to be utilized by the semiconductor industry. In addition, since a membrane filter is a continuous polymeric structure and can be made from chemically-resistant fluoropolymers, filter shedding is unlikely. However, these very efficient membrane filters do have a greater pressure drop than HEPA-type fibrous filters and hence are currently only being used to filter compressed gases utilized by the semiconductor industry. Typical installa-

tions range from a few square centimeters for low flow
rates at the point-of-use to tens of square meters in
central gas distribution systems.

PROCESS GAS FILTER DEVICES

The preceeding section has shown that particles can
be removed effectively by either fiberglass media ($<10^{-4}$
penetration) or microporous membrane media ($<10^{-9}$ penetration). Both types of filter media have been utilized
in filter devices suitable for POU applications. Figures
10-12 show photographs of various configurations commercially available.

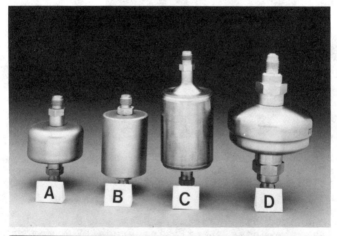

Fig. 10--POU process gas filters.

Fig. 11--Mini-POU process gas filters.

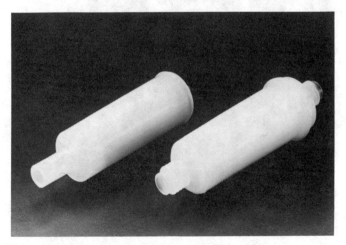

Fig. 12--All-PVDF inert gas filters.

Figure 10 shows four filter devices which can be used to filter at rates up to 300 liters/minute. Filter A uses a HEPA (high efficiency particulate air) type fiberglass filter sandwiched between two stainless steel support screens (8 cm^2 disc). Filter B also uses a fiberglass filter media but in a cylindrical configuration (25 cm^2). The end and seam seals are provided with an epoxy adhesive.

Filter C contains two layers of PTFE (polytetrafluoro-ethylene) membrane filter in pleated form (300 cm^2 filter

area).   The membranes are supported upstream and downstream using a non-woven, spun-bonded, polypropylene fiber
fabric, and the end caps also are polypropylene.

Filter D utilizes a stacked disc, microporous, membrane filter of PVDF (polyvinylidene fluoride) polymer.
The membrane is bonded to molded polysulfone plastic
discs, which are stacked up to provide a device with 300
cm^2 of filter area.   Filter D is also available with all
fluoropolymer materials of construction.   The discs are
molded from ECTFE (ethylene chlorotrifluoroethylene) and
the membrane is PTFE.

Figure 11 shows a recently developed POU process gas
filter designed to fit into equipment that has space
limitations (Fig. 13).   This filter consists of a membrane
support core molded from PFA-Teflon[R] (perfluoroalkoxy) and
a PTFE membrane (10 cm^2 area).   The filter easily handles
flow rates up to 30 liters/minute (50 cm/sec face velocity) with low pressure drop.

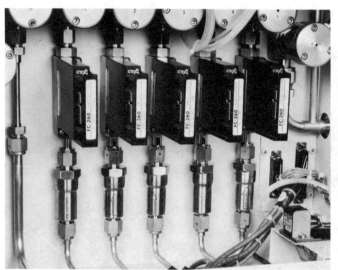

Fig. 13--Mini-POU process gas filter installation.

All of the above filters are fully enclosed in a welded
stainless steel housing and are available with a variety
of fittings utilized by the semiconductor industry.

Gas filters of all plastic construction are also
available for low pressure, inert gas applications, e.g.,
gas guns and gas-driven automated process equipment.
Figure 12 is a photograph of a 10 cm^2 membrane filter
constructed completely of PVDF (membrane and housing).
The filter is available with various end fittings.

Downstream Cleanliness

A filter device should not only remove the particles in the upstream gas but should also not contribute particles from the materials downstream of the filter media (i.e., filter supports, flow channels and fittings).

Particle shedding studies have been conducted at Millipore to characterize the downstream cleanliness of various POU gas filters. The results of these studies for filters A-D (Fig. 10) have been previously reported[14,15] and will be reviewed below.

The test protocol consisted of supplying filtered air to each type of filter with an inlet pressure of 69 kPa (10 psig). Particles shed during the first five minutes were ignored, since they could be attributed to filter handling and installation into the test system. The filters were then subjected to five minutes of mechanical vibration (30 raps/minute with a screwdriver handle) and then five minutes of pressure pulsing (10 to 1 psid, 20/minute). Each filter was then loaded with 10^{10} room air particles over a 20-hour period. During this period, no particles were detected above the typical background counts of the particle counter (LAS-X automatic optical counter, 0.09-3.0um range, PMS Corp., Boulder, Colorado).

After particle loading, the test filters were again fed with filtered air and the mechanical vibration and pressure pulsing phase of the test protocol repeated.

The results from the above test protocol are summarized in Figs. 14 and 15. The data represent the average of three of each of the four types of filters evaluated.

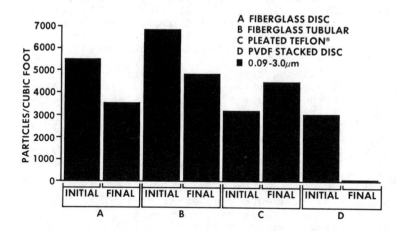

Fig. 14--Particle shedding due to mechanical shocking.

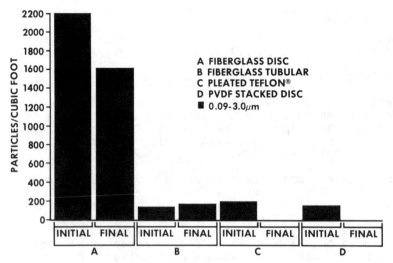

Fig. 15--Particle shedding due to pressure pulsing.

Figure 14 shows the average concentration of particles released by mechanical shocking of the various POU filters before and after loading the filter with 10^{10} particles. All of the filters initially released from 3000-7000 particles/ft^3 (>0.09um). After the filters were exposed to 20 hours of diluted room air, three of the filters (A-C) continued to shed particles during the final mechanical shocking. However, the stacked disc cartridge appears to be significantly cleaner after the 20-hour room air challenge test. The LAS-X optical particle counter essentially registered zero counts during this test period. Due to the LAS-X low sample flow rate, we can only have confidence in saying that the particle concentrations were<60 particles/ft^3 (>0.09um). These shedding results are similar to those obtained in the filter evaluation studies conducted by Duffin at Motorola[16].

Figure 15 shows the average concentration of particles released by pressure pulsing before and after particle loading. With the exception of filter A, the initially-released concentration of particles ranged from 100-200 particles/ft^3 (>0.09um). The concentration of particles released by filter A (~2000 particles/ft^3) may be due to the higher velocity through the smaller filter area, since all of the POU filters delivered approximately the same flow rate of 150-200 liters/minute at the 10 psi pressure drop. After the 20-hour room air challenge, filter A continued to shed a greater number of particles (~1600 particles/ft^3) than filter B (~200 particles/ft^3) during the final pressure pulsing mode. Filters C and D registered zero counts indicating a particle shedding concentration of <60 particles/ft3 (>0.09um).

Particle shedding from the Mini-POU devices shown in Fig. 11 has been recently measured at Millipore. The devices were tested using the system shown in Fig. 16. Particle-free compressed air flows through the test filters under controlled flow conditions and is combined with particle-free dilution air. The combined streams then flow into an isokinetic sampler at ambient conditions. Streams are drawn from the sampler and analyzed for particle concentration by both an optical particle counter (OPC) and a condensation nucleation counter (CNC). The OPC used is a PMS LPC525 which samples at 1.0 cfm (28 lpm) and is sensitive down to 0.2um. The CNC is a TSI model 3200 which samples at 0.01 cfm (0.3 lpm) and is sensitive down to 0.01um particles.

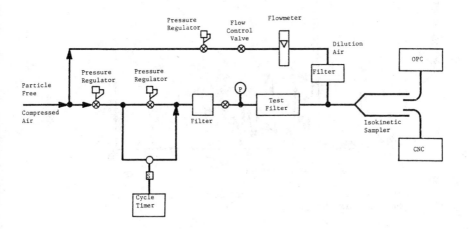

Fig. 16--Mini-POU process gas filter test system

The cartridges were tested under five sets of conditions selected to simulate use in the microelectronics industry:

- 10 liter/min steady-flow for 15 minutes
- 20 liter/min steady-flow for 15 minutes
- 30 liter/min steady-flow for 15 minutes
- pulsing flow from 0 to 30 liter/min at 60 cycles/min for 15 minutes
- 20 liter/min steady-flow for 30 minutes

The flow rate through the filter was controlled by a pressure regulator. The dilution air (60 liter/min) was necessary to insure that there was sufficient air in the sampler to prevent contamination of the filtrate by room air prior to sampling by the particle counters.

The shedding performance of the filters is given as a function of time in Fig. 17. The particle concentrations downstream of the test filter are corrected for the dilution air and system background (1-3 particle/ft^3). Concentrations were less than 20/ft^3 >0.20um at all times except during the first two minutes at 30 liter/min.

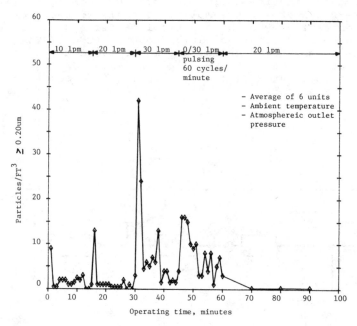

Fig. 17--Particle shedding performance of Mini-POU process gas filters.

The data obtained using the CNC cannot be accurately shown as a function of time because the instrument samples at such a low flow rate, and 1 particle/minute is equivalent to approximately 100 particles/ft^3. Therefore, only the average concentrations measured during each set of operating conditions are meaningful. These concentrations along with the average concentrations obtained with the OPC were as follows:

Test Conditions	Part/ft^3 ≥0.01um	Part/ft^3 ≥0.2um
10 liter/min	18.2	1.9
20 liter/min	7.2	1.6
30 liter/min	15.9	8.4
Pulsing	32.0	7.9
20 liter/min	1.2	0.3

In all cases except during the pulsing the average concentrations measured by the CNC were <20 particles/ft3 ≥0.01um.

CONCLUSIONS

The mechanisms of aerosol retention in both fiberglass and membrane type filter media predict a most penetrating particle size. Hence, filter performance ratings should be based on penetration measurements at this size range.

For fiberglass filter media the most penetrating particle size is typically between 0.1-0.2um and the fractional penetration of liquid DOP particles ranges between 10^{-4} and 10^{-5}.

The most penetrating particle size for membrane filter media is smaller than for fiberglass media and typically in the 0.05um range. The fractional penetration of solid NaCl particles of this size is less than 10^{-9}.

Particle shedding of POU gas filters appears to be dependent on both filter media and filter construction. All filters initially shed particles when subjected to mechanical and pressure pulses. However, fiberglass filters continue to shed particles during pulsing after 20 hours of use. On the other hand, membrane filters appear to clean up with use and indicate reduced particle shedding during pressure pulsing after 20 hours of use.

There appears to be a difference in shedding performance for membrane filter devices using pleated structures and filter devices using molded supports. The pleated filters continue to release particles when subjected to mechanical pulses whereas the stacked disc filters remain at instrument background levels.

Future studies will be aimed at identifying the source of the shed particles in order to aid in the development of even cleaner filter products for critical process gases.

ACKNOWLEDGMENTS

The authors are grateful for the careful and diligent experimental work performed by John Jaillet and Dr. Ramesh Hegde, and the help from Deborah Moschella and Frances Carlson in the preparation of this paper.

REFERENCES

[1] Mucha, J.A., "The Gases of Plasma Etching: Silicon-Based Technology", Solid State Technology, March 1985.

[2] Mitchell, J.W., "Chemical Analysis of Electronic Gases and Volatile Reagents for Device Processing", Solid State Technology, March 1985.

[3] Duffalo, J.M. and Monkowski, J.R., "Particulate Contamination and Device Performance", Solid State Technology, March 1984.

[4] Tolliver, D.C. and Schroeder, H.G., "Particle Control in Semiconductor Process Streams", Microcontamination, June/July 1983.

[5] Rubow, K.L., "Submicron Aerosol Filtration Characteristics of Membrane Filters", Ph.D. thesis, University of Minnesota, Mechanical Engineering Department, Minneapolis, MN (1981).

[6] Liu, B.Y.H., D.Y.H. Pui, and K.L. Rubow, "Characteristics of Air Sampling Filter Media", Aerosols in the Mining and Industrial Work Environment, V.A. Marple, B.Y.H. Liu (ed.) Vol. III, pp. 989-1038, Ann Arbor Science Publishers, Ann Arbor, MI (1983).

[7] Orr, C., (ed.) Filtration Principles and Practices, Parts I and II, Marcel Dekker, Inc., New York, N.Y. (1979).

[8] Shaw, D.T., (ed.) Fundamentals of Aerosol Science, John Wiley & Sons, New York, N.Y. (1978).

[9] Davies, C.N., Air Filtration, Academic Press, London (1973).

[10] Hinds, W.C., Aerosol Technology, John Wiley & Sons, Inc., New York, 1982.

[11] Liu, B.Y.H., Rubow, K.L. and Pui, D.L.H., "Performance of HEPA and ULPA Filters", Proceedings of 31st Annual Meeting of IES, Las Vegas, Nevada, April 29-May 2, 1985.

[12] Accomazzo, M.A., Rubow, K.L., and Liu, B.Y.H., "Ultrahigh Efficiency Membrane Filters for Semiconductor Process Gases", Solid State Technology, March 1984.

[13] Rubow, K.L. and Liu, B.Y.H., "Evaluation of Ultrahigh-Efficiency Membrane Filters", Microcontamination, March (1985).

[14] Accomazzo, M.A., "Particulate Retention and Shedding Characteristics from Point-of-Use Process Gas Filters", Technical Program Proceedings SEMICON/WEST, San Mateo, California, May 21-23, 1985.

[15] Accomazzo, M.A., "Particulate Retention and Shedding Characteristics of Membrane Filtration Devices for Semiconductor Process Gases and Liquids", SEMI Technology Symposium, Tokyo, Japan, December 11-12, 1985.

[16] Duffin, R.L., "Process Gas Filtration in Integrated Circuit Production", Microcontamination, December 1983/January 1984.

Subject Index

A

Acceptability, indoor air quality, 198–200

A. C. Fine and A. C. Coarse, Dustron results, 123

Aerodynamic particle sizer, 134–135

Aerosol filtration (See also Gas filtration)
efficiency measurements, 175–176
granular bed
flow fields, 47–49
mechanisms, 50–52
minimum efficiency, 54–57
most penetrating particle size, 52–54
mechanisms, 404–407
models for membrane filters
capillary tube, 76–83, 88–90
fibrous, for porous media, 83–86, 90
most penetrating particle size, 405–407

Aerosols
ASTM F21.20 Round Robin tests (See ASTM F21.20 Round Robin test procedure)
deposition in model filters, 71–72
latex, 142, 144–146
monodisperse, generation, 129–131
small-diameter, filter media and filters (DIN 24184 Standard), 219
test and atmospheric, table, 218
"worst case," 173–174

AFNOR NFX 44011 (uranin aerosol test), 218

Air cleaners, electrostatic, problems with, 226

Air cleaner testing (See also Filter testing)
SAE J726 Standard, 266–268
accuracy and initial efficiency, 270–272
recommendations for revisions, 273

reproducibility, 269–270
special adaptations, 272

Air cleanliness in cleanrooms
air filtration and, 386
air flow and pressurization, 399
definition and classification, 394–395
factors affecting, 397–398
Federal Standard 209B: 384
HEPA and ULPA filters, 386
HEPA filters and room integrity, 398–399
history, 391
instrumentation and methodology, 395–396
internal sources, 386
in microelectronics industry, 387–388
particle measurement methods, 384–386
in pharmaceutical, medical, and food industries, 388
room types and testing modes, 396–397
standards and specifications, 384
validation (certification), 392–393

Air permeability measurements, 105–108

Air pollutants, effect on gas turbines, 258–259

Air quality
acceptability criteria, 198–200
ambient, effect on gas turbines, 259–260
control
interaction of mini- and microenvironments, 209
methods, 196–198
strategies, 200–212
indoor, technical definition, 199

AIT 744 Lampblack, 231, 232, 235–237

Alumina dust test (British Standard 2831), 217

ARI Standard 850-84 (commercial and industrial air filter equipment), 217

ASHRAE Standards
  52-76: 194, 215-216, 219-226,
    229-237
  55-1981: 199
  62-73: 195
  62-1981: 194-196, 199
ASHRAE test dust, 116
ASTM F21.20 Round Robin test
    procedure, 149-151, 154
  filter media, 153
  participants, 154
  particle concentration and
    accumulation, 161
  percent penetration data
    analysis, 154-160
  pressure drop data analysis,
    161-162
  latex sphere challenge
    particles, 153-154
ASTM Standards
  D737-75: 105
  D757-75: 278
  D1682-64: 279
  D2176: 279
  D2176-69: 296
  D2905-81: 146
  D2906-85: 152
  D2986: 280
  D3467: 226
  D3786: 279
  D4029-83: 295
  F778: 163
  F778-82: 105-107, 109, 123,
    162, 216
Australian Standard 1132-1973
    (general ventilation
    filters), 219

B

Baghouses (See also Fabric
    filters; Fabric filtration)
  technology for utility opera-
    tions
    cleaning, 336-338
    filtration, 335-336
    sonic power augmentation, 338
  troubleshooting
    clay emissions, 285-286
    pressure drop, 287
    system upgrade, 286
  woven fiberglass
    need for new test methods,
      296-297
    test method research, 298-299
    test methods, 295-296

British Standards
  2831: 217
  3928: 217-218

C

Capillary tube model
  conventional membranes and,
    88-90
  geometry, 77-78
  particle deposition theories
    diffusion, 82-83
    direct interception, 80-81
    impaction and interception,
      81-82
    inertial impaction, 80
    by mechanism, table, 79
Carbon black, dust spot testing,
    231, 233, 235
Certification of cleanrooms (See
    Validation of cleanrooms)
Chemical analysis, fabric filter
    media, 280-281
Clay emissions, troubleshooting
    baghouses, 285-286
Cleanliness, air (See Air
    cleanliness)
Cleanrooms
  air cleanliness
    air flow and pressurization,
      399
    definition and classifica-
      tion, 394-395
    factors affecting, 397-398
    HEPA filters and room
      integrity, 398-399
    instrumentation and metho-
      dology, 395-396
    requirements, 384
    room types and testing modes,
      396-397
    validation (certification),
      392-393
  Federal Standard 209B: 384
  HEPA and ULPA filters, 386
  internal particle sources, 386
  in microelectronics industry,
    387-388
  particle measurement, 384-386
  in pharmaceutical, medical, and
    food industries, 388
Commercial and industrial air
    filter equipment (ARI
    Standard 850-84), 217
Condensation nucleus counter,
    133-134

Contaminants
  A. C. Fine and A. C. Coarse,
    Dustron results, 123
  gaseous, filter testing, 226
  gas turbine filters, 263-264
  personal exposure control,
    208-212
  preparation, Dustron, 116-117
Control strategies
  personal exposure, 208-212
  room ventilation, 201-208

D

Dendrite growth
  in granular bed filters, 64-65
  numerical modeling, 21-24
Deposition effect, particles on
  granular filters, 60-63
Diffusion, capillary tube particle deposition, 82-83
DIN 24184 (filter media and
  filters for small-diameter
  aerosols), 219
DIN 24185 (general ventilation
  filters), 219
Dioctyl phthalate/photometer
  technique, 127-128
Dustcake
  description and filtration
    mechanics, 339-340
  model development, 340-342
  removal methods
    reverse gas, 336-337
    shaking, 337-338
    sonic power augmentation, 338
Dust collectors, pulse-cleaned
  cartridge (See Pulse-
  cleaned cartridge dust
  collectors)
Dustron
  condition filters, 119
  contaminant preparation,
    116-117
  contaminant types, 123
  electronics, 116
  felt vs. wire surfaces and,
    121-122
  filter media, 123-125
  filter preparation, 117
  gravimetric analysis and data
    reduction, 119
  instrumentation and control,
    114-115
  pneumatics, 111-114
  procedure, fig., 117
  software, 116

  standard media for, 120-121
  system capabilities, 119-120
  test condition selection, 118
  test initiation, 118-119
Dusts, test, 123, 267-269
Dust-spot efficiency revisions
  (ASHRAE 52-76), 223-226
Dust spot testing
  carbon black feeding system,
    fig., 233
  development, 231-232
  test data summary, fig., 234
  test variables, 232, 235

E

Efficiency
  during dust collector season-
    ing, fig., 250
  fabric filtration, testing, 280
  filter fiber, 3-5
  fractional
    for ACF loading of heavy duty
      engine air cleaner, fig.,
      271
    pulse-cleaned cartridge dust
      collectors, 244
    standard test dusts and, 269
  HFATS, description, 349-351
  initial, 142
  NIOSH approach to updating,
    173-177
  particulate filter testing,
    169-170
  proposed specifications
    aerosol size, 346-347
    penetration rejection
      criterion, 347-348
    test material density, 348
  shortcomings of certification
    testing methods
    discriminating ability and
      reproducibility, 173
    environmental conditions, 173
    integrated vs. instantaneous
      monitoring, 171-172
    particle size, aerosol type,
      and flow rate, 172
  steady state, pulse-cleaned
    cartridge dust collectors,
    248-249
  test procedures and require-
    ments, 170-171
  test system, 175
Effluent concentration, and
  pressure drop in granular
  filters, fig., 61-62

Electric power generation
 baghouse configuration, 334-335
 baghouse operations, 335-338
 dustcake characteristics,
  338-342
 fiberglass fabric filters,
  292-300
  need for new test methods and
   research, 296-299
  test methods, 295-296
 process description, 333-334
Electrostatic air cleaners,
 problems with, 226
Electrostatic enhancement of
 fabric filtration
 methods
  external electric field,
   324-328
  particle charging, 321-324
 pressure reduction mechanisms,
  319-3221
Environment, thermal, acceptable,
 199
EUROVENT 4/5 (general ventilation
 filters), 218

F

Fabric filters (See also
  Baghouses)
 cleaning
  reverse gas, 336-337
  shaking, 337-338
  sonic power augmentation, 338
 electrical forces
  particle collection efficiency and, 318-319
  pressure drop and, 319
 fiberglass
  accelerated fabric testing,
   298
  bag failure analysis, 298
  bag standardization, 299
  bag testing, pilot scale, 299
  Manufacturers Advisory Group,
   299
  new, testing, 298-299
  test methods, 295-297
 lab testing objectives, 282-283
 quality assurance programs,
  283-285
 resistance, 303-306
 specific resistance coefficient
  experimental measurements,
   306-310
  in situ measurement, 312-314
  measurement, 311-312

 in modeling, 310-311
 troubleshooting
  baghouse system upgrade, 286
  clay emission collection,
   285-286
  pressure drop, 287
Fabric filtration
 electrostatic enhancement
  economic viability, 328-329
  by external electric field,
   324-328
  by particle charging,
   321-324, 328
  pressure reduction mechanisms, 319-321
 in utility power plants
  baghouse configuration,
   334-335
  baghouse operations, 335-338
  dustcake characteristics,
   338-340
  process, 333-334
Fabric testing
 laboratory
  objectives, 282-283
  usefulness, 288-290
 limitations, 289
 methods
  chemical analysis, 281
  efficiency testing, 280
  Flex Test, 279-280
  microscopic examination,
   280-281
  Mullen Burst, 279
  permeability, 278
  tensile strength, 278-279
  visual bag inspection, 281
 quality assurance programs,
  283-285
 recommendations, 289-290
 troubleshooting
  baghouse system upgrade, 286
  collecting clay emissions,
   285-286, 290
  pressure drop, 287, 290
Federal Specification F-F-310:
 216
Federal Standards
 209B: 384, 394, 396, 400
 209C: 396
Fiberglass filters
 Manufacturers Advisory Group,
  299
 research needs
  bag failure analysis, 298
  bag standardization, 299

fabric testing, 298–299

pilot-scale bag testing, 299
test methods, 296–297
test methods, 295–296
Fibrous filters, 407–408
clean, pressure drop, 19–21
efficiency, 3–5
filtration theory, 2–7
model for porous media, 83–86,
90–91
performance, 5–7
pressure drop, 2–3
theory, 2–7
Fibrous filtration
electrically enhanced,
numerical modeling
dendrite growth, 21–24
model formulation, 14–16
particle trajectories, 17–19
pressure drop in clean
filters, 19–21
pressure loss, mitered cylinder
model
matrix geometry, 30–37
slip effects, 40
theory, 29–30
thickness effects, 37–40
Filter media and filters for
small-diameter aerosols
(DIN 24184 Standard), 219
Filters
downstream particle shedding,
415–418
Dustron, 117–118
efficiency (See Efficiency)
fabric (See Fabric filters)
felt vs. wire surfaces, 121–122
fiberglass (See Fiberglass
filters)
fibrous (See Fibrous filters)
gas turbines, inlet
aging, 263
configuration, 262–263
contaminants, 263–264
self-cleaning, 264–265
testing, 261–262
granular (See Granular filters)
HEPA (See High efficiency
particulate air filters)
membrane (See Membrane filters)
model, aerosol deposition,
71–72
particulate, testing, 169–171
process gas, 412–414
ULPA (See Ultra-low penetration
air filters)
Filter testing (See also Air

cleaner testing)
AFNOR NFX 44011 (uranin aerosol
test), 218
ARI Standard 850-84 (commercial
and industrial air filter
equipment), 217
ASHRAE Standard 52-76 (See
ASHRAE Standards)
ASTM F-778-82 (See ASTM
Standards)
Australian Standard 1132-1973
(general ventilation
filters), 219
automated flat sample (See
Dustron)
automated for efficiency
measurements
aerodynamic particle sizer,
134–135
aerosol flow system, 132–133
aerosol generation of particles (0.5µm and
≥ 0.5µm, 129–130, 131
condensation nucleus counter,
133–134
filter holder assembly,
131–132
microcomputer system, 135–136
operation, 137
schematic, 130
test data, 137–138
variations, 139
British Standards (See British
Standards)
DIN 24184 (filter media and
filters for small-diameter
aerosols), 219
DIN 24185 (general ventilation
filters), 219
dioctyl phthalate/photometer,
127
dust spot technique for ASHRAE
52-76:
air flows, 235
concentration and supply
pressure, 235
development, 231–232
effects of carbon blacks, 235
number of holes, 235
test container, 232
test data, table, 234
EUROVENT 4/5 (general
ventilation filters), 218
Federal Specification F-F-310:
216
gaseous contaminant, 226
for gas turbines, 261–262

gravimetric techniques, 128
HFATS (See High flow alterna-
    tive filter test system)
MIL-STD-282 (thermal DOP test
    for HEPA filters), 216
sodium chloride/flame photo-
    meter, 128
UL 900 (flame resistance), 217
Filtration (See also Removal
    control)
    for cleanrooms, 386
    internal particle sources in
        cleanrooms, 386
    systems for gas turbines, 261
    theory for fibrous filters, 2-7
Fit testing, quantitative
    history, 182-186
    problems and research needs,
        186-188
Flame resistance (UL 900), 217
Flat sheet
    filter media, standard test for
        initial efficiency, 141-151
    procedure, 146-151
    test system and materials,
        144-146
    filtration performance test
        (See Dustron)
Flex test, 279-280
Flow (See also Inertial flow;
    Laminar flow; Slip flow;
    Turbulent flow)
    fields in granular beds, 47-49
    HEPA filters
        characteristics, 370-372
        standards for testing, 373-
            374
    mixed, in cleanrooms, 396
    and pressurization in clean-
        rooms, 399
Food industry, air cleanliness
    requirements, 388
Fractional efficiency (See
    Efficiency)

G

Gas filtration (See also
    Aerosol filtration)
    by fibrous media, 1-12
    process
        downstream cleanliness,
            415-418
        filters, 412-414
Gas turbines
    air filtration systems, 261
    air pollutant effects, 258-259

ambient air quality and,
    259-260
filters
    aging, 263
    configuration, 262-263
    contaminants, 263-264
    self-cleaning, 264-265
    testing, 261-262
Granular bed filtration
    flow fields, 47-49
    minimum efficiency, 54-57
    most penetrating particle size,
        52-54
    particle collection mechanisms,
        50-52
Granular filters
    dendrite growth, 64-65
    deposition effect, 60-63
    deposit morphology on model
        filters, fig., 65
    effluent concentration and
        pressure drop data, fig.,
        61-62
    performance
        experimental determination,
            69-71
        theory, 63-69
Gravimetric techniques for filter
    testing, 128

H

Heating, ventilating, and air
    conditioning systems
    conventional, fig., 194
    and minienvironment, 201
HEPA filters (See High effi-
    ciency particular air
    filters)
HFATS (See High flow alternative
    filter test system)
High efficiency particulate air
    filters, 2
    definition, 391
    efficiency test specifications
        aerosol size, 346-347
        penetration rejection
            criterion, 347-348
        test material density, 348
    flow
        characteristics, 370-372
        constants and physical
            characteristics, 368
        parameters and performance,
            368-369
        standards for testing,
            373-374

HFATS (See High flow alternative filter test system)
homogeneous and layered, 369-370
leakage, in cleanrooms, 398-399
pressure and temperature standards, 374
pressure drop performance assessment using
  inertial flow, 366-367
  slip flow, 367
  turbulent flow, 367-368
  viscous laminar flow, 367
quality assurance measures, 377
recommendations, 377-379
test method selection
  filter characterization/performance, 375-376
  pressure drop only, 374-375
  viscous laminar flow and, 376-377
thermal DOP test (MIL-STD-282), 216
High flow alternative test system
description, 349-351
performance
  damaged filter tests, 351-353
  intact filter tests, 353-357
  measurement uncertainty, 357-358
HVAC systems (See Heating, ventilating, and air conditioning systems)
HV Dustron (See Dustron)

I

IES-RP-CC-006-84-T: 393, 394, 396, 398, 400
Impaction, capillary tube particle deposition, 80-82
Indoor air quality procedure, 196
Inertial flow, HEPA filter pressure drop and, 366-367
Interception, capillary tube particle deposition by, 80-82

L

Laminar flow
in cleanrooms, 396
viscous, 367, 376-377

Latex
aerosols, 142, 144-146
spheres, results of ASTM Round Robin tests
  challenge particles, 153
  filter media, 153
  participants in, 154
  particle concentration and accumulation, 1612
  percent penetration analysis, 154-160
  pressure drop data analysis, 161-162
  procedure, 154
Liquid extrusion technique, 98

M

Media resistance, ASTM F-778-82: 216
Medical industry, air cleanliness requirements, 388
Membrane filters, 74-75, 409-412
capillary tube model, 76-83
  conventional membranes and, 88-90
  diffusion, 82-83
  direct interception, 80-81
  geometry, 77
  impaction and interception, 81-82
  inertial impaction, 80
  overall filter efficiency, 83
  particle deposition theories, 78-80
fibrous filter model, 83-86, 90-91
micrographs, fig., 76
Nucleopore, 86-88
Mercury porosimetry, 98
Methylene blue test (British Standard 2831), 217
Microelectronics industry
air cleanliness requirements, 387-388
process gas filter devices, 412-418
Microenvironment, personal exposure control, 208-212
Microscopy, fabric filter media, 280-281
Military Standard 282: 216, 348
Minienvironment and HVAC system, thermal and air quality control interaction, 201

Minimum bubble pressure, 99-105
Minimum collection efficiency,
    granular bed filter, 54-57
Mitered cylinder pressure loss
    model
  matrix geometry, 30-37
  single-layer screen models and,
    42-43
  slip effects, 40
  theory, 29-30
  thickness effects, 37-40
MIT Flex test, 279-280
Models
  numerical (See Numerical
    modeling)
  membrane filter particle
    collection
    capillary tube, 76-83
    fibrous, 83-86
Monodisperse aerosols, gener-
    ation, 129-131
Most penetrating particle size,
    52-54
  concept, 405-407
  membrane filter, fibrous model,
    90-92
Mullen Burst test, 279
Multichannel analyzer
  optical particle counter with,
    145
  for time-resolved measurements
    of pulse-cleaned cartridge
    dust collectors, 244-246

N

National Institute for Occupa-
    tional Health and Safety
  particulate filter testing,
    169-171
  research objectives for up-
    dating respirator filter
    testing, 173-177
  shortcomings of certification
    testing methods, 171-173
NIOSH (See National Institute
    for Occupational Health and
    Safety)
Nucleopore filter
  micrograph, fig., 76
  model, 86-88, 92
Numerical modeling of electri-
    cally enhanced fibrous
    filtration
  dendrite growth, 21-24
  model formulation, 14-16
  particle trajectories, 17-19

pressure drop in clean filters,
    19-21

O

Optical particle counter, 142,
    145-147, 244-246

P

Particle deposition (See also
    Aerosol deposition)
  effect on granular filters
    experimental determination,
      69-71
    loading phenomenon, 60
    theory, 63-69
  stochastic simulation, 66-69
  theories for capillary tube
    model, table, 79
Particles
  aerosols (See Aerosols)
  measurement methods for
    cleanrooms, 384-386
  shedding, downstream, 415-418
  trajectories with/without
    electric field, fig., 18
Particle size distribution
  filter resistance and, 303-306
  specific resistance coeffi-
    cient, 305-306
  experimental measurement,
    310
  in situ measurement, 312-314
  measurement, 311-312
  in modeling, 310-311
Particle-size-penetration test,
    development (ASHRAE 52-76),
    223-226
Particle sizer, aerodynamic,
    134-135
Particulate filters (See also
    High efficiency particulate
    air filters)
  respirator, NIOSH objectives
    for updating, 173-174
  testing, 169-171
Permeability
  air, measurement, 105-108
  fabric test method, 278
Pharmaceutical industry, air
    cleanliness requirements,
    388
Pollutants, air, effect on gas
    turbines, 258-259
Pore size distribution, 98-99

Pore throat size distribution, 99-105
Pore volume distribution (See Pore size distribution)
Pressure drop
  ASTM F21.20 Round Robin tests, 161-162
  across granular bed, 47-49
  baghouse, troubleshooting, 287
  in clean fibrous filters, 19-21
  during dust collector seasoning, fig., 250
  and effluent concentration in granular filters, fig, 61-62
  in electrostatically enhanced fabric filters, 319
    by external electric field, 324-328
    mechanisms, 319-320
    by particle charging, 321-324
  filter resistance, 303-306
  flat-sheet filters, 2-3
  HEPA filters
    flow characteristics, 370-372
    flow standards for testing, 373-374
    inertial flow and, 366-367
    slip flow and, 367
    test method selection, 374-376
    turbulent flow and, 367-368
    viscous laminar flow meters as standards, 376-377
  specific resistance coefficient
    experimental measurements, 306-310
    in situ measurement, 312-314
    measurement, 311-312
Pressure loss in fibrous filters
  matrix geometry of mitered cylinder model, 30-37
  slip effects, 40
  theory, 29-30
  thickness effects, 37-40
Pressurization, air flow in cleanrooms and, 399
Process gas
  downstream cleanliness, 415-418
  filters, 412-414
Pulse-cleaned cartridge dust collectors
  fractional efficiency, 244
  instrumentation, 244-246
  recovery from change, 249-251
  seasoning, 249
  short-time scale, 251-255

  steady state efficiency, 248-249
  time scales, 246-248
  uncertainty, 246
Pulse-cleaning, effect on collector life and efficiency, 251-255

Q

Quality assurance
  filter bags, 283-285
  HEPA filters, 377
Quantitative fit testing
  history, 182-186
  problems and research needs, 186-188

R

Removal control, 197, 207, 212
  acceptability criteria for contaminant exposure, 198-200
  air quality control and, 196-198
  strategies, 200-212
    personal exposure, 208-212
    room ventilation, 201-208
  ventilation control and, 194-196
Residential air filter equipment (ARI Standard 680-80), 217
Respirator filtration
  evaluation methods, 169-171
  NIOSH objectives for updating, 173-177
  shortcomings of certification testing methods, 171-173
Respirators, quantitative fit testing
  history, 182-186
  problems and research needs, 186-189
Reverse gas, dustcake removal with, 336-337
Room acceptability ratio, 202, 207
Room purifiers, problems with, 226

S

SAE J726 Air Cleaner Test Code
  accuracy and initial efficiency, 270-272
  dust loading with specific contaminants, 272

performance under wet/humid
    conditions, 272
recommendations, 273
standard test dust and
    reproducibility, 269-270
vibration and, 272
Seasoning, pulse-cleaned car-
    tridge dust collectors,
    249-250
Self-cleaning filters, 264-265
Shaking, dustcake removal by,
    337-338
Shedding, particle, 415-418
Slip flow, 40
    HEPA filter pressure drop and,
    367
Sodium chloride/flame photometer
    technique, 128
Sodium-flame test (British
    Standard 3928), 217-218
Sonic power augmentation, dust-
    cake removal by, 338
Specific resistance coefficient,
    305-306
    experimental measurement,
    306-310
    in situ measurement, 312-314
    measurement, 311-312
    in modeling, 310-311

T

Tensile strength, fabric test
    methods, 278-279
Test dusts (See Dusts, test)
Thermal DOP test for HEPA filters
    (MIL-STD-282), 216
Thermal environment, acceptable,
    199
Trajectories, particle, 17-19
Troubleshooting fabric filter
    bags, 285-287
Turbulent flow
    in cleanrooms, 396
    HEPA filter pressure drop and,
    367-368

U

UL 900 (flame resistance), 217
ULPA filters (See Ultra-low
    penetration air filters)
Ultra-low penetration air
    filters, 386
Uranin aerosol test (AFNOR NFX
    44011), 218

United States Department of
    Energy
    filter efficiency test speci-
        fications (proposed)
        aerosol size, 346-347
        penetration rejection
            criterion, 347-348
        test material density, 348
    HFATS
        description, 349-351
        measurement uncertainty, 357-
            358
        performance, 351-357

V

Validation of cleanrooms, 392-393
    air cleanliness
        definition and classifi-
            cation, 394-395
        factors affecting, 397-398
    air flow and pressurization,
        399
    Federal Standard 209B: 394-395,
        398
    HEPA filters and room integ-
        rity, 398-399
    IES-RP-CC-006-84-T: 393
    instrumentation and method-
        ology, 395-396
    room types and testing modes,
        396-397
Ventilation
    control, 194-196, 201-208
    filters, general
        Australian Standard 1132-
            1973: 219
        DIN 24185: 219
        EUROVENT 4/5 Standard, 218
Ventilation rate procedure, 196
Verification of cleanrooms (See
    Validation of cleanrooms)
Very large scale integration
    circuits
    design rules, table, 387
    gaseous impurity concentra-
        tions, 403
Viscous laminar flow, HEPA filter
    pressure drop and, 367
Visual inspection, fabric filter
    bags, 281
VLSI circuits (See Very large
    scale integration circuits)

W

"Worst case" aerosol, 173-174

Author Index

A

Accomazzo, M. A., 402
Agarwal, J. K., 127
Ariman, T., 13

D

Davis, W. T., 302
Donovan, R. P., 316

E

Engel, M. R., 241

F

Fain, D. E., 364
Felix, L. G., 292
Flannery, J. L., 390

G

Gehri, R. P., 292
Giovanni, D. V., 292
Grant, D. C., 402

H

Haley, L. H. Jr., 277
Henry, F. S., 13

J

Japuntich, D. A., 152
Johnson, B. R., 127
Johnson, T. W., 241

K

King, J. G., 383
Krafthefer, B. C., 193

L

Lee, K. W., 46
Liu, B. Y. H., 1, 74, 241

M

McDonald, B. N., 241
McKenna, J. D., 277

Miller, B., 97
Monson, D. R., 27
Moulton, J. E. III, 110
Moyer, E. S., 167
Murphy, D. J., 214
Myers, W. R., 181

N

Nicholson, R. M., 141, 266

P

Pontius, D. H., 332
Pui, D. Y. H., 241

R

Raber, R. R., 229
Remiarz, R. J., 127
Rivers, R. D., 214
Rubow, K. L., 1, 74

S

Schaller, R. E., 241
Scripsick, R. C., 345
Smith, W. B., 292, 332

T

Tatge, R. B., 257
Tien, C., 60
Tyomkin, I., 97

V

VanOsdell, D. W., 316

W

Walcroft, J. P., 390
Wehner, J. A., 97
Weisert, L. E., 266
Wilson, J. C. Sr., 110
Woods, J. E., 193

83 NFT